# THE ENVIRONMENT
## OF THE
## DEEP SEA

W. W. Rubey (1898–1974)

W.G. ERNST

and

J.G. MORIN

*Editors*

# THE ENVIRONMENT
# OF THE
# DEEP SEA

Rubey Volume II

Prentice-Hall, Inc., Englewood Cliffs, New Jersey 07632

*Library of Congress Cataloging in Publication Data*
Main entry under title:

The environment of the deep sea.

    (Rubey volume; 2)
    Lectures presented at UCLA, Apr.–June 1979, as the
Rubey Colloquium.
    Bibliography: p.
    Includes index.
    1. Deep-sea ecology–Congresses.  2. Abyssal zone–
Congresses.  3. Marine biology–Congresses.
4. Oceanography–Congresses.  5. Marine sediments–
Congresses.  I. Ernst, W. G. (Wallace Gary)  (date)
II. Morin, J. G.  III. Series.
QH541.5.S3E58  ·   574.5'2636      82–342
ISBN  0-13-282822-7           AACR2

Editorial/production supervision
    and interior design by Karen Skrable
Manufacturing buyer: John Hall

Printed in the United States of America

10  9  8  7  6  5  4  3  2  1

ISBN   0-13-282822-7

Prentice-Hall International, Inc., *London*
Prentice-Hall of Australia Pty. Limited, *Sydney*
Prentice-Hall of Canada, Ltd., *Toronto*
Prentice-Hall of India Private Limited, *New Delhi*
Prentice-Hall of Japan, Inc., *Tokyo*
Prentice-Hall of Southeast Asia Pte. Ltd., *Singapore*
Whitehall Books Limited, *Wellington, New Zealand*

# CONTENTS

# PREFACE

The Department of Earth and Space Sciences of the University of California, Los Angeles (UCLA), conducts an annual lecture series entitled "The Rubey Colloquium." Specialists from research institutions around the country are invited to present lectures on a central theme of broad interest to earth and/or space scientists. As appropriate, written and elaborated summaries of these presentations subsequently have been gathered together as a book; collectively, this series is known as the Rubey Volumes.

Both colloquia and book series are named in honor of the late W. W. Rubey (1898–1974), career geologist with the U. S. Geological Survey and Professor of Geology and Geophysics at UCLA. A brief sketch of Rubey's geologic accomplishments is contained in the preface to Rubey Volume I, *The Geotectonic Development of California* (Ernst, ed., 1981).

Rubey Volume II, *The Environment of the Deep Sea*, presents the results of a Rubey Colloquium that took place during April–June, 1979. This particular lecture series was envisioned by the late Thomas E. Ronan, Jr. (1946–1978), Assistant Professor of Paleobiology at UCLA. An ecologist and marine biologist, his tragic death while collecting invertebrates on a diving expedition off Santa Catalina Island on September 7, 1978, abruptly terminated the career of a promising young scientist. His warmth; boundless enthusiasm; and dedication to his family, his students, and his science are deeply missed.

Because of the timely nature of the subject, and the well-advanced detailed plans formulated by Ronan, we assumed responsibility for continuing this colloquium. The present Rubey Volume, which reflects that stimulating series of lectures, stands as a tribute to Tom Ronan and is dedicated to his memory. The volume is divided into two portions: Part I deals with the physical and chemical environment, whereas Part II treats biological aspects. This convenient distribution is somewhat misleading because of the interconnectedness of biological, chemical, and physical processes in the deep sea.

We express appreciation to UCLA for financial support of the colloquium, and to the individual contributors for their enthusiasm, diligence, and patience, both during their visits to the Earth and Space Sciences Department and later during preparation of their chapters. We are especially indebted to Tom Ronan, whose vision and energetic organization made possible the presentation of Rubey Volume II, *The Environment of the Deep Sea.*

W. G. ERNST
Department of Earth and Space Sciences
University of California, Los Angeles

J. G. MORIN
Department of Biology
University of California, Los Angeles

# THE ENVIRONMENT
## OF THE
## DEEP SEA

# I

# PHYSICAL AND CHEMICAL
# ENVIRONMENT
# OF THE DEEP SEA

J. R. Heirtzler
Woods Hole Oceanographic Institution
Woods Hole, Massachusetts 02543

# 1

# THE EVOLUTION OF THE DEEP-OCEAN FLOOR

## ABSTRACT

Over the last ten years marine scientists have come to understand how the ocean floor is created and why it has the shape that it does. They are turning their attention to details of the shape of the ocean floor in the major ocean provinces and to appreciating the interrelationship of the shape of the sea floor, the proximity of the coast, the chemistry and flow of water masses, and the plants and animals that inhabit the oceans.

## INTRODUCTION

Since the earliest recorded history of civilized man, the ocean's surface has been used as a means of transportation and as a place to fish for food. For thousands of years we had little serious interest in and no ability for exploring beneath its surface. It was only in the late 18th century that the first depth sounding in the deep ocean was made. This apparently was accomplished by a Captain Phipps, who recorded a depth of 683 fathoms (4098 ft or 1249 m) in the North Atlantic. In the early 19th century the first plankton was collected and in the mid-19th century Matthew F. Maury assembled wind and current information, which appreciably reduced the time of transit of sailing ships by taking advantage of favorable sea conditions. Most persons, however, consider the study of the deep-ocean floor to have begun with the expedition of the HMS *Challenger* during the years 1873–1876 when biological, geological, and a host of general oceanographic observations were made and subsequently published.

The first modern depth measurements by electronic echosounders were made by the German ship *Meteor* in 1924, and it was only then that we began to understand the actual shape of the sea floor. In 1935, Maurice Ewing started to get some idea of the gross structure beneath the sea floor when he began his extensive marine seismic studies. The first sea-floor cameras became operational during the Second World War when they were used to identify sunken ships. It is to be noted that the first exploitation of the oceans for oil appeared shortly thereafter. In 1948, the first oil wells were drilled in the Gulf of Mexico.

In 1954, however, a new generation of global studies began when it was then recognized, from the locations of earthquakes, that a mid-ocean ridge system encircled the world. From a study of magnetic field anomalies associated with this ridge system, it soon became clear that the mid-ocean ridge played a fundamental role in the evolution of all the surface expression of the earth. During the years of 1963–1966 the theories of sea-floor spreading and plate tectonics were established. Starting in 1968 the drillship *Glomar Challenger* began drilling into the deep-sea floor and recovering samples which not only confirmed sea-floor spreading but also advanced the new science of paleoceanography.

Starting in 1974, (Ballard and Van Andel, 1977) engineering developments made it possible to navigate precisely over a few square kilometers of the deep-ocean bottom and to undertake detailed geologic mapping; because of that, the study of ecosystem processes in the deep sea first became possible.

After the 1920s, echosounding in the deep sea proliferated. Bruce Heezen and Marie Tharp, of Lamont-Doherty Geological Observatory, assembled great quantities of echosounding data and published them in the form of "physiographic" maps that gave a general indication of the type of relief on the ocean floor in an attractive format (Fig. 1-1). Since they did not have data from every part of the world's oceans, they had to make careful guesses in a few areas to complete the maps. Nevertheless their maps, published by the National Geographic Society in the 1960s and '70s, still give the best "feeling" for the shape of the sea floor.

Some of the main features displayed by Heezen and Tharp's maps are the prominence of the mid-ocean ridge system; a rift valley along the axis of the ridge in the North Atlantic; fracture zones which offset the ridge axis; the lack of sediment at the axis of the mid-ocean ridge, but great expanses of flat abyssal plains away from the ridges; the existence of seamounts in many parts of the ocean; the prominence of continental shelves and slopes near the continents with canyons cutting the slopes; and the continuity of island arcs and trenches, especially around the edges of the Pacific Ocean.

These different features have been analyzed in detail by marine geologists and geophysicists. Some of the features are more abundant in one ocean than another. Table 1-1 shows the statistics.

From these numbers it is clear that ocean basin and mid-ocean ridge provinces are the most common types of features. Continental shelves and slopes are more important in the Arctic Ocean than elsewhere, whereas trenches occur much more in the Pacific. The zero percentages for the Arctic Ocean reflect how little we know about the ocean floor there and probably are not accurate.

A common graphic way of showing how the depths of the ocean are distributed is with the use of the so-called hypsometric curve (Fig. 1-2). This curve shows, for example, that most of the ocean floor is about 4 to 6 km deep. This depth distribution, coupled with the fact that about 70% of the earth's surface is covered with oceans, can be used to show that, if the earth's surface were completely flat, it would be covered with a uniform layer of seawater about 2 km deep! The mean depth of all the oceans is about 3730 m, compared to only 840 m for the mean elevation on land. The deepest trench in the oceans is about 11,000 m; in contrast, Mt. Everest, the highest mountain on land, has an altitude of 8845 m. It is clear from these comparisons that the sea floor has areas more rugged than those on land and has great, flat abyssal plains larger than the great plains or deserts on land.

The depth of most of the ocean prohibits direct exploration by man in today's submersibles (Heirtzler and Grassle, 1976). Although one bathyscaph can go to the deepest part of the ocean, a few others can go only to approximately 4000 m and the greater depths can only be photographed and sampled by instruments on the ends of cables from surface vessels (Heezen and Hollister, 1971). Our inability to get to most of the sea floor has not prohibited our studying the gross structure of the ocean, but it *has* inhibited, until quite recently, our ability to examine relatively small areas in a coherent way.

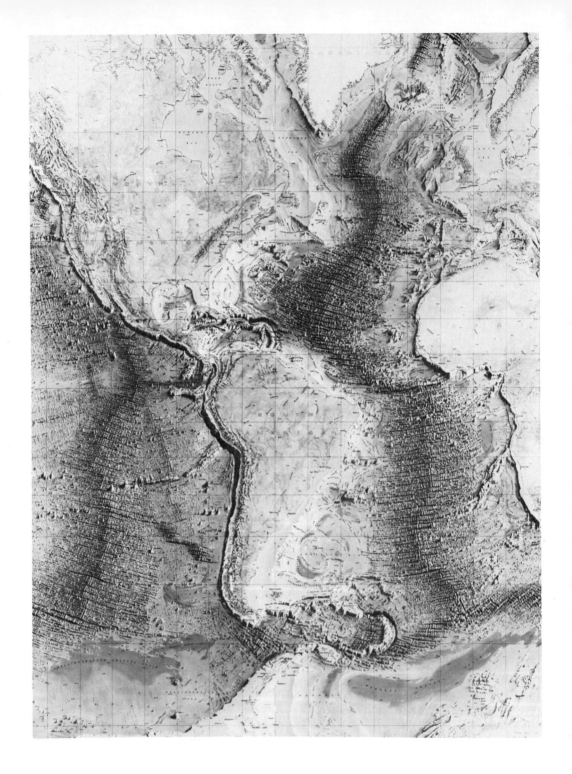

Fig. 1-1. A portion of the physiographic map World Ocean Floor published in 1977 by the U.S. Navy Office of Naval Research based upon similar maps by Heezen and Tharp (copyright by Marie Tharp). Lines are drawn each 10 degrees of latitude and longitude.

TABLE 1-1. Distribution of Provinces in Oceans and Adjacent Seas (in percent)[a]

| Oceans and Adjacent Seas | Mid-ocean Ridge | Ocean Basin | Other Ridges | Cont. Shelf and Slope | Island Arc and Trench | Cont. Rise |
|---|---|---|---|---|---|---|
| Pacific | 35.9 | 43.0 | 2.5 | 13.1 | 2.9 | 2.7 |
| Atlantic | 32.3 | 39.3 | 2.0 | 17.7 | 0.7 | 8.0 |
| Indian | 30.2 | 49.2 | 5.4 | 9.1 | 0.3 | 5.7 |
| Arctic | 4.2 | 0.0 | 6.8 | 68.2 | 0.0 | 20.8 |
| Percent of World Ocean | 32.7 | 41.8 | 3.1 | 15.3 | 1.7 | 5.3 |

[a]After Menard and Smith, 1966.

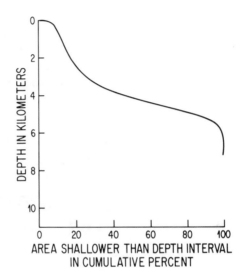

Fig. 1-2. Hypsometry of all ocean basins. Menard and Smith, 1966.

AREA SHALLOWER THAN DEPTH INTERVAL IN CUMULATIVE PERCENT

## THE ORIGIN OF THE SEA FLOOR

The structure of the sea floor appears to be much simpler than the structure of the continents. It seems to have two components — sediments and basaltic rock — that originate by different processes. The basaltic rock is entirely exposed near the axis of the mid-ocean ridge but is covered with an increasing thickness of sediment as one proceeds away from the ridge axis and towards the continents. Sediment thicknesses reach 10 or more km in some areas.

In the early 1960s it was discovered that the magnetic field measured on the sea surface includes an unusual magnetic anomaly associated with the axis of the mid-ocean ridge. The anomaly is caused by the way elementary particles of iron, which are in the minerals of the earth's crust, are magnetized. It was found that the magnetic anomaly over the axis of all parts of the mid-ocean ridge was similar, suggesting similar geologic structure. Not only was the anomaly elongated following the ridge axis, but there were paralleling and axially symmetric anomalies on both sides of the axis. The elongated magnetic anomalies, or magnetic anomaly stripes, were accounted for by a relatively simple but revolutionary theory put forward by Fred Vine and Drummond Matthews of Cambridge University and L. W. Morley of the Geological Survey of Canada in 1963. The so-called Vine-Matthews' theory indicates that the magnetic anomaly stripes are caused by strips of magnetic rock on the sea floor being magnetized normally (with a direction like today's magnetic field) or reversely (opposite to today's field). The rock presumably became magnetized as it welled up hot and then cooled through the Curie point as the material moved away from the axis in both directions. It cooled at the mid-ocean ridge axis as the earth's magnetic field alternated its polarity in past geologic ages. By knowing when the magnetic field reversed its polarity, one can determine the age of the rock; by knowing how far a particular magnetic stripe is from the axis, one can determine the velocity of sea-floor spreading (Vine and Matthews, 1963).

By plotting on a map all of the directions in which the sea floor is moving, it becomes apparent that large sections of the earth's surface move as a unit or as a plate (Fig. 1-3). These plates include both sea floor and continents (plate tectonics). It is clear that if all the plates are moving away from the ridges, they will be bumping into one another unless some of the plates override other plates. That seems to be happening around the edges of the Pacific Ocean where the continental parts of plates override the sea floor. A deep trench is found near continents where the sea floor turns down (is subducted). Many earthquakes are located at subduction zones (Fig. 1-4). A cross-sectional view would reveal that earthquake epicenters extend down in a plane under the continents to a depth of several hundred kilometers. This is in contrast to mid-ocean ridge epicenters, which are quite shallow, i.e., probably not more than a few tens of kilometers.

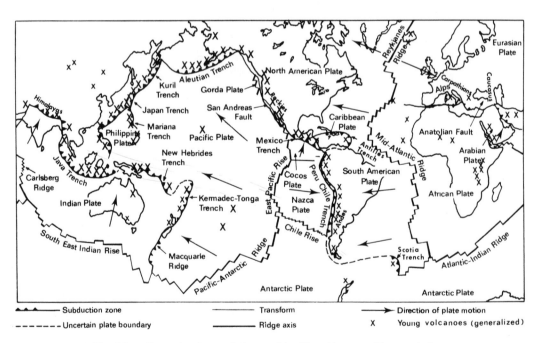

Fig. 1-3. Tectonic plates of the world with mid-ocean ridges and rises, fracture (or transform) faults, subjuction zones, and young volcanoes. After Uyeda, 1978.

THE EVOLUTION OF THE DEEP-OCEAN FLOOR

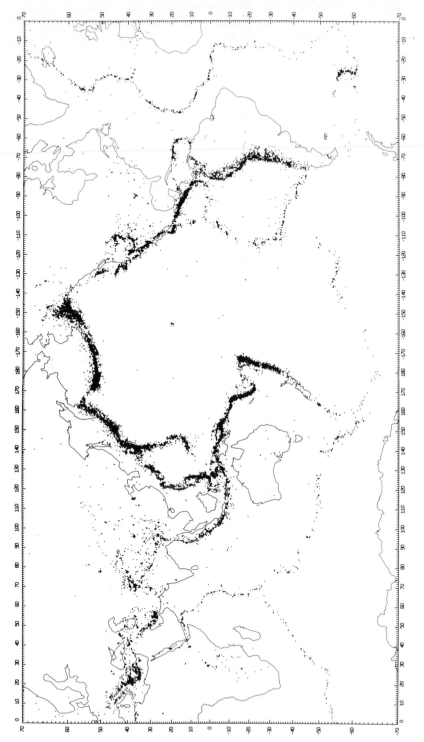

SEISMICITY OF THE EARTH, 1961-1967, ESSA, CGS EPICENTERS

DEPTHS 0-100 KM.

Fig. 1.4.    Location of shallow earthquakes of the world for the years 1961–1967. Barazangi and Dorman, 1969.

9

Thus, the plate tectonics theory provides us with a coherent picture of the generation of the oceanic basaltic basement. It does not indicate how the continents originated nor how sediments of the oceans came about.

The sediments of the oceans arise from several sources and are generally classified as terrigenous or pelagic. Terrigenous sediment comes from land either through rivers emptying into the ocean, through coastal erosion, through dust carried to sea by wind, or as rocks transported by icebergs. Pelagic sediments can originate as biproducts of life in the sea, especially planktonic life. How these sediments vary from place to place and how they are moved by ocean currents and so-called turbidity currents (heavy suspensions of sediments moving downslope with bottom water) are subjects that will be discussed in other chapters of this book. Although most sediments are relatively stable and, once deposited on basaltic basement, stay with that basement as the sea floor spreads, other surface sediments are moved about in a very energetic environment on the ocean bottom.

## THE MID-OCEAN RIDGE

Since the axis of the mid-ocean ridge is where the tectonic plates of the earth are created, it is a place of very special interest to earth scientists. Starting in 1971 a project to explore a small portion of the Mid-Atlantic Ridge was undertaken. This project, with the acronym FAMOUS, was planned to allow scientists to compare the sea floor to

Fig. 1-5.    A drawing of the topography across the 3000-m-wide inner rift valley of the mid-Atlantic Ridge (copyright by National Geographic Society). Heirtzler, 1975.

THE EVOLUTION OF THE DEEP-OCEAN FLOOR

places on land that were thought to be similar geologically. Accordingly, it was decided to employ manned submersibles for the first time in the deep ocean to make a detailed geologic map of an area a few kilometers on a side. One American and two French research submersibles were used in 1974. Prior to the dives, numerous surveys of the area were made, some with new and special instruments, so that the diving scientists could relate the features they saw on the sea floor to the larger picture of plate tectonics. During the course of these studies, considerable information about the ocean-bottom geology, water, and biota was obtained. Thus this is one area where some tentative comments can be made about the ocean floor as an ecosystem. Since the diving scientists were geologists and geophysicists rather than physical oceanographers or biologists, these observations are likely to be somewhat simplistic, but they do provide a first look and are suggestive of more thorough studies that could be done.

The area chosen was 900 km southwest of the Azores. The water depth was about 3000 m and the pressure on the sea floor about 360 kg/cm$^2$. The inner rift valley floor was found to be 3-5 km wide and bounded on the east and west by walls 500 m high (Fig. 1-5). This section of rift valley is intersected and offset by two fracture zones separated by about 50 km. The American dives, which form the basis of comments here, were made in the inner rift valley, between the two fracture zones. Some dives were also made in the south fracture zone. The inner rift valley is a field of exposed volcanic rock not unlike the field of solidified volcanic rocks that one finds in recent Hawaiian eruptions or elsewhere. On the sea floor the lava forms are elongated and shaped as if squeezed from a giant tube of toothpaste with an orifice a meter or so in diameter (Fig. 1-6). These forms are called pillow lava and some are piled into small hills 100 m

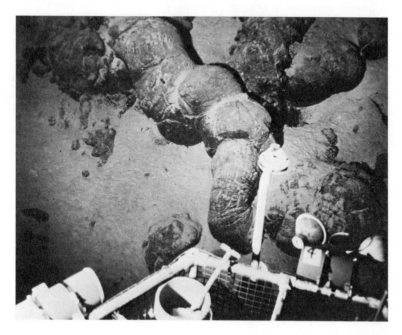

Fig. 1-6.    Pillow basalts in FAMOUS area of Mid-Atlantic Ridge. Heirtzler, 1975.

or so high and 100–200 m wide along the axis of the rift valley. The volcanic terrain is cracked by fissures paralleling the rift axis. The width of the fissures ranges from a few centimeters to a few meters. Studies suggest that the small volcanic hills are created over a period of about 10,000 years and that the age of inner rift floor is from less than 10,000 years near the exact axis to about 200,000 years near the walls. In some places between the individual pillows, thin sediments of up to a meter in thickness have collected. However, the average thickness of sediment is very small. Shallow, weak earthquakes with depths less than a few tens of kilometers occur frequently here, many times per hour within a distance of a few kilometers of the axis. During the 80 hours spent on the bottom by U.S. scientists, no earthquakes were known to have occurred in their range of vision and no ongoing volcanic activity or heated water was found.

The movement of water on the ocean floor was of great practical concern since the submersible could only move at a slow speed and would be out of control if regional or local currents of greater than a few km/h existed. Current meters were deployed and they showed that currents never exceeded one km/h, being mainly tidal and in phase with tides as measured in the Azores. Where the rift intersected the fracture zones, the water depth was slightly greater and a basin was formed. Here water of slightly different characteristics was apparently trapped, but some of it spilled over a sill into the rift valley. When the submersible approached nearly vertical walls, it would usually be pushed downward as if there were a slight waterfall down the face. Time did not permit exploring these phenomena to see if they existed over a large area or for how long they persisted. The driving force behind these water movements is not understood, although it is clear that there is a relationship between the shape of the sea floor and the movement of water.

Since the biota of the sea floor is entirely dependent on the local water conditions, it will naturally change as the character of the water changes from one part of the inner valley floor to another. To the nonspecialist most of the mobile forms of life here (fish, shrimp, sea urchins, octopus, sea cucumbers, etc.) seem very similar to shallow-water species. The nonmobile ones (corals, sponges, etc.) are much more dependent upon the local environmental conditions. They need a rock to which to attach themselves and a flow of water from which to derive nutrients. Corals, for example, were frequently found at the edges of small basins, such as near the fracture zone rift intersection where bottom water was spilling into the rift valley. Along the axis of the rift valley, the nonmobile forms were not randomly distributed over the rocks but seem to be collected in what euphemistically have been called "thickets" or "gardens." Some of these colonies contained several different types of species; some were flourishing, whereas others were entirely dead and starting to become covered with manganese oxide.

Since the first studies made on the Mid-Atlantic Ridge, other studies have been undertaken in the Caribbean, near the Galapagos, and near the Gulf of California. The two latter areas in the Pacific proved especially interesting because hot water was found issuing from the rocks of the mid-ocean ridge. These hot water springs support colonies of clams, crabs, tube worms, and other creatures. These colonies apparently obtain nutrients directly from the chemicals in the spring water and not from products of photosynthesis or plankton at the ocean surface.

Seamounts are a unique feature of the deep sea. They may occur in isolation or as members of a chain. One of the best examples of a chain in the Atlantic is the New England seamount chain which extends for more than 1000 km eastward from the coast of New England towards the Mid-Atlantic Ridge. They rise 2000-3000 m from a seabed that is up to 5500 m deep. They have a spacing of typically 50-100 km but Corner Rise is some 300 km further east than the nearest seamount. Preliminary exploratory dives by manned submersibles have taken place on the tops of eight of these peaks. The depth limit of the submersible prohibits its going deep along the sides. On seven of the eight seamounts, only a single dive has been made. Such limited direct viewing, plus the fact that they have not been adequately mapped by echosounders, point out that the initial observations made may not be typical of the seamount as a whole. These few dives did, however, permit direct observation of these deep-ocean features for the first time.

The most important general observation made was that the seamounts are not gently rounded forms, as is suggested by most bathymetric maps. They are all extremely rugged features that would have been difficult to climb had they been on dry land. A diverse group of rock forms is present—many are too large to photograph (i.e., larger than about 2 by 3 m). As one proceeds from Corner Rise on the southeast towards Mytilus Seamount on the northwest, the number of glacially transported rocks clearly increases, and in fact few, if any, such rocks could be seen south of Rehoboth Seamount at about 37°N latitude.

In places where the local slope was low enough to hold sediment, there are waves of a meter or two in wavelength in the sediment produced by currents (Fig. 1-7). In

Fig. 1-7. Sediment waves on a New England seamount. The wire mesh tray in the foreground is part of the submersible *Alvin*.

some areas, motions of the submersible showed that the direction of water current changed measurably over rather small distances. Many live coral were seen moving back and forth with the current, especially over ledges. Manganese oxide is ubiquitous as a thin coating or as slabs up to 10 cm thick (Fig. 1–8). There are large areas of dead coral that are being covered with manganese oxide, suggesting that the environmental conditions suddenly became unfavorable for them during the recent geologic past.

At least four of the New England seamounts are believed to have been at sea level at one time, but to have subsided. The tops of these four contain shallow-water coral that is of approximate Eocene (40 million years) age. None of these seamounts is within 1000 meters of the sea surface today.

Surface measurements with seismic instruments show Mytilus Seamount to have a 300-meter-thick transparent cap sitting on its summit at a depth of about 3000 m. The submersible easily recovered pieces of this coral. Mytilus was the only place where large spongelike forms were observed. The forms were white and had a diameter and height of 50 cm or more. The interior, when viewed from the top, greatly resembled a huge swirl of whipped cream. These forms have not yet been identified.

## ABYSSAL PLAINS

The rugged igneous rock forms that are created at the mid-ocean ridges become covered by an increasing thickness of sediment as the sea floor spreads away from the axis. In some areas, where the source of sediment from land or sea is plentiful, the thickness of the sediments on older sea floor may be 10 or more kilometers. If there is a biogenic component to the sediment, if the conditions of temperature and pressure are right, and if the sediments are sufficiently porous but with occasional nonporous caprock, then gas or petroleum might form in commercial quantities deep within the sediments near the continents — but such deposits have not yet been confirmed off the continental shelf.

As the sea floor moves away from the mid-ocean ridge axis, the outer layers of the crust slowly cool and shrink, causing the sea-floor rocks and their sediment cover to be deeper than the sea floor in the axial region. By the time the sea floor has reached an age of about 100 million years, the sediments are sufficiently thick to blanket nearly all relief. Except for scattered seamounts, plateaus, or fracture zones, the sea floor, which is 5 to 6 km deep, is now thought to be monotonously flat between the mid-ocean ridges and the continental margins.

These abyssal plains are too deep to be reached by manned submersibles. Our knowledge of the small-scale features has been obtained from bottom cameras suspended from cables and from any bottom material or solid specimens that have been dredged up. Accordingly, our knowledge of small-scale features can best be described as primitive for these parts of the ocean floor.

At latitudes higher than about 40 degrees in both Northern and Southern hemispheres, one finds glacially transported rocks of up to a meter or so in size resting on the ocean floor. These rocks were deposited by melting icebergs about 14,000 years ago, so they have accumulated a small "dusting" of sediment since that time. The frequency with which one finds these rocks increases as one proceeds landward toward the

Fig. 1-8. Manganese-covered biogenic material on Manning seamount of the New England seamount chain. The area of the photograph is about 1.00 by 1.25 m.

source areas of the rocks. At about 40°N off New England, for instance, it is not unusual to find such a rock each few tens of meters.

In places where appreciable and persistent water currents extend to the sea floor, waves or furrows in the sediment are found. These features, measuring a few meters across, a few tens of centimeters in depth, and many kilometers or tens of kilometers in length have been found in the western and northeastern North Atlantic. New side scanning sonar mapping devices are revealing additional bottom wave forms caused by the action of ocean currents. New studies are seeking to understand the transport of bottom sediments.

Near ships' regular traffic routes one now frequently photographs man-made artifacts, such as glass or plastic bottles, cloth or paper, and even objects made of rapidly biodegradable materials such as wood, on the ocean floor. It is now routine to see man-made debris floating on the ocean surface far from land — a condition that did not exist a few years ago — so it is not surprising that we find some of the same material on the ocean floor.

On the abyssal plain floor one also finds manganese nodules or pavement in certain areas. Why they are found in some areas and not others is not entirely clear. An area southeast of Hawaii and another off Florida are well-known places where this material is found, but it is found in lesser concentrations in most parts of the oceans (see Chap. 6, this volume, by Heath).

The huge expanses of the abyssal plains are modified slightly by biological activity —either bottom-dwelling or swimming creatures. Larger creatures dig holes a few meters in size and smaller creatures dig, eat, excrete, or otherwise churn up the bottom to depths of a few tens of centimeters each few days or years.

If we can take a lesson from recent manned dives on the mid-ocean ridge, we will likely find small-to-intermediate scale ecosystems on at least certain parts of the floor of the abyssal plains when we are able to go there and observe first-hand.

## CONTINENTAL MARGINS

The transitional regime between the deep abyssal plains and the shoreline is called the continental margin. It is generally divided into three regions: the continental shelf, the continental slope, and the continental rise. The continental shelf is closest to shore and is typically no deeper than 200 m. There is a marked drop to greater depths at the seaward edge of the shelf. The area beyond the "shelf break" is called the continental slope. The continental slope approaches the abyssal plain floor in a region called the continental rise. Beneath the sea floor, the continental shelf and slope are structurally part of the continent. The continental shelves are of special interest to commercial fishermen who fish on the bottom for lobsters or other creatures, and to petroleum companies in some areas.

The continental shelves and slopes occupy about 15% of the oceans of the world and because of their potential economic significance have played a key role in Law of the Sea negotiations in the United Nations. Countries such as those bordering the deep ocean trenches around the Pacific generally have very narrow continental shelves, but others, such as New Zealand, have extensive shelves.

Fig. 1-9. Typical sea floor photo in sediment covered areas showing track left by bottom dweller and occasional hole by burrowing organisms. Water depth is 2500 m and photograph covers about 0.5 by 1.0 m.

THE EVOLUTION OF THE DEEP-OCEAN FLOOR

A few of the larger continental shelves include islands but most are quite feature-less (Fig. 1-9). There are some canyons cutting shelves where rivers empty onto them. These canyons usually continue across the shelf and incise the continental slope. In other places along the slope, some canyons seem to be unrelated to any adjacent river and their origin is obscure.

Throughout geologic ages, sea level is believed to have changed by several hundred meters. When sea level was at a lower stand than at present, some of the present sea floor near shore was exposed. Accordingly, beach rocks and even some of the remains of man-made artifacts (piles of shells, housing structures, etc.) are found on the ocean floor on near-shore parts of the continental shelf. The uplift or subsidence of the land near shore due to geologic forces may also contribute to retreat or advance at the shore-line during geologic time. For example, about 150 million years ago when the Atlantic was a young and narrow sea, the shoreline was several hundred kilometers further east than at present off eastern North America. At that time an extensive coral bank existed on the shelf, but it has since been covered with a thick sedimentary layer. In other areas the most recent sediment has been removed to expose very old sedimentary layers. In very few places have the continental shelf and slope been drilled to recover actual samples of the sub-bottom material, but commercial, governmental, and scientific drilling have taken place at some locations on the shelf. Drilling on the continental slope is expected to take place in the next few years.

It is clear from the brief descriptions given here of the different types of deep-sea areas that the ocean still contains many unexplored features. Marine scientists have come to recognize the floor of the ocean as a place with diverse characteristics that are just beginning to be studied as an ecosystem.

# REFERENCES

Ballard, Robert D., and Van Andel, Tjeerd H., 1977, Project FAMOUS: Operational techniques and American submersible operations: *Geol. Soc. Amer. Bull.*, v. 88, p. 495–506.

Barazangi, M., and Dorman, J., 1969, Word Seismicity Map Compiled from ESSA Coast and Geodetic Survey Epicenter Data, 1961–1967: *Seismol. Soc. Amer. Bull.*, v. 59, p. 369–380.

Heezen, Bruce C., and Hollister, Charles D., 1971, *The Face of the Deep*: Oxford University Press, New York, 657 p.

Heirtzler, J. R., 1975, Man's First Voyages down to the Mid-Atlantic Ridge. Where the earth turns inside out: *National Geographic*, v. 147, p. 586–603.

——, and Grassle, F., 1976, Deep Sea Research by Manned Submersible: *Science*, v. 194, p. 294–299.

Menard, H. W., and Smith, S. M., 1966, Hypsometry of ocean basin provinces: *J. Geophys. Res.*, v. 71, p. 4305–4325.

Uyeda, Seiya, 1978, *The New View of the Earth*: San Francisco, Freeman and Co., p. 194.

Vine, Fred J., and Matthews, Drummond H., 1963, Magnetic Anomalies over Oceanic Ridges: *Nature*, v. 199, p. 947–949.

Stanley V. Margolis,
Peter M. Kroopnick,
and William J. Showers
Department of Oceanography
University of Hawaii
Honolulu, Hawaii 96822

# 2

# PALEOCEANOGRAPHY: THE HISTORY OF THE OCEAN'S CHANGING ENVIRONMENTS

Paleoceanography is an exciting new branch of earth sciences that deals with how the oceans have evolved to their present configuration in terms of circulation patterns, temperature structure, chemical composition, and organic productivity. Studies of this nature have been made possible by the recovery of long, well-preserved deep-sea sediment cores containing calcareous remains of oceanic plankton and benthos. The Deep Sea Drilling Project and the U.S. oceanographic community are responsible for the efforts and initiative in obtaining these cores.

Recent advances in technology and in our understanding of the distribution of the stable isotopes of oxygen and carbon in the hydrosphere, atmosphere, and biosphere have enabled paleoceanographers to quantify their interpretations of the environmental conditions that existed in the oceans during critical times in earth's climatic history. Detailed biostratigraphic age dating and paleomagnetic stratigraphy have permitted correlations across and between ocean basins so that we can obtain a picture of what the oceans were like at any given time in the past, as well as allowing us to look at the changes that have occurred over tens of millions of years. Isotopic and ecological techniques now are available to determine surface- and bottom-water temperatures, global temperature gradients, thermocline depth, salinity, oxygen content, and organic productivity rates in the ancient oceans by analysis of marine sediments.

The evolution of the present oceanographic and climatic regime appears to be controlled by a number of factors including the degree of buildup of polar ice, especially in the Antarctic; plate tectonics and sea-floor spreading; and extraterrestrial events during certain critical periods. This chapter will discuss the significant paleoceanographic changes that occurred at the Cretaceous/Tertiary boundary (65 mya), Eocene/Oligocene boundary (38 mya), the early-middle Miocene (14 mya), and the late Miocene (6 mya). Each is associated with major changes in world climate and the oceanic and terrestrial biosphere. We will show how an understanding of paleoceanographic trends will help identify changes occurring in the present-day oceans that will affect sea level, world climatic patterns, biogeographic distributions, and ultimately humanity's future.

# INTRODUCTION

During the last decade of oceanographic and geological investigations, it has become quite evident that the world's oceans have undergone a series of progressive changes that have altered their physical boundaries, circulation patterns, and thermal structure, changes which in turn have affected their chemical and biological regimes. With our current efforts to understand the causes of climatic changes on the long and short time scale, out of obvious self-interest, we are turning more to the oceans to obtain information about the interactions of the air and sea that produce our present climate, and to ocean sediments to obtain a record of past climatic trends. The goal of this type of research is to identify the various stages that led up to our present global oceanic and atmospheric climatic system so that we can use past trends to predict future changes.

STANLEY V. MARGOLIS, PETER M. KROOPNICK, WILLIAM J. SHOWERS

This chapter will discuss the materials and techniques developed thus far in our attempts to reconstruct what the ancient oceans were like and will summarize our present understanding of the major events that have affected the evolution of the world's oceans. We will also take glimpses of physical and chemical changes that were occurring in the oceans during certain critical intervals in earth history. Finally, we will show that the ice ages were not isolated events but rather a manifestation of a series of paleocean-ographic events that began in the Early Cenozoic with the glaciation of Antarctica and the concomitant cooling of the world's oceans.

The term paleoceanography, initially coined by W. R. Reidel in the early 1960s, was used to describe the interpretation of data gathered from the distribution of fossil plankton from piston cores in terms of species diversity and latitudinal temperature zonations (Funnel and Reidel, 1971). Studies of this kind were limited by the availability of piston cores of sufficient length and age. The Deep Sea Drilling Project, however, has greatly expanded the opportunities for study with its coverage of the oceans, both in terms of oceanic regions and biostratigraphic age. This, together with our increasing knowledge of the evolution of the ocean basins by plate tectonics and sea-floor spreading, has enabled us to enter a new phase of quantitative paleoceanography. The materials that are used for these studies are the planktonic and benthic components of deep-sea sediments as well as the authigenic and detrital mineral constituents. The initial data base for such studies consists of the Deep Sea Drilling Project reports and the research papers published on these cores. The techniques are classical biostratigraphy for age determinations and correlations; quantitative micropaleontology for paleo-environment and changing watermass characteristic identification (see Sachs *et al.*, 1977 for a review); and oxygen and carbon isotope analyses for paleotemperature, paleoproductivity, and isotope signature characterization for chronology and correlations.

## METHODS

The time period covered in this chapter will be the last 75 million years of earth history in general, and the Cenozoic in greater detail. The data that will be used to reconstruct the temperature, productivity, and circulation patterns of the past will be those of stable oxygen and carbon isotopes of calcium-carbonate-secreting plankton and benthic fossils. This is because a record of the fluctuations in these environmentally sensitive isotopes in the world's oceans is found in the fossil remains of benthic and planktonic foraminifera and nannofossils. The oxygen isotopes record surface- and bottom-water paleotemperatures, changes in salinity, and degree of buildup of continental ice sheets. A quantification of the relative amounts of each of these parameters is dependent upon an understanding of the time period and oceanic area in which the sediments were deposited, as will be discussed in more detail below. Carbon isotopes are believed to reflect changes in the global carbon budget, paleocirculation, and surface-water productivity. During the last 30 years, determining stable isotope compositions of biogenic calcium carbonate has become a standard tool for paleoceanographic and paleoclimatic studies. Stable isotope paleoceanography began with the classical work of Urey (1957), in which it was suggested that the oxygen isotope ($^{18}O/^{16}O$) composition of calcium carbonate or the $\delta^{18}O$, defined as $[R\ Sample/R\ Standard - 1] \times 10^3$, where $R = {}^{18}O/^{16}O$ or $^{13}C/^{12}C$, would reflect the ratio of the solution in which it was growing, which varies as a function of temperature. Experimental work by Urey et al. (1951), Epstein et al. (1951), and others, later confirmed this suggestion and demonstrated its applicability to carbonate-secreting biota and its potential use as a paleotemperature tool. One of the first questions the oxygen isotope paleotemperature technique was used to answer was that of the temperature regime during the mass extinctions at the Cretaceous/Tertiary time boundary (Lowenstam and Epstein, 1954; Urey et al., 1951). An up-to-date discussion of this and the latest data on this interesting time interval will be presented below.

The $\delta^{18}O$ of fossil calcium carbonate can be determined with a precision of 0.1 per mil (parts per thousand), which corresponds to an estimated paleotemperature accuracy of about 0.5°C (Savin, 1977). A discussion of the mass spectrometric and analytical techniques used to determine $\delta^{18}O$ and $\delta^{13}C$ can be found in McCrea (1950) and more recently in Shackleton et al. (1973). Sample preparations vary with the types of organisms studied. The basic assumption of the technique is that the organisms under consideration will secrete their calcium carbonate in isotopic equilibrium with seawater or with the medium in which they are growing. In order to determine the temperature during growth, it is also necessary to know the isotopic composition of the seawater or growth media. It must also be assumed that the original isotopic composition of the fossil shells remains unaltered, i.e., that it was not subject to recrystallization, exchange, or crystal overgrowths that would change the original isotopic composition. With regard to the first assumption, it has been demonstrated that some species of marine mollusks and foraminifera deposit their calcium carbonate at isotopic equilibrium (Emiliani, 1966; Epstein et al., 1953).

## OXYGEN AND CARBON ISOTOPES AS PALEOCEANOGRAPHIC INDICATORS

### Oxygen Isotopes of Planktonic Foraminifera

Emiliani (1954, 1955, 1958, 1966, 1971, 1972) was the first to study the oxygen and carbon isotopic composition of benthic and planktonic foraminifera in order to conduct paleoceanographic investigations of Pleistocene sediments. Assuming equilibrium in certain species of planktonic foraminifera that inhabit surface waters, Emiliani determined that the $\delta^{18}O$ of the tests reflected a combination of temperature and seawater isotopic composition as mentioned above, with two thirds of the glacial/interglacial $\delta^{18}O$ difference being due to temperature change, and one third due to changing $\delta^{18}O$ of seawater caused by buildup of ice caps and consequent enrichment of the oceans in $^{18}O$. More recently Shackleton and Opdyke (1973) and others have argued that two thirds of the $\delta^{18}O$ difference is due to ice volume change and only one third is due to temperature. This topic is still the subject of much debate, as summarized by Savin (1977).

Another problem with paleotemperature determinations of planktonic foraminifera is that these plankton can change their depth habitat as temperature and other water column hydrographic properties change, by adjusting their density to stay within a water mass, thus optimizing feeding or other vital functions (Emiliani, 1954). Uncertainties of depth habitat further complicate paleotemperature estimates, especially those of extinct species. Studies of planktonic foraminifera from plankton tows and sediment traps indicate that some species do not deposit their tests in isotopic equilibrium (Fairbanks, et al., 1980; Kahn, 1979; Williams et al., 1979a, b). Recent core top analyses have also shown disequilibrium when isotopic values are compared to the hydrography of the overlying water column (Berger et al., 1978b; Shackleton and Vincent, 1978; Williams and Healy-Williams, 1980). Several studies have investigated the "vital" effect in foraminifera and several explanations for the isotopic disequilibrium have been advanced. While differing in their conclusions, most workers agree with the observations discussed below.

Ontogenetic changes in metabolic rates produce different $\delta^{18}O$ values for different size fractions of specimens of the same species (Curry, 1977; Curry and Mathews, 1979). Buchard and Hansen (1977) and Erez (1978) have proposed incorporation of metabolic $CO_2$ in foraminiferal calcite due to the presence of symbiotic zooxanthellae, which may account for isotopically lighter tests. Fairbanks et al. (1980) found that nonspinose, symbiont-barren planktonic foraminifera are close to oxygen isotope equilibrium while spinose symbiont-bearing forams have an average -0.35‰ oxygen depletion. This agrees with Angell's (1979) conclusion that the calcium and carbon ions used for calcification in the symbiont-barren benthic foraminifera Rosalina are extracted from seawater and are not found in the protoplasm prior to calcification. In symbiont-bearing foraminifera, Muller (1978) has shown that up to 10% of the carbon fixed by algal photosynthesis in the symbiotic benthic foraminifera Amphistegina may be recycled between symbiotic and host. Assuming a metabolic $CO_2 \cdot {}^{13}C$ of -14‰, Craig (1957) estimates 30% of carbon of planktonic foraminiferal calcite is metabolic, while Williams et al. (1977) using -25‰ for $\delta^{13}C$ metabolic $CO_2$ calculates 4-18% metabolic $CO_2$ incorporation. Similarly, Kahn (1979) calculates 3-4% incorporation. The exact percentage of metabolic $CO_2$ incorporated in planktonic foraminiferal calcite cannot be precisely determined until they can be successfully cultured. It is possible that the percentage of metabolic $CO_2$ incorporated could change with increasing age or depth habitat. Other explanations for disequilibrium are: ontogenetic sinking, which would cause the final isotope value to reflect the average depth where most of the $CaCO_3$ was secreted; and seasonal temperature cycles, during which certain foraminifera record seawater temperatures of only a particular season (Deuser, 1979; Williams et al., 1979a).

In the absence of laboratory data for cultured planktonic foraminifera, field studies have attempted to document isotopic disequilibrium by analyzing foraminifera collected in the water column and at the top of sediment cores. In the past these studies have usually suffered from the lack of synchronous hydrographic data. Recent plankton work such as Williams et al. (1979a,b) and Fairbanks et al. (1980) indicates that the $\delta^{18}O$ depth stratification of certain planktonic species of foraminifera as proposed by Emiliani (1954, 1971) and Savin and Douglas (1973) is essentially correct if seasonal fluctuations are taken into account. However, when $\delta^{13}C$ is plotted against $\delta^{18}O$ of Holocene core top foraminifera (Berger et al., 1978b), no apparent correlation occurs with the inferred depth habitat (from $\delta^{18}O$ values). There does appear to be a trend of increased variability in shallow-water species and an ontogenetic change in oxygen-isotopically-light forms. The wide scatter of $\delta^{13}C$ values in recent foraminifera indicates that $\delta^{13}C$ is not precipitated in equilibrium with seawater at all times, and that this non-equilibrium $^{13}C$ precipitation does not appear to have a constant offset in foraminifera.

## Carbon Isotopes in Planktonic Foraminifera

Parker (1964) and Smith and Epstein (1970) suggested that animals reflect the isotopic composition of their food. Several other investigators have suggested that carbon isotopes may be controlled by trophic relationships, productivity, or upwelling (Berger et al., 1978a; Degens, 1979; Kahn, 1979; Killingley and Berger, 1979; Weiner, 1975; Williams et al., 1977); however no measurement of productivity or the isotopic composition of other planktonic trophic levels were made in previous plankton studies. Equilibrium water-isotope values are generally calculated from sparse hydrographic data and then correlated with the isotopic value of the planktonic samples to quantify isotopic disequilibrium. Species-dependent biogenic fractionation of $\delta^{18}O$ is not as great a problem as $\delta^{13}C$ biogenic fractionation (Kroopnick et al., 1977). This may be because $\delta^{13}C$ depth profiles in the water column are known to vary over space and time as a result of local conditions and water mass mixing. Calculated equilibrium water-isotope values do not take this variation in account.

In the present-day oceans, the $\delta^{18}O$ of dissolved inorganic carbon varies from about +2‰ at the surface to below -0.5‰ in the oxygen minimum (500 m) of the North Pacific. Bottom-water values are generally between -0.2 and +0.2‰ (Kroopnick, 1974a, 1980). The isotopic composition of dissolved inorganic carbon in surface- and near-surface water is controlled by several processes: (1) interaction with atmospheric $CO_2$, (2) in situ production of $CO_2$, (3) oxidation of organic matter and dissolution of carbonate, and (4) overall rate of deep-water inorganic carbon introduced into surface water. Because of the preferential uptake of isotopically light carbon due to metabolic activities of phytoplankton in the upper surface water, a $\delta^{13}C$ minimum coincides with the dissolved oxygen minimum in the water column, whereas maximum $\delta^{13}C$ values occur in the surface layer (Kroopnick, 1974a,b; Kroopnick et al., 1977; Fig. 2-1).

Although the carbon isotopic compositions of planktonic calcareous organisms are modified by vital effects, they are ultimately dependent on the $\delta^{13}C$ of the dissolved inorganic carbon of the ambient seawater. Figures 2-2 and 2-3 show the latitudinal variation of surface waters for $\Sigma CO_2$ and $\delta^{13}C$ of the $\Sigma CO_2$ along 150°W during the HUDSON 1970 Expedition (Kroopnick et al., 1977). They show that the dynamics of the equatorial system (upwelling and downwelling) have a profound influence on the distribution of $\delta^{13}C$ (note the dip in $\delta^{13}C$ at the equator and the increases at 10°N and 10°S). Correcting these raw data for the changes in sea surface temperature and constructing a model that allows equilibration between sea surface $CO_2$ and atmospheric $CO_2$ produces the curve shown in Fig. 2-4. At high latitudes, the measured $\delta^{13}C$ of atmospheric $CO_2$ is enriched by up to 2‰ in relation to the calculated equilibrium value. At these latitudes a net transfer of $CO_2$ into the ocean is occurring. $^{12}C$ must then be entering the sea in preference to $^{13}C$.

STANLEY V. MARGOLIS, PETER M. KROOPNICK, WILLIAM J. SHOWERS          23

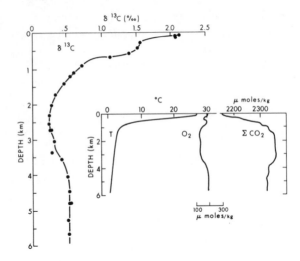

Fig. 2-1.    Composite of $\delta^{13}C$ of $\Sigma CO_2$ and temperature, dissolved oxygen and $CO_2$ versus depth for antipode 15 Station 6 (17°S-172°W). From Kroopnick *et al.*, 1977.

This uptake of $^{12}C$ most likely occurs during photosynthesis by phytoplankton which preferentially remove $^{12}C$ from the $HCO_3^-$ pool, reflecting a fractionation factor of approximately 20-25‰. In the mid-latitude areas photosynthesis is low and the net biological effect is for $^{13}C$-depleted organic carbon to be added to the atmosphere by respiration. Thus, the carbon isotope data indicate that invasion and evasion of $CO_2$ to and from the sea is at least partially controlled by biological productivity (Kroopnick *et al.*, 1977).

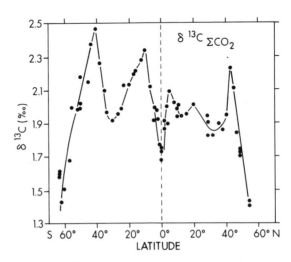

Fig. 2-2.    $\delta^{13}C$ of $\Sigma CO_2$ along 150°W in April and May 1970. From Kroopnick *et al.*, 1977.

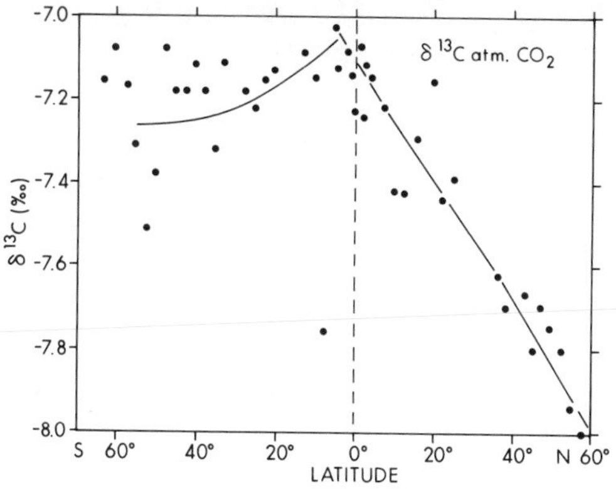

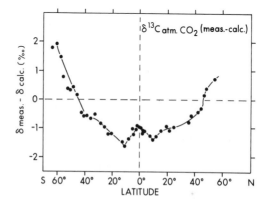

Fig. 2-3. $\delta^{13}C$ of atmospheric $CO_2$ along 150°W in April and May 1970. Kroopnick et al., 1977.

Fig. 2-4. $\delta^{13}C$ measured $- \delta^{13}C$ calculated for atmospheric $CO_2$ along 150°W in April and May 1970. From Kroopnick et al., 1977.

## Oxygen and Carbon Isotopes of Calcareous Nannofossils

The $\delta^{18}O$ and $\delta^{13}C$ of coccoliths (phytoplankton) have recently been investigated (Anderson and Cole, 1975; Berger et al., 1978b; Douglas and Savin, 1975; Lindroth et al., 1977; Margolis et al., 1975). Margolis et al. (1977) and Kennett et al. (1979) show that when nannofossils (coccoliths) are compared with planktonic and benthonic foraminifera, more information about paleoceanographic conditions can be obtained. The depth distribution of planktonic foraminifera, as discussed above, may fluctuate with changing marine conditions, but as much as coccoliths are primary producers, their depth distriburion is confined to the surface by light penetration. Unlike planktonic foraminifera, coccoliths can be grown in culture in the laboratory. The oxygen-isotopic composition of cultured coccoliths, although not in isotopic equilibrium, shows a definite temperature dependence which closely parallels that of equilibrium precipitation of calcium carbonate (Dudley, 1976; Dudley and Goodney, 1979; Dudley et al., 1980). Oxygen and carbon isotopes of calcareous nannofossils from deep-sea sediments similarly reveal that their isotopic values depart from thermodynamic equilibrium with oceanic surface-water temperatures; however, there is a good correlation with values for surface-dwelling planktonic foraminifera, indicating that they are both responding to the same variations in seawater temperature and isotopic composition.

## Interpretation of Carbon and Oxygen Isotopes
## of Calcareous Plankton

Several workers have proposed that $\delta^{18}O$ values of planktonic foraminifera generally reflect their depth habitat. Theoretically, if one could find a suite of calcareous organisms that calcified in equilibrium with respect to carbon and oxygen isotopes at various depths throughout the water column, then a paleo-$^{13}C$ profile and a paleothermocline could be reconstructed. However, the assignment of depth habitats to extinct species is not an easy task (see Berger, 1969; Goodney et al., 1980; Savin and Douglas, 1973; Shackleton et al., 1973). Temperature changes and food availability cause vertical migration in zooplankton, which along with horizontal advection could alter such depth rankings. By analyzing several species from the same core, reliable depth habitats have been determined. However, the $\delta^{13}C$ data for planktonic foraminifera can show large, species-dependent non-equilibrium fractionation effects (Goodney, 1977; Savin and Douglas, 1973; Shackleton et al., 1973). Large fractionations have also been observed recently by several workers within a single species as a function of size or age.

There is a relationship between the $\delta^{13}C$ of foraminiferal calcite and the $\delta^{13}C$ of $\Sigma CO_2$ in the water column, although foraminifera apparently do not grow in isotopic equilibrium. Recent calcareous nannoplankton similarly depart from carbon isotopic equilibrium, but their variations in $\delta^{13}C$ are also closely related to the $\delta^{13}C$-$CO_2$ of surface waters. Calcareous nannofossils from Cenozoic deep-sea cores, however, show values from +1.5-4.0‰, which are closer to isotopic equilibrium with respect to $\delta^{13}C$-$\Sigma CO_2$ than planktonic foraminifera. The differences in $\delta^{13}C$ values of these plankton are probably related to mechanisms of calcification and utilization of photosynthetically derived and metabolically mediated carbon reservoirs. Because some surface-dwelling planktonic foraminifera contain symbiotic photosynthetic zooxanthellae, they therefore would show $\delta^{13}C$ of $CaCO_3$ similar to coccoliths. Alternatively, both forams and coccoliths may use carbon from the same reservoirs, depending upon variations in environmentally controlled growth conditions, which affect productivity in a given region of the ocean.

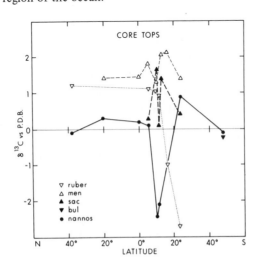

Fig. 2-5.   Plot of $\delta^{13}C$ versus latitude for piston core tops from Goodney et al. (1980); ruber = *G. ruber*, men = *G. menardii*, sac = *G. sacculifer*, bul = *G. bulloides*, nannos = nannofossils.

Productivity controls the $\delta^{13}C$ of the $HCO_3^-$ reservoir. When the $\delta^{13}C$ of foraminifera and coccoliths from Recent sediments are plotted with respect to latitude, an inverse relationship appears (Fig. 2-5). This relationship is expected if the $\delta^{13}C\text{-}CO_2$ of the ocean reservoir is to some extent affected by phytoplankton productivity. In areas of low productivity, phytoplankton and coccolith productivity would not appreciably alter the $\delta^{13}C$ of seawater, so that foraminiferal $CaCO_3$ would only be slightly enriched in $\delta^{13}C$. On the other hand, high phytoplankton productivity would greatly deplete the inorganic carbon reservoir in $^{12}C$ relative to $^{13}C$ by the incorporation of metabolic $CO_2$ into the coccolith $CaCO_3$ plates; consequently the foraminiferal $CaCO_3$ would become enriched in $^{13}C$. Therefore, positive $\delta^{13}C$ excursions in forams, and negative ones in coccoliths, indicate high productivity when they occur together in the present-day oceans. The reverse situation indicates low levels of productivity. Local upwelling conditions and selective solution will, of course, also complicate this interpretation.

## Oxygen and Carbon Isotopes of Benthic Foraminifera

According to Savin (1977), benthic foraminifera, like calcareous plankton, exhibit departures from isotopic equilibrium in deep-sea sediments (Duplessy *et al.*, 1970; Shackleton *et al.*, 1973; Vinot-Bertouille and Duplessy, 1973). Smith and Emiliani (1968), however, found that the $\delta^{18}O$ of Recent shallow-water benthic foraminifera were consistent with bottom-water temperatures. Shackleton (1974) found that at least one genus, *Uvigerina*, grows its test at or near equilibrium, and that even those benthic genera that were in disequilibrium were reasonably consistent in their fractionation, therefore permitting correlations within cores of different taxa (Savin, 1977). Woodruff *et al.* (1980) found that of 14 species of benthic foraminifera sampled from five equatorial Pacific stations, only *Uvigerina* sp. closely approached oxygen isotope equilibrium with bottom waters, although there were species-specific isotopic variations within the genus. None of the taxa of benthic foraminifera studied, however, exhibited $\delta^{13}C$ of their $CaCO_3$ in equilibrium with the dissolved $HCO_3^-$ found in bottom waters (Woodruff *et al.*, 1980).

Thus we see that although planktonic and benthic foraminifera and calcareous nannofossils all show species dependent "vital" effects that cause non-equilibrium precipitation of oxygen and carbon isotopes in their calcareous tests, there is still much valuable paleoceanographic information that can be obtained by analyzing calcareous deep-sea sediments.

## HISTORY OF OCEAN TEMPERATURES

Having looked at the present-day oceans, we now look at the fossil record preserved in deep-sea cores to see what we can learn about the oceans. Figure 2-6 summarizes oxygen isotope data for benthic and planktonic foraminifera and calcareous nannofossils for the past 75 million years based on analysis of deep-sea cores. During the time represented by this figure, there has been a general long-term cooling of the oceans; however, there are several periods of warming, followed by rapid temperature drops. These data

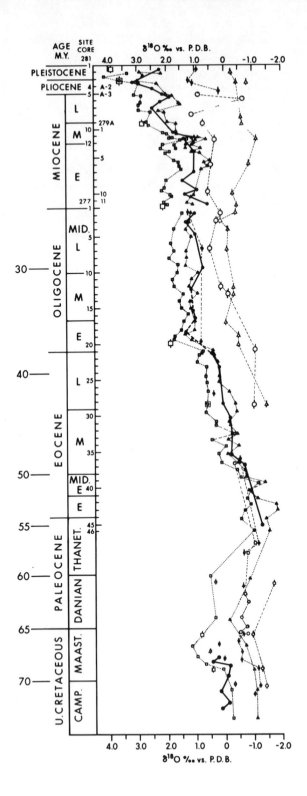

Fig. 2-6. Summary plot of oxygen isotope data for benthic foraminifera (squares), planktonic foraminifera (triangles), and nannofossils (circles and dots). Data are from: Douglas and Savin (1975), Margolis *et al.* (1975), Saito and Van Donk (1974), Savin *et al.* (1975). Solid dots with dark line in Late Cretaceous are nannofossils from E55-26. (See facing page.)

Site 305 Nannofossils, Shatsky Rise
Site 47 Nannofossils, Shatsky Rise
Site 167 Nannofossils, Megallan Rise from Douglas and Savin (1975)
Site 167 Planktonic foraminifera
Site 167 Benthic assemblages

Benthic and planktonic foraminifera data older than 60 million years are from Savin *et al.* (1975) and Saito and Van Donk (1974). The carbon and oxygen isotope record of calcareous biogenic compounds contained in sediments from DSDP Sites 277, 279, and 281 are from Shackleton and Kennet (1975b) using monospecific foraminiferal samples, and Margolis *et al.* (1975a) using mixed assemblages of nannofossils. Those data for Sites 47, 167, and 305 are from Douglas and Savin (1975), using mixed assemblages of benthic and planktonic foraminifera and calcareous nannofossils. From Margolis *et al.*, 1977.

have been reviewed recently by Margolis *et al.* (1977), Savin (1977), Kennett (1977), and Arthur (1979).

## Late Cretaceous-Paleocene

The bottom water paleotemperature record from the northern Pacific and Circum-Antarctic region can be summarized as follows: In the North Pacific after a late Maastrichtian cooling, a warming trend of about 5°C occurred across the Cretaceous/Tertiary time boundary. This was followed by a continued warming of surface and bottom waters of another 5°C throughout the rest of the Paleocene. In the Circum-Antarctic region, the Late Cretaceous cooling of deep water was synchronous with, and similar in magnitude to that occurring in the North Pacific. Equatorial waters, however, exhibited a much smaller change during this interval. According to Saito and van Donk (1974), Douglas and Savin (1975), and Savin (1977), the paleotemperature histories of the South Atlantic and North Pacific show similar trends across the Cretaceous/Tertiary boundary. In DSDP sediments in both of these areas, surface and bottom waters dropped a few degrees in early-middle Maastrictian time (Fig. 2-5) followed by a warming through early Paleocene time. The temperature drop near the Cretaceous/Tertiary boundary is small compared to other coolings in the Cretaceous (Douglas and Savin, 1975). The data presented by Margolis *et al.* (1977) show that the Cretaceous/Tertiary boundary appears actually to fall during a time of significant global warming of bottom and surface waters by about 5°C. Since the isotope data shown in Fig. 2-5 represent a composite of several DSDP sites, the exact time interval of this warming cannot be estimated. Recent isotope work on foraminifera by Boersma *et al.* (1979) on Cretaceous and Paleocene sediments from several Atlantic Ocean DSDP sites confirm the significant warming across the Cretaceous/Tertiary boundary. Site 384 in the western North Atlantic, which contains a relatively complete carbonate section across the Cretaceous/Tertiary boundary according to Boersma *et al.* (1979), shows the temperature rise to have actually occurred before the basal Tertiary zone, so that 3°C of the 5°C temperature rise in the early Paleocene must have occurred suddenly, during the transition. This temperature rise was more pronounced in bottom waters and in samples from higher latitudes, according to Boersma *et al.* (1979). Two other rises in surface-water temperature between 59 and 62 mya, were also found in the Paleocene of the Atlantic by Boersma *et al.* (1979). Temperatures during the later Paleocene were generally cooler, with no significant change across the Paleocene/Eocene boundary.

## Eocene-Oligocene

In the North Pacific and Circum-Antarctic region, the Early Cenozoic warming trend carries through until the mid-early Eocene (Fig. 2-5). Surface-water paleotemperatures at middle and high latitudes show a similar warming trend. Equatorial surface waters, however, exhibited a much smaller change during this interval. During the remainder of the Eocene, a steady parallel decrease occurred in both bottom- and surface-water temperatures. Close to the Eocene/Oligocene boundary, in subantarctic waters, there was at first a slight warming in bottom-water paleotemperatures and then a sudden cooling in the earliest Oligocene. The temperature drop of about 4–5°C occurred during a period inferred to be about 100,000 years on the basis of sedimentation rates at DSDP Site 277 (Kennett and Shackleton, 1976). A parallel drop in surface-water temperature also occurred within the subantarctic, as is evident from oxygen isotope ratios of planktonic foraminifera and calcareous nannofossils (Margolis *et al.*, 1975).

In the equatorial and northern Pacific, the early Oligocene temperature drop in bottom waters appears to have been equally abrupt, though suitable data are not yet available because of wider sampling intervals (Savin, 1977). In contrast, equatorial surface-water paleotemperatures do not show a corresponding dramatic temperature drop. Therefore the early Oligocene cooling seems to have been primarily a high-latitude, surface phenomenon. Kennett and Shackleton (1976) have argued that this drop in bottom temperature occurred when cold Antarctic bottom waters began to dominate deep-water oceanic circulation in a similar manner as in the present-day oceans. Although there is a slight warming of bottom waters at times during the early-middle Miocene, at no time in the Cenozoic did bottom waters warm to pre-Oligocene values (Fig. 2-5). Kennett and Shackleton (1976) stated that the late Eocene-early Oligocene isotopic temperature drop reflects a decrease in Antarctic surface-water temperatures to near-freezing and the production of extensive sea ice, although isotopic temperatures of bottom water at this time (5-6°C) at Site 277 (present water depth 1200 m) were well above freezing. They also state that oxygen isotopic evidence indicates no significant buildup of Antarctic ice at this time.

There is, however, geological evidence for at least limited glaciation of western Antarctica during the Eocene and Oligocene (Le Masurier, 1970; Margolis and Kennett, 1971). Stump *et al.* (1980) report subglacially erupted volcanic rocks potassium/argon dated to be of early Miocene age from East Antarctica, suggesting the presence of an ice sheet since at least early-Miocene (~18 mya) time. There is also other evidence that supports the possibility that the oxygen isotope change near the Oligocene/Eocene boundary may be partially due to continental ice-induced changes in the isotopic composition of seawater, and not just due to temperature. The fact that the increase in $\delta^{18}O$ is found in both bottom- and surface-water indicators (Margolis *et al.*, 1975) argues for a possible change in seawater isotopic composition or a sudden refrigeration of the entire ocean. Corliss (1979) states that the lack of benthic foraminiferal faunal change after the bottom-water temperature drop suggests that the species present must have had fairly wide environmental tolerances, "based on the assumption that the isotopic event reflects a temperature decrease, as previously suggested, and not a change in the oceanic isotopic composition. No faunal change would be expected if the isotopic event reflected a change in oceanic isotopic composition."[1] This lack of significant faunal change could also mean that part of the isotopic change was due to ice volume change. Matthews assumes an ice-free world prior to 50 mya, but with the progressive isolation of the Antarctic continent by sea-floor spreading, the development of an Antarctic ice cap would produce "the $\delta^{18}O$ positive shift and contribute to the severity of regression observed in continental margin seismic stratigraphy."[2]

O'Keefe (1980) reports that at the time of the terminal Eocene event, the earth may have been surrounded by a ring similar to those around Saturn. The ring, presumably composed of tektites, which have been found widely scattered around the world in sediments of this age, would have reduced insolation in the temperate zone of the Northern Hemisphere and resulted in an enormous temperature decrease of $20°C$ for one to several million years. This would explain the climatic and oceanic cooling, the buildup of Antarctic ice, and the floral and faunal extinctions.

Other, less dramatic mechanisms that have been suggested for the sudden oxygen isotopic change near the Eocene/Oligocene boundary include:

1. Isolation of Antarctic surface waters resulting from northward spreading movement of Australia and the opening of the Drake Passage (Kennett and Shackleton, 1976).
2. Increase in Antarctic alpine glaciation due to uplift of mountain ranges above the snow line and related paleoclimatic and atmospheric circulation changes.
3. A slight change in the tilt of the earth's axis that would change the seasons so as to cool the polar regions without drastically changing equatorial climates.

According to Margolis et al. (1977), data in currently available literature indicate that surface- and bottom-water paleotemperatures during the remainder of the Oligocene through the early Miocene in the Circum-Antarctic oceans were more stable. Savin (1977), however, notes that there were a series of paleotemperature fluctuations in low-latitude bottom waters during the late Oligocene. It is not known at this time if these temperature fluctuations also occurred in the subantarctic region because there is a gap in the published high-latitude isotope record.

## Miocene

The next significant change in paleotemperatures occurred about 18 mya during late-early Miocene time, when there was a warming of subantarctic and equatorial surface and bottom waters (Fig. 2-5). This was followed by an extremely sharp drop in $\delta^{18}O$ values of high-latitude surface water and oceanic bottom waters at all latitudes during early-middle Miocene time (14–15 mya). The magnitude of this isotopic change is similar

[1] Corliss, 1979, p. 64.
[2] Matthews, Personal communication, 1979.

to that near the Eocene/Oligocene boundary, but there is uncertainty as to what portions of this isotopic drop are due to temperature change or to seawater oxygen isotopic compositional changes caused by accumulation of Antarctic ice (Savin, 1977). From middle Miocene time onward, this uncertainty complicates estimates of paleotemperatures from isotopic data (Savin, 1977; Shackleton and Kennett, 1975a). Although middle Miocene high-latitude and bottom-water isotopic changes occurred, low-latitude surface temperatures did not undergo as extreme a change (Savin et al., 1975). In any event, the early-middle Miocene was a time during which oceanic circulation patterns and resultant latitudinal thermal gradients underwent a significant change (Savin, 1977). Beginning at this time there was also decoupling of high-latitude from low-latitude temperature fluctuations (Savin, 1977). Benthonic foraminiferal isotopic changes between middle Miocene and middle-late Miocene time mostly reflect changes in the size of the Antarctic ice sheet (Shackleton and Kennett, 1975a). Shackleton and Kennett (1975b) have interpreted subantarctic oxygen isotopic data to indicate that the Antarctic ice cap had reached its present size by early-late Miocene time; however, Savin et al. (1975) did not feel it was possible to separate the ice effect from the temperature effect during that time interval. Oxygen isotopic values for planktonic foraminifera and calcareous nannofossils which are indicative of surface-water conditions also responded in a similar manner. However, during middle-late Miocene time, there apparently was a warming in subantarctic surface waters with no coinciding change in deep-water temperatures. According to Shackleton and Kennett (1975b), at this time glaciation of Antarctica was extensive enough that the ice sheet was an essentially permanent feature. Ice-volume changes occurred, but the stability of the ice sheet was not affected by warmer periods, although again it is difficult to separate the ice volume from the temperature effect strictly on isotopic composition (Savin, 1977).

Matthews[3] alternatively explains the middle Miocene (14–15 mya) positive excursion in $\delta^{18}O$ to be caused by the cessation of production of warm saline bottom waters from the Mediterranean and the production of colder, less saline bottom waters. In this manner Matthews[4] "ascribes the $\delta^{18}O$ shift to bottom-water temperature rather than change in ice volume."

Thus we see that there is much speculation and controversy concerning the interpretation of the oxygen isotope data both at the Oligocene/Eocene boundary and during the middle Miocene, all being critical events in the history of the Cenozoic oceans. Events in the Mediterranean have also had great influence on global paleoceanography during the latest Miocene (Messinian, 6.2-5.2 mya) according to Arthur (1979). Isolation of the Mediterranean from the North Atlantic may have occurred about 6.2 mya, resulting in the formation of a major evaporite basin described by Hsü et al. (1977). The oxygen isotope record for this interval at several sites shows a positive $\delta^{18}O$ excursion that could reflect either an increase in Antarctic ice volume or a decrease in temperature. The increase in ice volume could have produced a lowering of sea level, resulting in the isolation and dessication of the Mediterranean (van Couvering et al., 1976); however, isotopic analysis of a well-dated sequence of DSDP cores by Cita and Ryan (1980) show that the salinity crisis preceded the positive $\delta^{18}O$ shift in the late Miocene (Arthur,

[3] Personal communication, 1979.

[4] Ibid.

1979). A more detailed discussion of paleoceanographic events of the late Miocene will be presented below under the heading of carbon isotopes.

### Pliocene-Pleistocene

The next critical event in the paleoceanographic history with respect to oceanic temperatures is the initiation of Northern Hemisphere glaciation, which according to Shackleton and Kennett (1975b) occurred about 3 mya, based on oxygen isotope data from the southwestern Pacific Ocean at DSDP Site 284. There is evidence, however, for a precipitous drop in surface-water isotopic values during the early Pliocene (Margolis et al., 1977). More recently, Margolis and Herman (1980) have found evidence of glacial ice-rafting, and for the development of sea-ice cover in sediment cores from the central Arctic Ocean of early Pliocene (4.5 mya) to Recent age. Apparently, the development of both mountain glaciers extending to sea level in the Northern Hemisphere and Arctic Sea ice cover preceded development of a major Northern Hemisphere ice cap by several million years (Herman and Hopkins, 1980).

This chapter will not present a discussion of the paleoceanographic and paleoclimatic changes that have occurred and are still occurring during the Quaternary. A detailed presentation of the Pleistocene isotope record can be found in Shackleton and Opdyke (1973), Climap (1976), and Savin (1977).

# CARBON ISOTOPE HISTORY

Having already looked at the carbon budget in the present day oceans in an earlier section, let us now turn to the fossil record to see what we can learn about productivity and the carbon budget in the ancient oceans during the times of major climatic changes. In looking at the time series plot of $\delta^{13}C$ for the Cenozoic and Late Cretaceous (Fig. 2-7), we see that the range of values for benthic foraminifera and surface-dwelling plankton is between –0.5 and +0.5‰ for benthic forams and +1.0 and +4.0‰ for surface dwellers. The $\delta^{13}C$ of the $\Sigma CO_2$ in the present day ocean varies from about +2.0‰ at the surface to around –0.5‰ in bottom waters (Kroopnick, 1974). In both this time series plot, and the next of DSDP Site 284 from the South Pacific (Fig. 2-8), coccolith values of $\delta^{13}C$ are consistently larger than planktonic foraminifera, which is consistent with our model of coccoliths being closer to isotopic equilibrium with surface water $CO_2$, mainly $HCO_3^-$.

### Pliocene-Pleistocene

In the examination of the carbon budget throughout time, we will start in the Pleistocene where we ended our discussion of oxygen isotopes and go progressively back in geological time to look at $\delta^{13}C$ changes. At Site 284 in the South Pacific, which contains a record of the last 5 million years (Fig. 2-8), there is a large divergence in foraminifera and coccolith $\delta^{13}C$ values during the Pleistocene that is coincident with a positive excursion in both surface- and bottom-water oxygen isotopic values (Fig. 2-9). This indicates that either cooling or increased ice buildup stimulated oceanic circulation and

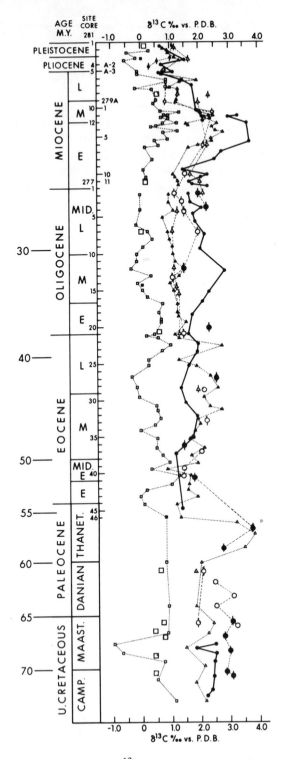

Fig. 2-7.  Summary plot of $\delta^{13}C$ for benthic foraminifera, planktonic foraminifera, and nannofossils. Symbols, sites, and references are the same as for Fig. 2-6. From Margolis *et al.*, 1977.

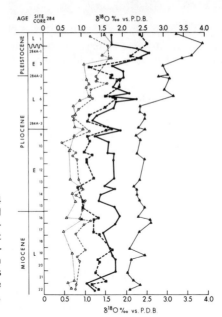

Fig. 2-8.  Plot of $\delta^{18}O$ for DSDP Site 284 (40°31'S, 167°41'E, 1066-m depth), solid triangles are benthic foraminifera (*Uvigerina*) from Shackleton and Kennett (1975a). Solid line with dark dots are calcareous nannofossils. Solid squares with dashed line are *G. bulloides*, open triangles are *G. quinqueloba*, and open circles are *N. pachyderma*. From Kennett *et al.*, 1979.

enhanced surface-water productivity. This higher production and concurrent higher preservation rate for calcareous plankton buries more of the heavier $^{13}C$ in the form of pelagic $CaCO_3$, while the $^{13}C$-depleted organic matter is recycled by oxidation and bacterial activity and returned to the surface water, resulting in a negative trend in $\delta^{13}C$ for the Pleistocene. Berger *et al.* (1978) also report a divergent $\delta^{13}C$ trend between the foraminiferal and coccolith size fractions in box cores from the Ontong-Java Plateau, which also indicates increased productivity during certain periods of the Pleistocene.

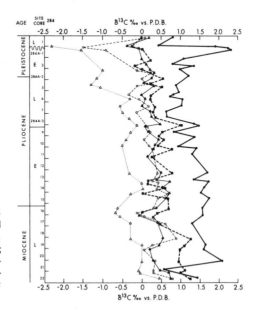

Fig. 2-9.  $\delta^{13}C$ versus PDB for Site 284 from Kennett *et al.*, 1979. Nannofossil isotope data (●) are from Margolis *et al.* (1975) and new data from this study; planktonic (△) and benthic (□) foraminiferal data are from Shackleton and Kennett (1975a), and (■) are *G. bulloides* from Loutit and Kennett (1979).

STANLEY V. MARGOLIS, PETER M. KROOPNICK, WILLIAM J. SHOWERS          35

Much of the Pliocene shows variable $\delta^{13}C$ values with increasing divergence between surface- and bottom-water values, as the surface to bottom temperature gradient increased and circulation was enhanced, resulting in still higher productivity.

## Late Miocene Negative $\delta^{13}C$ Excursion

The next significant change in $\delta^{13}C$ was a pronounced negative shift of about 0.5‰ in both benthic and planktonic foraminifera and calcareous nannofossils of late Miocene age. This negative shift is now documented in sediment cores from all the world's oceans (Bender and Keigwin, 1979; Loutit and Kennett, 1979; Vincent and Berger, 1980; Vincent et al., 1980), and by other workers in the Cenozoic paleoceanography program. The late Miocene negative shift in $\delta^{13}C$ (Fig. 2-10) has now been precisely dated by a combination of nannofossil first-appearance data and magnetostratigraphy to have occurred between 6.1 and 5.9 mya (Haq et al., 1980). It is believed that this shift was a global event, reflecting perhaps a change in the cycling of metabolic $CO_2$ related either to a change in upwelling rates or global oceanic circulation patterns (Bender and Keigwin, 1979). Vincent et al. (1980) believe that a lowering of sea level during the late Miocene caused an increase in the delivery rate of organic carbon from continents and shallow shelves. In either case, any process that enriched $^{12}C$ relative to $^{13}C$ could be responsible. Again, organic carbon $\delta^{13}C$ is around -23‰, while pelagic carbonates are between +1 and +4‰, so that either bringing in carbon from the continents, or releasing it by recycling more organic carbon or burying more pelagic carbonate, could cause the negative $\delta^{13}C$ shift.

A detailed comparison of coccolith and foraminifera oxygen and carbon isotopes across the $\delta^{13}C$ shift at DSDP Site 158 in the eastern equatorial Pacific Ocean (Keigwin and Margolis, in preparation), shows that the difference between benthic foraminifera and coccolith $\delta^{13}C$ values range from about 1.9‰ before the shift to 2.8‰ after the shift. Most of this change is a decrease in benthic $\delta^{13}C$ values, rather than in $\delta^{13}C$ of surface plankton. Therefore any explanation for the negative $\delta^{13}C$ excursion must also explain why it affected bottom $\delta^{13}C$ values approximately twice as much as surface-

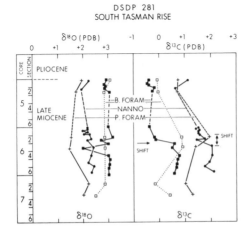

Fig. 2-10. Oxygen and carbon isotope time series plots for DSDP Site 281 (48°S-147°46'E, 1591-m water depth) at the South Tasman Rise. See Fig. 2-9 for symbols.

water values. At DSDP Site 158, the negative $\delta^{13}$C shift is associated with a positive $\delta^{18}$O excursion, which could have resulted from either a cooling of surface and bottom waters or a further buildup of polar ice caps. Elsewhere, at DSDP Site 278 in the Southern Ocean, the $\delta^{13}$C shift is evident in benthic foraminiferal data but not in data from nannofossils. The $\delta^{13}$C shift at this site occurs in the middle of a period of major northward migration of the *Antarctic Convergence* or Polar Front (Fig. 2–11), as indicated by a positive excursion in coccolith $\delta^{18}$O values. There is also an increase in benthic $\delta^{18}$O values exactly at the negative $\delta^{13}$C event, indicating either cooling of bottom waters or Antarctic ice buildup. The long duration $\delta^{18}$O positive shift in nannofossils, however, is not accompanied by a corresponding shift in benthic foraminiferal $\delta^{18}$O values; this implies that the $\delta^{18}$O shift might be due to a decrease in surface-water temperature rather than a change in the isotopic composition of seawater, which would equally affect both surface and bottom waters. Since it is not known if either the benthic foraminifera or the calcareous nannofossil data shown in Fig. 2–11 reflect a true equilibrium with respect to temperature and $\delta^{18}$O, we cannot determine exact paleotemperatures at this site. The $\delta^{18}$O values above and below the nannofossil $\delta^{18}$O shift, however, are quite similar, indicating a well-mixed water column with a vertical thermocline

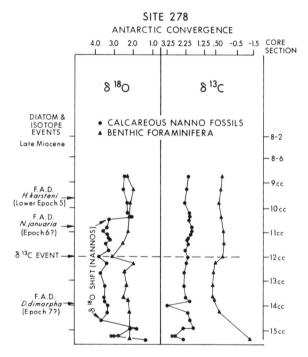

Fig. 2-11.    Oxygen and carbon isotopes of calcareous nannofossils and benthic foraminifera from DSDP Site 278 (56°33'S, 160°04'E, 3675-m water depth) near the *Antarctic Convergence* (Polar Front). Isotope data are from the Cenozoic Paleoceanography program. Investigators as follows: Nannofossil isotopes are from Margolis and Kroopnick, Univ. of Hawaii; benthic foraminifera isotope data are from Keigwin, Bender, and Kennet, Univ. Rhode Island; and diatom data are from L. Burckle (personal communication), Lamont-Doherty Geological Observatory.

structure similar to what is presently found at this site, located at the *Antarctic Convergence*. The isotope data thus confirm the observation of Margolis (1975) that DSDP Site 278 contains a record of the fluctuating position and intensity of the *Antarctic Convergence* caused by changes in climate. The distribution of ice-rafted sand and rock fragments over the interval shown in Fig. 2-11, according to Margolis (1975), indicates very high concentrations of glacially derived sediments to the bottom of core 8. Below core 8, glacially derived quartz grains decrease significantly from tens of thousands of grains to below 200 per 10-cc sample as the sediment changes from a diatom ooze to a nannofossil ooze and the sedimentation rate decreases. In cores 9–15, the number of quartz grains fluctuates between zero and 180 per 10-cc sample, and those present are around 1-2 mm in diameter. No evidence of any ice-rafted sediment was found below core 15. Therefore, in this interval we see a major change in the delivery rate of ice-rafted sediment, along with the shift in the position of the *Antarctic Convergence* and the late Miocene negative $\delta^{13}$C event.

At DSDP Site 281, further north on the South Tasman Rise (Fig. 2-10), the negative $\delta^{13}$C shift is demonstrated in both benthic foraminifera (-1.0‰ shift) and nannofossil (-0.5‰ shift) values. Planktonic foraminifera show an increase in $\delta^{13}$C over the same interval, indicating that phytoplankton productivity must have been high at this site at that time, as previously discussed. The benthic and planktonic $\delta^{18}$O record, with the exception of one negative spike in benthic values just before the carbon shift, is remarkably uneventful over this time interval. The northward shift in the *Antarctic Convergence* found at Site 278 further to the south apparently did not effect surface-water temperatures in other regions.

In summary, the negative $\delta^{13}$C shift has now been documented in benthic and planktonic foraminifera, and nannofossils (Bender and Keigwin, 1979; Savin and Sommer, 1979; Haq *et al*.; Vincent *et al*., 1980). The evidence seems to indicate that it is associated with either a cooling or an increase in global ice storage. Its timing is also critical relative to the Messinian Crisis in the Mediterranean, although the timing of these events needs to be documented further by detailed biostratigraphic and isotopic analysis of well-preserved marine sections.

## Positive $\delta^{13}$C Excursions, Miocene, and Paleocene

(Schlanger and Jenkyns, 1976; Scholle and Arthur, 1979). The positive $\delta^{13}$C excur-positive excursions (Fig. 2-7), the largest of which occurs in the late-early to early-middle Miocene (~14–18 mya), roughly just before the early-middle Miocene climatic event at about 14 mya. This is also the time of a steepening of the global latitudinal temperature gradients, and the equatorial thermocline, as well as of an increase in bottom-water circulation (Savin, 1977). Benthic and planktonic $\delta^{13}$C values are heavier within this interval than anytime else in the Cenozoic, with the exception of a positive excursion in planktonic values only during the late Paleocene (Fig. 2-7). Positive excursions in $\delta^{13}$C in both calcareous benthos and plankton are caused by a removal of $^{12}$C-rich organic carbon from the oceanic system by burial in sediments. Thus the carbon reservoir available for calcifying organisms is enriched in $^{13}$C and recorded as a positive isotopic excursion (Kroopnick *et al*., 1977). These two positive $\delta^{13}$C Tertiary

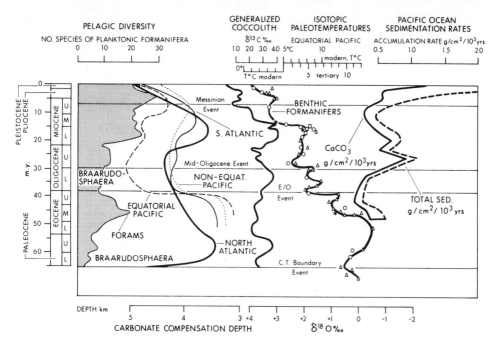

Fig. 2-12. Composite Cenozoic summary plot of pelagic diversity, carbonate compensation depth, generalized coccolith $\delta^{13}$C, isotopic temperatures, and Pacific sedimentation rates (modified from Arthur, 1979, and references therein). Pelagic diversity from Fisher and Arthur (1977); calcite compensation depth after van Andel *et al.* (1977); Pacific Ocean sedimentation rates after Worsley and Davies (1979); oxygen isotopes after Savin (1977); carbon isotopes of coccoliths after Margolis *et al.* (1977).

excursions have been identified in sediments from all the world's oceans and after more detailed biostratigraphic studies will become isotopic event markers much like the late Miocene negative $\delta^{13}$C shift at 6 mya discussed above. Burial of organic carbon during times like these would make strata of this age potential source rocks for petroleum (Schlanger and Jenkyns, 1976; Schoole and Arthur, 1979). The positive $\delta^{13}$C excursions in late-early Miocene and late Paleocene time are both also associated with shoaling of the calcite compensation depth (CCD) in the North Atlantic (Fig. 2-12). The production of North Atlantic deep water serves as a major means of introducing $CO_2$ and particulate organic carbon into the deep waters of all the oceans by mixing to form common water around the Antarctic (Kroopnick, 1974a, b). The duration of the positive $\delta^{13}$C excursion is several million years, with a gradual transition to "normal," more negative $\delta^{13}$C values which precede and follow the $\delta^{13}$C excursion (Fig. 2-12). In order to maintain $^{13}$C enriched conditions in the oceans for this long time period, there must have been a continued high input of carbon from the continents and selective burial of $^{12}$C in the form of organic carbon in sediments, leaving $^{13}$C enrichment in the $HCO_3^-$ available for calcifying plankton. Such conditions of increased burial of $^{12}$C existed until the mid-Miocene cooling and/or buildup of polar ice. The late-early Miocene positive $\delta^{13}$C excursion is associated with warm temperatures and relatively high stands of sea level (Fig. 2-12). The late Paleocene positive $\delta^{13}$C excursion appears to have similar associations. Sluggish bottom-water circulation may result in a deep oxygen minimum and the deposition of sediments rich in organic carbon. Such conditions have been described for times of relatively high sea level and warmer climates during the Mesozoic and Cenozoic (Arthur, 1979; Berger, 1979; Schlanger and Jenkyns, 1976). There is a positive $\delta^{13}$C excursion in calcareous nannofossils in the mid-Oligocene also associated with high stands of sea level and warm temperatures (Fig. 2-12). It correlates in age with the mid-Oligocene injection event, where less saline waters from previously isolated basins were proposed by Thierstein and Berger (1978) to have entered the oceans. Thus far the mid-Oligocene carbon excursion has only been identified in two cores (Kroopnick et al., 1977).

The late Paleocene positive $\delta^{13}$C excursion has been identified by Saito and van Donk (1974) in the South Atlantic and by Douglas and Savin (1975) in the North Pacific. These and other data for this time period, as summarized by Margolis et al. (1977) and Kroopnick et al. (1977), show a 2 per mil increase in the difference between surface- and bottom-water $\delta^{13}$C values between 56 and 58 mya, similar in magnitude to the late-early Miocene positive $\delta^{13}$C excursion described above. More recently, Boersma et al. (1979) found a similar excursion in the late Paleocene in the North and South Atlantic in their extremely well-documented study. These authors explain the increase in $\delta^{13}$C gradient between surface and bottom waters as being due to a major change in the oceanic carbon reservoir, associated perhaps with an increase in productivity of surface plankton as suggested by Broecker (1971, 1975). These increased rates of productivity would also intensify the oxygen minimum zone, and cause a rise in the calcium carbonate compensation depth (Boersma et al., 1979), all features similar to those described for the late-early Miocene positive $\delta^{13}$C excursion. Both events are also associated with warming trends in surface and bottom waters. Possible explanations for the late Paleocene positive $\delta^{13}$C excursion are increased intensity of bottom-water circulation (Boersma et al., 1979) producing a greater upwelling rate of nutrients, or a higher delivery rate of nutrients from shallow-water areas expanded by the transgression shown by many authors for this time period.

## Negative $\delta^{13}$C Excursion at Cretaceous/Tertiary Boundary

There are several smaller $\delta^{13}$C excursions in the mid-Oligocene, in the mid-Eocene, near the Paleocene/Eocene boundary, and near the Cretaceous/Tertiary boundary (Fig. 2-7, 2-12). All but the last have not been found as yet in a sufficient number of cores to warrant their being considered global events. Boersma et al. (1979) show a negative excursion in $\delta^{13}$C occurring in surface-dwelling plankton at the Cretaceous/Tertiary boundary, concurrent with an increase in $\delta^{13}$C for deeper dwelling planktonic foraminifera, and then a decrease in $^{13}$C for benthic foraminifera. This unusual $\delta^{13}$C profile, according to Boersma et al. (1979), indicates either (1) that the oceans at the time had quite the opposite $\delta^{13}$C profile than that found today, (2) that subsurface waters were warmer and more saline than surface water, or (3) that the extinct planktonic species analyzed had unusual depth habitat and/or non-equilibrium deposition of oxygen and carbon isotopes during the first three million years of the Paleocene.

Isotopic analysis of benthic foraminifera and calcareous nannofossils from a relatively complete, shallow-water shelf section across this same time interval in the vicinity of Braggs in Southern Alabama, described by Worsley (1974), similarly shows a significant warming (about $7°C$) in bottom- and surface-water isotopic temperatures across the Cretaceous/Tertiary boundary. Nannofossil (surface-water) $\delta^{13}C$ values were also more negative than values for benthic foraminifera at the boundary (Margolis et al., in preparation). Analysis of carbon and oxygen isotopes of calcareous nannofossils from the samples described by Boersma et al. (1979), as discussed below, may clarify the interpretation of the isotope data from the earliest Paleocene, since the depth habitat of the nannofossils is within the photic zone.

Thierstein and Berger (1978) have isotopically analyzed coccoliths, using the techniques of Margolis et al. (1975), from samples across the Cretaceous/Tertiary boundary at DSDP Site 356 in the South Atlantic. They found that coccolith $\delta^{18}O$ values across the boundary exhibit a 2 per mil negative excursion, which if calculated purely in terms of an increase in isotopic temperature would be equivalent to about a $10°C$ warming of surface temperature. Thierstein and Berger (1978) believe that part of this excursion is due to an injection of low-salinity water from the Arctic, which previously was tectonically isolated from the world's oceans. A lowering of oceanic surface-water salinity by 10 to 25‰ would result in the observed -2.0‰ change in coccolith $\delta^{18}O$ values. An even more remarkable observation of Thierstein and Berger (1978) is a 3‰ negative shift in coccolith $\delta^{13}C$. The range of coccolith $\delta^{13}C$ values for the Tertiary are between +1 and +4 per mil (Margolis et al., 1977) with present-day surface-water $\delta^{13}C$ values being +2 per mil. According to Thierstein and Berger (1978), in the basal Danian sediments at Site 356, coccolith $\delta^{13}C$ values drop to values between -0.5 and -1.5‰. The only reported coccolith carbon values similar to these are from eastern equatorial Pacific, and equatorial Indian Ocean surface sediments of Recent age shown in Table 2-1 (Goodney et al., 1980), and in late Pleistocene sediments (Kennett et al., 1979; Margolis et al.). We see in Table 2-1 that in present-day equatorial open-ocean normal salinity environments, depth stratification by oxygen isotopic temperature rankings of the planktonic foraminifera holds true. However, $\delta^{13}C$ shows considerable variation in surface waters. Both G. ruber and nannofossils show negative $^{13}C$ values

TABLE 2-1.   Foraminifera and Coccoliths of Recent Age, from Piston Core Tops [a]

| Core | Latitude | Longitude | Depth (m) | Fossil Type | $\delta^{18}O$‰ PDB | $\delta^{13}C$‰ PDB | Average Summer Temperature Surface Water °C |
|---|---|---|---|---|---|---|---|
| Eastern equatorial Pacific | | | | | | | |
| KK71–104 | 12°05'S | 110°37'W | 3087 | G. sacculifer | -3.57 | 1.44 | 25.5 |
| | | | | G. ruber | -3.25 | -0.85 | – |
| | | | | Coccoliths | -3.24 | -2.07 | – |
| | | | | G. menardii | -0.27 | 2.04 | – |
| Indian Ocean | | | | | | | |
| V–19–178 | 8°07'N | 73°15'E | 2188 | Coccoliths | -2.24 | -0.55 | 28.0 |
| V–19–185 | 6°42'N | 57°20'E | – | Coccoliths | -2.45 | -0.75 | 28.5 |

[a]From Goodney et al., 1980.

(-0.85 and -2.07‰) while *G. sacculifer* shows a $\delta^{13}C$ of +1.44‰, close to the present day value of +2.0‰. It has been shown in Recent sediment that in areas of high productivity, coccoliths reflect more negative values of $\delta^{13}C$ of $\Sigma CO_2$, as do symbiote-bearing planktonic foraminifera (Goodney *et al.*, 1980). Certain planktonic foraminifera, however, such as shallow-dwelling *G. ruber* and deeper dwelling *G. menardii*, exhibit more positive $\delta^{13}C$ values in areas of high productivity. There also appears to be a direct relationship between $\delta^{18}O$ and $\delta^{13}C$ of coccoliths, although the temperature dependence of $\delta^{13}C$ fractionation is small (Goodney, *et al.*, 1980). More negative $\delta^{13}C$ values are associated with warmer temperatures, and more positive values with cooler temperatures (Goodney *et al.*, 1980). This relationship may be due to differences in surface-water $\delta^{13}C$ of $\Sigma CO_2$, or due to species diversity changes as a function of water mass characteristics, nutrient levels, and other oceanographic environmental factors besides temperature.

When we look at nannofossil and foraminiferal oxygen and carbon isotopes of basal Paleocene age from DSDP Site 356 (Boersma *et al.*, 1979; Thierstein and Berger, 1978) in Table 2-2, we note the following: the nannofossils are the shallowest dwelling (most negative, and warmest) of the calcareous plankton according to oxygen isotopic temperature ranking, with the four planktonic foraminifera species showing progressively deeper habitats. There is also a clear trend of more positive $\delta^{13}C$ values with depth, the reverse of what is found in today's oceans, and during the majority of the last 70 million years of ocean history, as previously indicated by Boersma *et al.* (1979). However, the uncertainty of the depth habitat of the extinct species of planktonic foraminifera, with perhaps *G. cretacea* occupying a deeper habitat in waters of more negative $\delta^{13}C$, could be a problem for correct interpretation. The evidence from surface-dwelling photosynthetic coccoliths indeed demonstrates that the coccoliths are about 0.8 per mil more negative in $\delta^{18}O$ and therefore occupied significantly shallower (and warmer) water depths than the shallowest dwelling foram. The coccoliths also exhibit $\delta^{13}C$ values about 0.5 per mil more negative, further emphasizing the anomalously negative $\delta^{13}C$ characteristics of surface waters of the early Paleocene. Later on in the early Paleocene in Zone P-1 (Table 2-2,B) this trend is further accentuated, with nannofossils showing more negative (warmer) oxygen isotopes, and more negative carbon isotopes. According to Boersma *et al.* (1979) this unusual condition existed for the first three or four million

TABLE 2-2. Foraminifera and Nannofossil Isotopes of Basal Paleocene Age, *G. eugubina* Zone[a]

| DSDP Site | Core | Latitude | Longitude | Depth (m) | Fossil Type | $\delta^{18}O‰$ PDB | $\delta^{13}C‰$ PDB |
|---|---|---|---|---|---|---|---|
| A. 356 | 29-3 33 cms | 41°05'W | 28°17'S | 3175 | Coccoliths | -3.2 | -0.3 |
| | | | | | *Guembelitria cretacea* | -2.49 | +0.20 |
| | | | | | *Chiloguembelina midwayensis* | -1.78 | +0.80 |
| | | | | | *P. eugubina* and *P. gringa* | -1.36 | +1.50 |
| | | | | | Mixed benthics | +0.13 | +1.28 |
| | | | | *Zone P-1* | | | |
| B. 356 | 29-2 108 cms | | | | Coccoliths | -3.90 | -1.0 |
| | | | | | *G. cretacea* | -1.48 | +0.20 |
| | | | | | *S. pseudobulloides* | -1.13 | +1.45 |

[a]Data from Boersma *et al.* (1979) and from Thierstein and Berger (1978).

years of the Paleocene. Several million years later (Fig. 2-7) this grades into the late Paleocene positive $\delta^{13}$C excursion described above.

## DISCUSSION

### The Oceans During the Cretaceous/Tertiary Transition

There have been many explanations recently of exactly what happened at the Cretaceous/Tertiary boundary, and the carbon and oxygen isotope evidence presented above has been used as support for many of the theories proposed. Using the experience gained from looking at isotopes in the present-day oceans, let us try to put all these theories in proper oceanographic perspective. Gartner and Keaney (1978) and Thierstein and Berger (1978) invoke injections of brackish water from the Arctic Ocean as the cause of the extinctions of plankton, and of the oxygen and carbon isotope negative excursions at the Cretaceous/Tertiary boundary. This argument has been weakened by biostratigraphic and physical oceanographic evidence (Perch-Nielsen et al., 1979; Surlyk, 1980). The oxygen isotope data presented can also be reasonably accepted in terms of temperature alone, without invoking drastic salinity variations. Isotopic temperatures at Site 356 during the early Paleocene, as indicated by surface-dwelling forams and coccoliths, are equivalent to present-day equatorial temperatures, and according to Worsley (1974) this site was located within the equatorial zone at that time. The problem is not the magnitude of the basal Paleocene warming, but the suddenness of 3°C global oceanic warming of surface and bottom waters across the transition zone, followed by a continued temperature rise throughout the early Paleocene. Any theory must explain both the sudden and then the more gradual temperature rise in the oceans, although they may be due to two entirely different phenomenon.

Extraterrestrial phenomena such as asteroidal and cometary impacts have been suggested for the temperature rise, the mass extinctions, the platinum metal enrichments in the transitional clay layer, and the unusual carbon isotope profiles (Alvarez et al., 1980; Hsü, 1980). Looking at the carbon and oxygen isotope data alone from the Cretaceous/Tertiary transition zone, any theory must explain the sudden warming and the 1-3‰ negative excursion in $\delta^{13}$C. The only temperature effect discussed by Alvarez et al. (1980) during impact of a 6-14 km asteroid upon the earth would be a "temporary absence of sunlight which would effectively shut off photosynthesis."[5] This would have been caused by injection of the asteroid impact material into the stratosphere. Based on calculations from the Krakatoa eruption, after the proposed asteroidal impact 65 mya "day could have been turned into night for a period of several years" (Alvarez et al., 1980).[6] This would have resulted in a cooling, but because of its short duration, this event would probably not be recorded in the isotopic record preserved in marine sediments. There is evidence however, of a brief cooling immediately before the Cretaceous/Tertiary boundary (Margolis et al., 1977; Savin, 1977) in the late Maastrichtian in deep-sea sediments from the Atlantic and Pacific Oceans, and also from the Braggs Alabama shallow-shelf section discussed above (Margolis et al., in preparation). Geochemical trace metal analysis of these sediments would have to be undertaken

[5] Alvarez et al., 1980, p. 1106.

[6] Ibid.

to determine if the cooling was associated with anomalously high iridium, osmium, and other platinum metal contents before the cooling could be positively attributed to the proposed asteroidal impact event of Alvarez *et al.* (1980). If such an asteroid impacted in the ocean and had a diameter of 10 km or twice the average ocean depth, a crater on the sea floor would be produced, and there would be a huge outpouring of sediment, rock, and seawater according to Alvarez *et al.* (1980). The oceanographic effect would be a tsunami of enormous magnitude and of world wide scale, resulting both from the initial impact and from the return-flow of water. One would expect that there would be extensive evidence of this event on the continents, even if the ocean crater were subducted. In addition, there would be a heating of the atmosphere and the oceans created by expenditure of kinetic energy and by secondary chemical explosions (Hsü, 1980).

Hsü (1980) presents estimations of the effects if the impacting body were a comet. Similar estimates could be made for an asteroidal impact, but Hsü states that it is difficult to give exact figures for the magnitude of the atmospheric and ocean heating. In addition to the temperature increase, another physical effect would be the mixing of water mass properties or general disruption of established density stratification and circulation patterns. This would have the following effects on the oxygen isotopes of calcareous marine organisms in the oceans at that time, if it had occurred:

1. Sudden warming of surface and bottom waters from both heating and mixing processes; the relative warming would be greater at high latitudes and in deeper waters because of intrusion and homogenization of the warm water sphere into the colder bottom water; this would result in a reduction of the surface-to bottom-water thermal gradient; all of which are evident at Site 384 (Boersma *et al.*, 1979) and also at the Braggs, Alabama section (Margolis *et al.*, in preparation.

2. Deepening of the depth habitat of planktonic foraminifera, since many of them follow isotherms in their oceanic distribution (Williams *et al.*, 1977); those hardy and heat-tolerant calcareous nannofossils that survived the event would remain in the photic zone and record warmer (more negative) $\delta^{18}O$ values, as illustrated in the nannofossil isotope data of Thierstein and Berger (1978).

The sudden 1-3‰ negative excursion in $\delta^{13}C$ at the Cretaceous/Tertiary boundary has been attributed by Thierstein and Berger (1978) to a decrease in plankton productivity and therefore to the rate of removal of organically rich and $^{12}C$-depleted materials from the surface water. In addition, these authors state that the surface density stratification caused by injection of less-saline arctic waters would result in isolation of the surface layer and would produce a shallow, strong oxygen minimum. They calculate that these mechanisms were sufficient to produce the observed –2.0‰ change in $\delta^{13}C$. Boersma *et al.* (1979), however, have shown that there was also a 1.5‰ negative excursion in benthic forminifera at DSDP Site 384 so that an explanation must also be found for introduction of negative carbon into bottom waters. The basal Paleocene strata at several DSDP sites, and at the Braggs, Alabama locality all show a very small $\delta^{13}C$ gradient from surface to bottom waters. This can be explained by low productivity and by the mixing of oceanic waters, associated with a catastrophic event such as the proposed cometary or asteroidal impact. Hsü (1980) uses a large input of cometary carbon with an assumed $\delta^{13}C$ of -25‰ as the source of the -1.5‰ $\delta^{13}C$ excursion in biogenic carbonate. The addition of cometary $CO_2$ to the earth would also increase $CO_2$ in the ocean and the atmosphere and result in a "greenhouse" effect, which would explain the continued warming of surface ocean waters beyond the initial thermal event at the boundary. It would also result in the catastrophic rise in the calcite compensation depth proposed by Tappan (1968) and by Tappan and Loblich (1971), Worsley (1974), and many others (Hsü, 1980).

It is not necessary, however, to invoke an extraterrestrial source for the negative carbon excursion, as there is sufficient carbon on earth of the proper carbon isotopic composition in the organic carbon reservoirs on land and in the oceans (Table 2-3, from Shackleton, 1977). Various estimates place the destruction of life at the Cretaceous/Tertiary boundary on land and in the oceans at around 75% of the species living at that time, and estimates of the duration of the extinction interval vary from 1-100 years to as much as a million years (McLean, 1978). Whatever mechanism is proposed for the sudden warming at the Cretaceous/Tertiary boundary (increase in solar activity, nearby supernova, asteroidal or cometary impact), it may have resulted in the mass mortality and extinctions. If there were massive forest and brush fires on land this would have liberated $CO_2$ with a $\delta^{13}C$ of -25 per mil into the atmosphere. In addition to further enhance the warming trend, this $CO_2$ with its negative $\delta^{13}C$ would eventually get into the surface layer of the ocean and alter its isotopic composition. The present-day oceans contain $3.5 \times 10^{19}$ g of carbon with an average $\delta^{13}C$ of 0‰. If $10^{18}$ g carbon from the terrestrial biosphere with a $\delta^{13}C$ of -25‰ made its way into the oceans, it would decrease the $\delta^{13}C$ of oceanic $CO_2$ by about 0.7 per mil (Shackleton, 1977). The magnitude of this change would be partially decreased by the increased dissolution of biogenic carbonate if there was sufficient time, but not if it occurred in a matter of a few tens of years. The denudation of the forests would allow more dissolved and particulate organic carbon from the terrestrial biosphere and from the shallow-water shelf areas also to make its way into the ocean carbon reservoir by normal aqueous transport mechanisms. If the spike of $^{12}C$-rich $CO_2$ were concentrated in the surface layer of the oceans to a depth of 100, rather than dispersed equally throughout the oceans, it could easily result in a -2.00‰ $\delta^{13}C$ shift in surface values from interaction with the atmosphere, and a -1.5‰ shift in bottom waters from particulate organic carbon sedimented to the sea floor. Waters deeper than 100 m and above the bottom, would exhibit more positive or normal $\delta^{13}C$ as indicated by Boersma et al. (1979) for Sites 356 and 384. Hsü (1980)

TABLE 2-3.   Carbon Reservoirs: Their Size and $^{13}C$ Values[a]

| Reservoir | $\delta^{13}C$ | Mass (as Carbon) |
| --- | --- | --- |
| Living plant biomass | -25‰ | $8.4 \times 10^{17}$ g |
| Humus | -25‰ | $12 \times 10^{17}$ g |
| Marine biomass | -23‰ | $1.8 \times 10^{15}$ g |
| Ocean dissolved $CO_2$ | 0‰ | $3.5 \times 10^{19}$ g |
| Atmospheric $CO_2$ | -7‰ | $6 \times 10^{17}$ g |
| Fossil fuel $CO_2$ (year 2000 cumulative) | -24‰ | $2 \times 10^{17}$ g |

[a]From Shackleton (1977), p. 405, and references cited therein.

reported recently that the 1–3‰ decrease in $\delta^{13}C$ of ocean waters would be "equivalent to a change produced if the whole of the terrestrial biosphere were put into the oceans."[7] This is possible, although it probably represented varying inputs of plant and animal carbon from land, humus, and marine organic carbon released by the event at the Cretaceous/Tertiary boundary. The negative carbon shift at the boundary can therefore be explained without invoking an additional source of $^{12}C$-enriched $CO_2$ from a comet.

Carbon dioxide enrichment of the oceanic surface layer would result in undersaturation with respect to $CaCO_3$ and this together with high surface-water temperatures could have resulted in destruction and extinction of much of the calcareous plankton, including the majority of calcareous nannofossils and planktonic foraminifera possessing symbiotic algae. Some planktonic foraminifera perhaps were able to survive in deeper, cooler, less corrosive water.

The Cretaceous/Tertiary boundary is marked in most sedimentary sections by a widespread dissolutional hiatus (Worsley, 1974) with the exception of a few DSDP sites in the deep Atlantic such as Sites 356 and 384 discussed above. Land sections at the Cretaceous/Tertiary boundary in Spain, Italy, Denmark, and Alabama are marked by a reduction in carbonate content with an increase in clay and other authigenic minerals typical of reducing conditions, poor circulation, and low productivity. Those stress-tolerant coastal nannoplankton such as *Braarudosphaera* sp., *Thoracosphaera* sp., and *Cruciplacolithus* sp. that survived the boundary "event," perhaps by forming spores or cysts, underwent a series of blooms in the Early Danian, and are exceptionally well preserved in mid-gyre Atlantic sites (Hsü, 1980; Thierstein and Berger, 1978). Less stress-tolerant open-ocean coccoliths stop calcifying or die under conditions of high temperature, absence of sunlight, low nutrients, or the presence of higher than average pelagic levels of trace metals. These conditions would probably have a similar effect on the noncalcareous members of the marine phytoplankton, resulting in a collapse of the marine food chain and extinction of many marine species. The reduction of the calcareous plankton as a sink for $CO_2$ in the oceans would further enhance the shoaling of the calcite compensation depth, the negative characters of $\delta^{13}C$ of $\Sigma CO_2$ in the surface layer of the ocean, and the buildup of $CO_2$ in the atmosphere, which would produce the continued warming trend in the early Paleocene by the "greenhouse" effect.

The oxygen and carbon isotopic data discussed here are therefore compatible with an extraterrestrial thermal event as a triggering phenomenon 65 mya, followed by second-order atmospheric and oceanographic responses which lasted for several million years. The oxygen and carbon isotope data by themselves, however, cannot differentiate between the various mechanisms proposed, i.e., cometary, asteroidal, supernova, or solar hyperactivity. We leave this to the astrophysicists and to detailed studies of the geochemistry of trace elements such as iridium and osmium from transitional sections, which no doubt are presently in progress. We plan such a study of the Braggs Alabama section (Margolis *et al.*, in preparation) as well as isotopic analysis of well-preserved calcareous nannofossils and foraminifera from other transitional sections. Our aim is to insure that various scenarios proposed for the Cretaceous/Tertiary boundary event and other

[7]Hsü, 1980, p. 201.

significant paleoceanographic events in the earth's history are viewed in proper oceanographic prospective and fit the framework of our knowledge of the present-day oceans, and of our rapidly expanding understanding of the ancient oceans.

## $\delta^{13}$ C Trends in the Pleistocene and Present-Day Oceans

There is also a trend toward negative $\delta^{13}$C values in calcareous plankton in the late Pleistocene (Figs. 2-7 and 2-8). Shorter term negative excursions in $\delta^{13}$C are found during glacial-interglacial transition periods of low sea level and aridity, which greatly reduced the world's forests (Shackleton, 1977). Organic carbon from the deforestation rapidly becomes oxidized, and is transferred to the oceans in the form of $^{12}$C-enriched $CO_2$, which is manifested in the oceans as a rise in the calcite compensation depth and in negative $\delta^{13}$C values for calcareous plankton and benthic foraminifera (Shackleton, 1977). The activities of humans since the discovery of fire and the advent of an agricultural society may have contributed to the late Holocene transference of $CO_2$ and $^{12}$C to the oceans, as large-scale burning has been used by primitive and modern man for clearing forest lands for farming since long before the industrial era and his accelerated use of fossil fuels for energy and heat. The late Holocene negative $\delta^{13}$C excursion and dissolution event noted by Shackleton (1977) and other workers may reflect these effects as well as a period of dryer global climate.

It is easy to draw an analogy between the Cretaceous/Tertiary boundary event and the present from the data and discussion presented above. McLean (1978) has done so, and predicts that anthropogenically generated $CO_2$ could serve as a trigger event for dramatic climatic warming that not only would have a direct and dire thermal effect on life, but also would produce a surging or collapse of the polar ice caps that would cause a rise in sea level, and innundation of many coastal areas.

There is a similarity in the $CO_2$ buildup and the transference of $^{12}$C to the oceans between the Cretaceous/Tertiary boundary and the present; however, on further examination the analogy breaks down. First of all, oceanic circulation at the present time is driven by a strong equator to pole temperature gradient and by thermohaline-controlled bottom-water production at the poles. This has created a strong density stratification (pycnocline) in the oceans, and a well-developed thermocline in equatorial waters which separates the warm and cold water spheres. Mixing and upwelling of cold nutrient-rich bottom waters occurs in zones of upwelling at the polar fronts, the equator, and along continental margins. Entry of $CO_2$ into surface waters that communicate directly with deeper waters and spread throughout the ocean basins occurs mainly in polar regions where the thermocline is vertical. Such a circulation pattern was not present during the Late Cretaceous, prior to development of polar ice caps (Margolis *et al.*, 1977). Life presently on land and in the oceans has evolved and adapted to this period of rapidly changing climates. Very probably, evolution was accelerated by these climatic changes. Productivity in the present-day oceans, in particular that of calcareous plankton, is higher than estimates of sedimentation rates from the Late Cretaceous. Productivity is stimulated by vigorous oceanic circulation and upwelling. The present-day oceans therefore probably have a greater capacity to absorb $CO_2$ than the Cretaceous oceans because of their colder temperatures and higher $CaCO_3$ production rates.

Man is now introducing $CO_2$ into the atmosphere at a rate such that, if it resulted in a "greenhouse" effect-induced 1°C-increase in surface temperature in regions of deep-water formation, such as the Southern Ocean, it would cause atmospheric $CO_2$ content to further increase by 4.2% because of the reduced $CO_2$ storage capacity. There is a far greater amount of $CO_2$ stored in deep, cold ocean water than that produced yearly by the burning of fossil fuels (McLean, 1978). The fossil fuel $CO_2$-induced warming may then result in a release to the atmosphere of the $CO_2$ stored in the deep ocean. This could then even further enhance the "greenhouse" effect warming, and might result in catastrophic melting of the polar ice cap, ice surge, and rapid rise of sea level such as has been proposed to have occurred between 10,000 and 12,000 years ago (McLean, 1978). It is hoped that the ocean will respond to the increase in $CO_2$ in the atmosphere either by absorbing it into the biogenic $CaCO_3$ production system, by the dissolution of aragonite from shallow-water tropical shelves as suggested by Hay and Southam (1977), or by dissolution of deep-sea carbonate sediments as suggested by Broecker and Takahashi (1977). Perhaps also the reduced rate of increase in the burning of fossil fuels noted since the post–1974 increase in oil prices (Abelson, 1980) will continue for the next two decades and thus halt the exponential buildup of anthropogenic $CO_2$ that has occurred since 1860 (Rotty, 1977). Only careful measurements of the $CO_2$-$\delta^{13}C$ system in the present-day oceans, and understanding of how the ocean-atmospheric systems have responded in the past to natural terrestrially and astronomically induced changes, will help us understand how the system will respond to the human-induced changes that are occurring today and that may occur in the future. It is hoped that the buffers of the productive biogenic $CaCO_3$ system and the polar ice caps will prevent a recurrence of the catastrophic extinctions that occurred at the end of the Mesozoic. There may, however, be more severe second-order effects on sea level and the biosphere.

We have thus come full-circle back to the present day and have seen how the current oceanic temperature and circulation system has been established by a series of critical events over the last 75 million years of earth history. Most of these events appear to be relatively sudden on a geological time scale. With the trend towards neo-catastrophism in the last two years, scientists have often invoked astronomical and sometimes farfetched theories to explain events in earth history. We caution researchers to make certain that these theories do not require the oceans to respond in a fashion inconsistent with what we currently understand to be their nature, or to have external sources as the cause of fluctuations which have been within the range of the oceans natural responses throughout time.

## ACKNOWLEDGMENTS

This research was supported by National Science Foundation grants OCE78-25283, OCE76-81961, and OCE77-21099. The authors thank all the CENOP investigators, especially J. P. Kennett and R. K. Matthews for their advice and discussions. The authors also profitted from discussions with S. Schlanger and S. Savin on this subject.

Abelson, P. H., 1980, The global 2000 report: *Science,* v. 209, p. 761.

Alvarez, L. W., Alvarez, W., Asaro, F., and Michel, H. V., 1980. Extraterrestrial cause for the Cretaceous-Tertiary extinction: *Science,* v. 208, p. 1095-1108.

Anderson, T. F., and Cole, S. A., 1975, The stable isotope geochemistry of marine coccoliths—a preliminary comparison with planktonic foraminifera: *J. Foram. Res.,* v. 5, p. 188-192.

Angell, R. W., 1979, Calcification during chamber development in *Rosalina floridana*: *J. Foram. Res.,* v. 9, no. 4, p. 341-353.

Arthur, M. A., 1979. Paleoceanographic events—recognition, resolution and reconsideration: *Review of Geophysics and Space Physics,* v. 17, no. 7, p. 1474-1494.

Bender, M. L., and Keigwin, L. D., 1979, Speculations about the Upper Miocene change in abyssal Pacific dissolved bicarbonate: *Earth Planet. Sci. Lett.,* v. 45, p. 383-393.

Berger, W. H., 1969, Ecological patterns of living planktonic foraminifera: *Deep-Sea Res.,* v. 16, p. 1-24.

——, 1979, Impact of deep-sea drilling on paleoceanography, *in* Talwani, M., and Ryan, W. B. F., eds., *Results of Deep-Sea Drilling in the Atlantic Ocean, Vol. 2*: Proc. of the Second Maurice Ewing Symposium, American Geophysical Union.

——, Diester-Haas, L., and Killingley, J. S., 1978a, Upwelling off Northwest Africa—the Holocene decrease as seen in carbon isotopes and sedimentological indicators: *Oceanol. Acta,* v. 1, p. 3-7.

——, Killingley, J. S., and Vincent, E., 1978b, Stable isotopes in deep-sea carbonates—box core ERDC-92 western equatorial Pacific: *Oceanol. Acta,* v. 1, p. 8-21.

Boersma, A., Shackleton, N., Hall, M., and Given, Q., 1979, Carbon and oxygen isotope records at DSDP Site 384 (North Atlantic) and some Paleocene paleotemperatures and carbon isotope variations in the Atlantic Ocean, *in* Tocholke, B., Vogt, P., *et al., Initial Reports of the Deep Sea Drilling Project*: U.S.G.P.O., Washington, D.C., v. 43, p. 695-717.

Broecker, W. S., 1971, Calcite accumulation rates and glacial to interglacial changes in oceanic mixing, *in* Turekian, K. K., ed., *The Late Cenozoic Glacial Ages*: Yale Univ. Press, New Haven, Conn., p. 239-265.

——, 1975, Climatic change—are we on the brink of a pronounced climatic change? *Science,* v. 189, p. 460-463.

——, and Takahashi, T., 1977, Neutralization of fossil fuel $CO_2$ by marine calcium carbonate, *in* Anderson, N. R., and Malahoff, A., eds., *The Fate of Fossil Fuel $CO_2$ in the Oceans*; New York, Plenum, p. 295-322.

Buchardt, B., and Hansen, H. J., 1977, Oxygen isotope fractionation and algae symbiosis in benthic foraminifera from the Gulf of Elat, Israel: *Bull. Geol. Soc. Denmark,* v. 26, p. 185-194.

Cita, M. B., and Ryan, W. B. F., 1980, Late Neogene paleoenvironments—interpretation of the evolution of the ocean paleoenvironment, *in* Ryan, W. B. F., von Rad, U., *et al., Initial Reports of the Deep Sea Drilling Project*: U.S.G.P.O., Washington, D.C., v. 47, p. 1003-1036.

Climap Project Members, 1976, The surface of the ice-age earth: *Science,* v. 191, p. 1131-1137.

Craig, H., 1957, Carbonates and carbon dioxide: *in* Revelle, R., and Fairbridge, R. W.,

Geol. Soc. Amer. Mem. 67, Treatise on Ecology and Paleoecology (J. Hedgepeth, ed.), v. 1, p. 274–275.

Corliss, B. H., 1979, Response of deep-sea benthonic foraminifera to development of the psychrosphere near the Eocene/Oligocene boundary: *Nature*, v. 282, p. 63–65.

Curry, W. B., 1977, Carbon and oxygen isotopic variation within and among species of planktonic foraminifera: *Amer. Geophys. Union Trans., Abstr. Prog.*, v. 58, p. 415.

——, and Matthews, R. K., 1981, Paleo-oceanographic utility of oxygen isotopic measurements of planktic foraminifera: Indian ocean core-top evidence. *Palaeogeogr., Palaeoclimatol. Palaeoecol.* v. 33, p. 173–191.

Degens, E. T., 1979, Carbon in the sea: *Nature*, v. 279, p. 1372.

Deuser, W. G., Ross, E. H., Hemleben, C., and Spindler, M., 1981, Seasonal changes in species composition, numbers, mass, size, and isotopic composition of planktonic foraminifera settling into the deep Sargasso Sea: *Palaeogeogr. Palaeoclimatol. Palaeoecol.*, v. 33, p. 103–127.

Douglas, R. G., and Savin, S. M., 1975, Oxygen and carbon isotope analyses of Tertiary and Cretaceous microfossils from Shatsky Rise and other sites in the North Pacific Ocean, *in* Larson, R. L., Moberly, R., *et al., Initial Reports of the Deep Sea Drilling Project*: U.S.G.P.O., Washington, D.C., v. 32, p. 509–520.

Dudley, W., 1976, Paleoceanographic applications of oxygen isotope analyses of calcareous nannoplankton growth in culture: Ph.D. dissertation, Univ. Hawaii, 168 p.

——, Duplessy, J. C., Blackwelder, P. L., Brand, L. E., and Guillard, R. R. L., 1980, Coccoliths in Pleistocene/Holocene nannofossil assemblages: *Nature*, v. 285, p. 222–223.

——, and Goodney, D. A., 1979, Oxygen isotope analysis of coccoliths grown in culture: *Deep-Sea Res.*, v. 26, p. 495.

Duplessy, J. C., Chenouard, L., and Reyss, J. L., 1974, Paleotemperatures isotopiques de l'Atlantique Equatorial: *Colloques Internationaux du Centre National de la Recherche Scientifique*, v. 219, p. 251–258.

Emiliani, C., 1954, Depth habitats of some species of pelagic foraminifera as indicated by oxygen isotope ratios: *Amer. J. Sci.*, v. 252, p. 149–158.

——, 1955, Pleistocene temperatures: *J. Geol.*, v. 63, p. 538–578.

——, 1958, Pleistocene temperatures: *J. Geol.*, v. 66, p. 264–275.

——, 1966, Paleotemperature analysis of Caribbean cores P6304–8 and P6304–9 and a generalized temperature curve for the past 425,000 years: *J. Geol.*, v. 74, p. 109–126.

——, 1971, The amplitude of Pleistocene climatic cycles at low latitudes and the isotopic composition of glacial ice, *in* Turekian, K. K., ed., *Late Cenozoic Glacial Ages*: Yale Univ. Press, New Haven, Conn., p. 183–197.

——, 1972, Quaternary paleotemperatures and duration of the high-temperature intervals: *Science*, v. 178, p. 398–401.

Epstein, S., Buchsbaum, R. A., Lowenstam, H. A., and Urey, H. C., 1951, Carbonate-water isotopic temperature scale: *Geol. Soc. Amer. Bull.*, v. 62, p. 417–426.

——, Buchsbaum, R. A., Lowenstam, H. A., and Urey, H. C., 1953, Revised carbonate-water isotopic temperature scale: *Geol. Soc. Amer. Bull.*, v. 64, p. 1315–1326.

Erez, J., 1978, Vital effect on stable-isotope composition seen in foraminifera and coral skeletons: *Nature*, v. 273, p. 199–202.

Fairbanks, R. G., Wiebe, P. H., and Be, A. W. H., 1980, Vertical distribution and isotopic composition of living planktonic foraminifera in the western north Atlantic: *Science*, v. 207, p. 61–63.

Fischer, A. G., and Arthur, M. A., 1977, Secular variations in the pelagic realm, *in* Cook, H. E., and Enos, P., eds., *Deep Water Carbonate Environments*: Soc. Economic Paleontol. and Mineral., Spec. Publ. No. 25, p. 19–50.

Funnell, B. M., and Riedel, W. R., 1971, *The Micropaleontology of the Oceans*: Cambridge Univ. Press, London, 828 p.

Gartner, S., and Keaney, J., 1978, The terminal Cretaceous event—a geological problem with an oceanographic solution: *Geology*, v. 6, p. 708–712.

Goodney, D. E., 1977, Non-equilibrium fractionation of the stable isotopes of carbon and oxygen during precipitation of calcium carbonate by marine phytoplankton: Ph.D. dissertation, Univ. Hawaii, 146 p.

——, Margolis, S. V., Dudley, W. C., Kroopnick, P., and Williams, D. F., 1980, Oxygen and carbon isotopes of Recent calcareous nannofossils as paleoceanographic indicators: *Mar. Micropaleo.*, v. 5, p. 31–42.

Haq, B. U., Worsley, T. R., Burckle, L. H., Bender, M., Keigwin, L. D., Milstone, B. A., Savin, S., and Vincent, E., 1980, The Late Miocene carbon-isotopic shift and the synchroneity of some planktonic biostratigraphic datums: *Geology*, v. 9, p. 427–431.

Hay, W. W., and Southam, J. R., 1977, Modulation of marine sedimentation by the continental shelves, *in* Anderson, N. R., and Malahoff, A., eds., *The Fate of Fossil Fuel* $CO_2$ *in the Oceans*: New York, Plenum, p. 569–604.

Hsü, K. J., 1980, Terrestrial catastrophe caused by cometary impact at the end of the Cretaceous: *Nature*, v. 285, p. 201–203.

——, Montadert, L., Bernoulli, D., Cita, M. B., Erickson, A., Garrison, R. E., Kidd, F., Melieres, F., Muller, C., and Wright, R. H., 1977, History of the Mediterranean salinity crisis: *Nature*, v. 267, p. 399–403.

Kahn, M. I., 1979, Non-equilibrium oxygen and carbon isotopic fractionation in tests of living planktonic foraminifera: *Oceanol. Acta*, v. 2, no. 2, p. 195–208.

Keigwin, L. D., Jr., Bender, M. L., and Kennett, J. P., 1979. Thermal structure of the deep Pacific Ocean in the early Pliocene: *Science*, v. 205, p. 1386–1388.

——, 1980. Paleoceanographic change in the Pacific at the Eocene-Oligocene boundary: *Nature*, v. 287, p. 722–725.

——, and Margolis, S. V. (in preparation), Details in oxygen and carbon isotope stratigraphy across the late Miocene carbon shift at Site 158, Eastern Equatorial Pacific.

Kennett, J. P., 1977, Cenozoic evolution of Antarctic glaciation, the circum-Antarctic Ocean, and their impact on global paleoceanography: *J. Geophys. Res.*, v. 82, no. 27, p. 3843–3860.

——, and Shackleton, N. J., 1976, Oxygen isotope evidence for the development of the psychosphere 38 m.y. ago: *Nature*, v. 260, p. 513–515.

——, Shackleton, N. J., Margolis, S. V., Goodney, D. E., Dudley, W. C., and Kroopnick, P. M., 1979, Late Cenozoic oxygen and carbon isotopic history and volcanic ash stratigraphy—DSDP Site 284, South Pacific: *Amer. J. Sci.*, v. 279, p. 52–69.

Killingley, J. S., and Berger, W. H., 1979, Stable isotopes in mollusk shell—detection of upwelling events: *Science*, v. 205, p. 186–188.

Kroopnick, P., 1974a, The dissolved $O_2$–$CO_2$–$^{13}C$ system in the eastern equatorial Pacific: *Deep-Sea Res.*, v. 21, p. 211–227.

——, 1974b, Correlations between $^{13}C$ and $CO_2$ in surface waters and atmospheric $CO_2$, *Earth Planet. Sci. Lett.*, v. 222, no. 4, p. 397–403.

——, 1980, The distribution of carbon-13 in the Atlantic Ocean: *Earth Planet. Sci. Lett.*, v. 49, p. 469–484.

——, Margolis, S. V., and Wong, C. S., 1977, $\delta^{13}C$ variations in marine carbonate sediments as indicators of the $CO_2$ balance between the atmosphere and oceans, *in* Anderson, N. R., and Malahoff, A., eds., *The Fate of Fossil Fuel* $CO_2$ *in the Ocean*: New York, Plenum, p. 295–321.

Le Masurier, W. E., 1970, Volcanic evidence for Early Tertiary glaciations in Marie Byrd Land: *Antarctic J.*, v. 5, p. 154–155.

Lindroth, K. J., Miller, L. G., Durazzi, J. T., McIntyre, A., and von Donk, J., 1977, Coccoliths as isotopic temperature indicators–a preliminary investigation: Unpublished manuscript.

Loutit, T. S., and Kennett, J. P., 1979, Application of carbon isotope stratigraphy to late Miocene shallow water marine sediment, New Zealand: *Science*, v. 204, p. 1196–1199.

Lowenstam, H. A., and Epstein, S., 1954, Paleotemperatures of the Post-Aptian Cretaceous as determined by the oxygen isotope method: *J. Geol.*, v. 62, p. 207–248.

Margolis, S. V., 1975, Paleoglacial history of Antarctica inferred from analysis of Leg 29 sediments by scanning electron microscopy, *in* Kennett, J. P., Houtz, R. E., *et al.*, *Initial Reports of the Deep Sea Drilling Project*: U.S.G.P.O., Washington, D.C., p. 1039–1048.

——, and Herman, Y., 1980, Northern Hemisphere sea-ice glacial development in the Late Cenozoic: *Nature*, v. 286, p. 145–149.

——, and Kennett, J. P., 1971, Cenozoic paleoglacial history of Antarctic recorded in subantarctic deep-sea cores: *Amer. J. Sci.*, v. 271, p. 1–36.

——, Kroopnick, P. M., Goodney, D. E., Dudley, W. C., and Mahoney, M. A., 1975, Oxygen and carbon isotopes from calcareous nannofossils as paleoceanographic indicators: *Science*, v. 189, p. 555–557.

——, Kroopnick, P. M., Goodney, D. E., 1977, Cenozoic and late Mesozoic paleoceanographic and paleoglacial history contained in circum-Antarctic deep-sea sediments: *Mar. Geol.*, v. 25, p. 131–147.

——, Worsley, T., and Kroopnick, P. M. (in preparation), Oxygen and carbon isotopes across the Cretaceous/Tertiary boundary from a transitional sequence at Braggs, Alabama.

Matthews, R. K., 1979, An alternative scenario for the Tertiary: Manuscript draft sent to CENOP investigators, September 25, 1979, 6 p.

McCrea, J. M., 1950, On the isotopic chemistry of carbonates and a paleotemperature scale: *Science*, v. 18, no. 6, p. 849–857.

McLean, D. M., 1978, A terminal Mesozoic "greenhouse"–lessons from the past: *Science*, v. 201, p. 401–406.

Muller, P. H., 1978, $^{14}$Carbon fixation and loss in a foraminiferal-algal symbiont system: *J. Foram. Res.*, v. 8, p. 35–41.

O'Keefe, J. A., 1980, Earth's Saturnlike rings during the Eocene/Oligocene boundary: *Nature*, v. 285, p. 369.

Parker, P. L., 1964, The biogeochemistry of the stable isotopes of carbon in a marine bay: *Geochim. Cosmochim. Acta*, v. 28, p. 1155–1164.

Perch-Nielsen, K., 1979, *in* Birkelund, T., *et al.*, eds., *Cretaceous/Tertiary Boundary Events, Vol. 1*: Univ. Copenhagen, p. 115–135.

Rotty, R. M., 1977, Global carbon dioxide production from fossil fuels and cement, A.D. 1950–A.D. 2000, *in* Anderson, N., and Malahoff, A., eds., *The Fate of Fossil Fuel* $CO_2$ *in the Oceans*: New York, Plenum, p.167–180.

Sachs, H. M., Webb, T. III, and Clark, D. R., 1977, Paleoecological transfer functions, *in* Donath, F. A., *et al.*, eds., *Ann. Rev. Earth and Planet. Sci.*, v. 5, p. 159–178.

Saito, T., and van Donk, J., 1974, Oxygen and carbon isotope measurements of Late Cretaceous and Early Tertiary foraminifera: *Micropaleont.*, v. 20, p. 152–177.

Savin, S. M., 1977, The history of the earth's surface temperature during the past 100 million years: *Ann. Rev. Earth Planet. Sci.*, v. 5, p. 319–355.

——, and Douglas, R. G., 1973, Stable isotope and magnesium geochemistry of Recent planktonic foraminifera from the South Pacific: *Geol. Soc. Amer. Bull.*, v. 84, p. 2327–2342.

——, Douglas, R. G., and Stehli, F. G., 1975, Tertiary marine paleotemperatures: *Geol. Soc. Amer. Bull.*, v. 86, p. 1499–1510.

Schlanger, S. O., and Jenkyns, H. C., 1976, Cretaceous anoxic events; causes and consequences: *Geologie en Mijnbouw*, v. 55, p. 179–186.

Scholle, P. A., and Arthur, M., 1979 (in press), Carbon isotopic fluctuations in pelagic limestones—Potential stratigraphic and petroleum exploration tool: *A.A.P.G. Bull.*

Shackleton, N. J., 1974, Attainment of isotopic equilibrium between ocean water and the benthonic foraminifera genus *Uvigerina*—isotopic changes in the ocean during the last glacial: *Paris, CNRS Colloque*, v. 219, p. 203–209.

——, 1977, Carbon-13 in *Uvigerina*—tropical rainforest history and the equatorial Pacific carbonate dissolution cycles, *in* Anderson, N. R., and Malahoff, A., eds., *The Fate of Fossil Fuel* $CO_2$ *in the Oceans*: New York, Plenum, p. 401–427.

——, and Kennett, J. P., 1975a, Paleotemperature history of the Cenozoic and the initiation of Antarctic glaciation—oxygen and carbon isotope analyses in DSDP Sites 277, 279, and 281, *in* Kennett, J. P., Houtz, R. E., *et al.*, *Initial Reports of the Deep Sea Drilling Project*: U.S.G.P.O., Washington, D.C., v. 29, p. 743–755.

——, and Kennett, J. P., 1975b, Late Cenozoic oxygen and carbon isotopic changes at DSDP Site 284—Implications for glacial history of the Northern Hemisphere and Antarctica, *in* Kennett, J. P., Houtz, R. E., *et al.*, *Initial Reports of the Deep Sea Drilling Project*: U.S.G.P.O., Washington, D.C., v. 29, p. 801–806.

——, and Opdyke, N. D., 1973, Oxygen-isotope and paleomagnetic stratigraphy of equatorial Pacific core V28–238—oxygen isotope temperatures and ice volumes on a $10^5$ year and $10^6$ year scale: *Quat. Res.*, v. 3, p. 39–55.

——, and Vincent, E., 1978, Oxygen and carbon isotope studies in recent foraminifera from the southwest Indian Ocean: *Mar. Micropaleont.*, v. 3, p. 1–13.

——, Wiseman, J. D. H., and Backley, H. A., 1973, Non-equilibrium isotopic fractionation between seawater and planktonic foraminiferal tests: *Nature*, v. 242, p. 177–179.

Smith, B. N., and Epstein, S., 1970, Biogeochemistry of the stable isotopes of hydrogen and carbon in salt marsh biota: *Plant Physiol.*, v. 46, p. 738–742.

——, and Emiliani, C., 1968, Oxygen isotopic analysis of Recent tropical Pacific benthonic foraminifera: *Science*, v. 160, p. 1335–1336.

Stump, E., Sheridan, M. F., Borg, S. G., and Sutter, J. F., 1980, Early Miocene subglacial basalts, the East Antarctic ice sheet, and uplift of the Transantarctic Mountains: *Science*, v. 207, p. 757–759.

Surlyk, F., 1980, The Cretaceous/Tertiary boundary event: *Nature*, v. 285, p. 187–188.

Tappan, H., 1968, Primary production, isotopes, extinctions and the atmosphere: *Palaeogeogr., Palaeoclimatol., Palaeoecol.*, v. 4, p. 187–210.

——, and Loeblich, A. R., 1971, *Geol. Soc. Amer. Spec. Publ.*, v. 127, p. 247–340.

Thierstein, H. R., and Berger, W. H., 1978, Injection events in earth history: *Nature*, v. 276, p. 461–464.

Urey, H. C., 1947, The thermodynamic properties of isotopic substances: *Chem. Soc. J.*, p. 562–581.

——, Lowenstam, H. A., Epstein, S., McKinney, C. R., 1951, Measurements of paleotemperatures and temperatures of the Upper Cretaceous of England, Denmark, and the Southeastern U.S.: *Geol. Soc. Amer. Bull.*, v. 62, p. 399–416.

Van Andel, Tj. H., Thiede, J., Sclater, J. G., and Hay, W. W., 1977, Depositional history of the South Atlantic Ocean during the last 125 million years: *J. Geol.*, v. 85, p. 651–698.

Van Couvering, J., Berggren, W. A., Drake, R. E., Aguirre, E., and Curtis, G. A., 1976, The terminal Miocene event: *Mar. Micropaleont.*, v. 1, p. 263–286.

Vincent, E., and Berger, W. H., 1980, The carbonate record of deep-sea biostratigraphy versus chemostratigraphy: *Oceanol. Acta.*

——, Killingley, J. S., and Berger, W. H., 1980 (in press), The magnetic epoch 6 carbon shift—a change in the ocean's $^{12}C/^{13}C$ ratio 6.2 million years ago: *Mar. Micropaleont.*, v. 5, p. 185–194.

Vinot-Bertouille, A. C., and Duplessy, J. C., 1973, Individual isotopic fractionation of carbon and oxygen in benthic foraminifera: *Earth Planet. Sci. Lett.*, v. 18, p. 247–252.

Weiner, S., 1975, The carbon isotopic composition of the eastern Mediterranean planktonic foraminifera *Orbulina universa* and the phenotypes of Globigerinoides ruber: *Palaeogeogr., Palaeoclimatol., Palaeoecol.*, v. 17, p. 149–156.

Williams, D. F., Be, A. W. H., and Fairbanks, R., 1979a, Seasonal oxygen isotopic variations in living planktonic foraminifera off Bermuda: *Science*, v. 206, p. 447–449.

——, Be, A. W. H., and Fairbanks, R., 1979b, Seasonal oxygen isotope variations in living planktonic foraminifera off Bermuda: *Science*, v. 206, p. 447–449.

——, and Healy-Williams, N., 1980, Oxygen isotopic-hydrographic relationships among Recent planktonic foraminifera from the Indian Ocean: *Nature*, v. 283, p. 848–852.

——, Sommer, M. A., and Bender, M. L., 1977, Carbon isotopic compositions of Recent planktonic foraminifera of the Indian Ocean: *Earth Planet. Sci. Lett.*, v. 36, p. 391–403.

Woodruff, F., Savin, S. M., and Douglas, R. G., 1980, Biological fractionation of oxygen and carbon isotopes by recent benthic foraminifera: *Mar. Micropaleont.*, v. 5, p. 3–11.

Worsley, T., 1974, The Cretaceous/Tertiary boundary event in the oceans, *in* Hay, W. W., ed., *Soc. Econ. Paleont. Mineral. Spec. Publ. 20*, p. 94–125.

Worsley, T. R., and Davies, T. A., 1979, Sea level fluctuations and deep-sea sedimentation rates: *Science*, v. 203, p. 455–456.

J. Kirk Cochran*
Department of Geology and Geophysics
Yale University
New Haven, Connecticut 06520

# 3

# THE USE OF NATURALLY OCCURRING RADIONUCLIDES AS TRACERS FOR BIOLOGICALLY RELATED PROCESSES IN DEEP-SEA SEDIMENTS

*Present Address: Woods Hole Oceanographic Institution
Woods Hole, Massachusetts 02543

## ABSTRACT

Naturally occurring radionuclides serve as chronometers for deriving rates of certain biologically related processes in the deep-sea environment. Criteria for selecting appropriate nuclides as tracers include chemical behavior and half-life relative to the rate of the process in question. Nuclides which are scavenged by particles in seawater are potential tracers for biological mixing of deep-sea sediments. Near-interface gradients in shorter-lived nuclides like $^{210}$Pb and $^{32}$Si demonstrate that deep-sea sediments are mixed on time scales of $10^2$ y with particle mixing coefficients of $(1 - 13) \times 10^{-9}$ cm$^2$/s. The distributions of longer-lived nuclides in the deep-sea sediment column indicate that continuous particle mixing extends from 7 to 15 cm below the sediment water interface. Nuclides like $^{228}$Ra and $^{210}$Pb incorporated into the calcium carbonate skeletons of deep-sea corals and mollusks yield growth rates for these organisms. The distribution of $^{228}$Ra in different size fractions of a deep-sea deposit feeding bivalve indicates that the clam grows to a size of ~8 mm in ~100 y. The activity ratio of $^{228}$Th to $^{228}$Ra in a large vesicomyid clam from the Galápagos hot spring area gives an age for the 22-cm specimen of ~7 y. Application of the $^{228}$Ra and $^{210}$Pb chronometers to a population of deep-sea corals shows that a coral size of ~3 cm is reached in 6 or 60 y, depending on the growth function chosen.

## INTRODUCTION

A number of processes occurring near the deep-sea sediment-water interface can be attributed directly to the activities of animals living in this ecologic niche. Of particular interest in this context are the rate and depth of mixing of deep-sea sediments by organisms and the rate of growth of animals which secrete a hard skeleton or shell (e.g., of CaCO$_3$). In the near-shore regime, rates of these processes often are rapid and can be determined by direct observation. In the deep sea, direct observation is difficult and rates of metabolic processes tend to be slower (Jannasch and Wirsen, 1973; Smith and Teal, 1973). Thus there is the need for chronometers or clocks that record and time the process in question. Such chronometers are found in various naturally occurring radionuclides which are input to the ocean in a variety of ways. Those which will be considered in the present discussion include:

1. Members of the naturally occurring $^{238}$U, $^{235}$U, and $^{232}$Th decay series. These three primordial isotopes have long half-lives and decay to produce a number of daughter nuclides that are useful as chronometers (see Fig. 3-1).
2. Cosmogenic nuclides, which are produced by the interaction of cosmic rays with the atmosphere. Cosmogenic nuclides useful in the study of deep-sea, near-interface processes are given in Table 3-1.

The usefulness of any of the nuclides in the above groups for studying sediment mixing or animal growth rates depends on properties each nuclide possesses. Specifically, the radioactive nature of each nuclide makes it a chronometer, and its characteristic chemical behavior links it to the process in question.

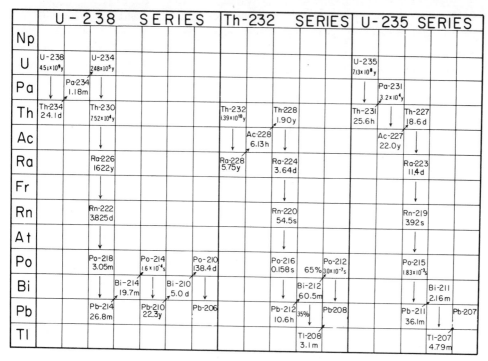

Fig. 3-1. $^{238}$U, $^{232}$Th, and $^{235}$U decay series.

TABLE 3-1. Cosmogenic Nuclides Useful in Studying Deep-Sea Processes

| Nuclide | Half-life |
|---------|-----------|
| $^{32}$Si | 300 y |
| $^{14}$C | 5730 y |
| $^{10}$Be | $1.5 \times 10^6$ y |
| $^{26}$Al | $0.75 \times 10^6$ y |

## THE NATURE OF RADIOACTIVITY

Radioactive decay is the spontaneous change of an atom of one nuclide into that of another. Such decay occurs at a characteristic rate, which is proportional to the number of atoms of the nuclide present at any given time. Thus this rate of change may be expressed

$$\frac{dN}{dt} = -\lambda N \qquad (1)$$

where $N$ = number of atoms of a particular nuclide at time $t$

$\lambda$ = constant of proportionality, termed the decay constant, which is peculiar to the nuclide in question

The quantity $dN/dt$ is termed the radioactivity or simply the activity of the nuclide, and the negative sign indicates that this quantity is decreasing (or decaying) with time.

Integrating Eq. (1) reveals the nature of this decrease:

$$N = N_0 e^{-\lambda t} \tag{2}$$

where $N_0$ = number of atoms present at time $t = 0$. Equation (2) indicates that the number of atoms decreases in an exponential fashion such that the number of atoms is reduced by half in a characteristic time, $t_{1/2}$. This time is referred to as the half-life and is related to the decay constant, $\lambda$ as $t_{1/2} = .693/\lambda$.

Many radioactive nuclides decay to stable, nonradioactive nuclides. However, Fig. 3-1 shows that although $^{238}$U, $^{235}$U, and $^{232}$Th do ultimately decay to stable isotopes of lead, they produce a variety of intermediate nuclides (termed "daughters") which are themselves radioactive. Thus a decay series is established in which all members but the last are radioactive. If the series is isolated long enough (generally four to five half-lives of the longest lived daughter), a state of radioactive equilibrium is reached in which all activities in the decay series are equal.

Because the various nuclides in the series correspond to different elements, they have different chemical behaviors in the marine realm. Such disparate behaviors allow for perturbations in the state of radioactive equilibrium. These perturbations create radioactive disequilibria, and the attempt to return to a state of radioactive equilibrium makes certain of the nuclides useful as clocks. Thus chemical behavior and half-life are keys to which nuclides will be useful in any given situation.

## SUPPLY AND REMOVAL OF RADIONUCLIDES IN THE OCEANS

Both uranium and thorium are chemically mobilized during weathering (Scott, 1968), but thorium has a much greater affinity for particles. Thus a significant fraction of $^{238}$U, $^{234}$U, and $^{235}$U added to the oceans arrives dissolved in rivers, and $^{232}$Th is transported dominantly in association with particles. The dissolved U in the oceans (as well as $^{228}$Ra) provides an *in situ* production of several Th isotopes: $^{234}$Th, $^{230}$Th, and $^{228}$Th. Like $^{232}$Th, these isotopes of Th are particle-reactive and the extent of their removal from seawater depends on the rate of removal onto particles compared with the rate of radioactive decay. The long half-life of $^{230}$Th virtually insures its scavenging by particles and deposition in deep-sea sediments (Fig. 3-2). The shorter half-lives of $^{234}$Th and $^{228}$Th allow for some fraction of these nuclides to remain in the water column, and precise evaluation of this "fraction" enables chemical residence times of these nuclides to be calculated.

In the deep-sea sediment column, $^{230}$Th is largely unaccompanied by any $^{234}$U and decays with time (or depth in the sediment column), according to Eq. (2). $^{226}$Ra, produced by $^{230}$Th decay, and its daughter $^{222}$Rn are considerably less particle-reactive than $^{230}$Th, and during their production they are mobilized into the porewater of the sediment. From there, they migrate into the overlying water column (Fig. 3-2). $^{228}$Ra is mobilized to sediment porewater in a manner analogous to $^{226}$Ra, even though $^{232}$Th is carried to deep-sea sediments dominantly in detrital phases. The migration of $^{226}$Ra and $^{228}$Ra from sediments to the overlying water is the dominant supply of these nuclides to the oceans. Their oceanic distributions are controlled by circulation time scales of the oceans relative to their half-lives. $^{226}$Ra has a half-life (1622 y) comparable to the residence time of the deep ocean ($\sim 10^3$ y) and thus is mixed throughout the water column. $^{228}$Ra, with a shorter half-life (5.75 y), is confined to near-bottom waters in the deep sea and to surface waters which have been in contact with shallow, coastal sediments. Ra is involved in the biological cycle in the oceans (Broecker *et al.*, 1976; Chan *et al.*, 1976; Ku *et al.*, 1970), and both Ra isotopes are incorporated into calcium carbonate skeletons such as those of corals and mollusks. Both Ra isotopes decay to more reactive daughters: $^{226}$Ra ultimately to $^{210}$Pb, and $^{228}$Ra to $^{228}$Th. Some of the $^{210}$Pb and $^{228}$Th so produced is scavenged by particles and removed to the sediments. This process operates in both the near-shore and deep-sea regimes (Fig. 3-2).

Another important route by which radionuclides enter the ocean is through the atmosphere. $^{222}$Rn (a noble gas) emanating from continental rocks and soils decays in the atmosphere to $^{210}$Pb, which, entrapped in rain, falls on both land and ocean.

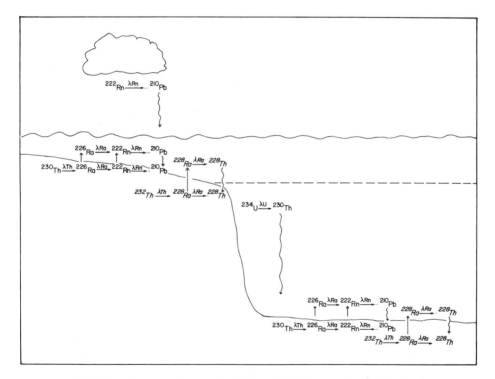

Fig. 3-2.   Supply and removal of U and Th decay series nuclides in the oceans. Horizontal arrows signify radioactive decay. Solid vertical arrows correspond to diffusive fluxes out of sediments into overlying water. Wavy arrows represent removal of nuclide onto particles, followed by deposition. Similar processes take place in the near-shore and deep-sea regimes, although the greater water depths of the open ocean allow the addition of excess $^{230}$Th (derived from dissolved $^{234}$U) to deep-sea sediments.

By this pathway some $^{210}$Pb is supplied to the surface of the oceans, where it can react with particles and be incorporated into sediments. The various cosmogenic nuclides are also supplied to the ocean surface from the atmosphere. There they behave chemically like their stable forms. Thus $^{14}$C is incorporated into C-containing material such as organic matter or shells made of calcium carbonate and $^{32}$Si is transferred from solution to the shells of silica-secreting organisms (e.g. diatoms, radiolaria, some sponges). $^{7}$Be and $^{10}$Be behave somewhat like Th and are scavenged by suspended particles. As is the case for Th, the extent of removal of Be from the water column depends on the rate of scavenging relative to the rate of decay.

From this brief description of the properties of the various nuclides, it is possible to delineate some processes they can be used to track. For example, (1) the removal from seawater of nuclides like $^{230}$Th, $^{228}$Th, $^{210}$Pb, $^{14}$C, and $^{32}$Si onto sediment particles makes them useful for determining rates of sediment associated processes like accumulation or biological mixing, and (2) the incorporation of $^{226}$Ra, $^{228}$Ra, $^{210}$Pb, and $^{14}$C into biological materials, calcareous shells in particular, enables the determination of growth rates or ages of calcareous skeletons.

## PARTICLE-MIXING RATES IN DEEP-SEA SEDIMENTS

Photographs of the deep-sea bottom (e.g. Fig. 3-3) show abundant evidence of bioturbation by bottom-dwelling fauna. However, a photograph alone does not convey any real information regarding how fast this mixing takes place, and thus we search for tracers

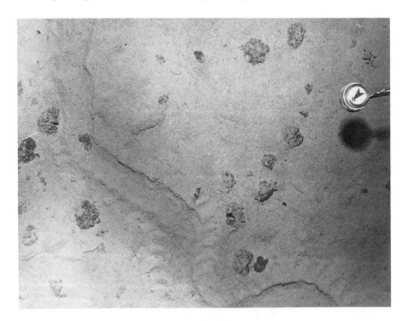

Fig. 3-3.    Photograph of the deep-sea floor southeast of Hawaii (5000-m depth). The dark objects are ferromanganese nodules about 8 cm in diameter. Disruption of the surface sediment by the burrowing activities of animals is evident. Photo courtesy of NOAA.

THE USE OF NATURALLY OCCURRING RADIONUCLIDES AS TRACERS

which can integrate mixing effects over long periods of time. As described in the previous section, our choice of a tracer depends on the degree to which it is associated with the process in question and its half-life. If we wish to obtain time scales for the mixing of particles, we must choose a nuclide which is strongly bound to the particulate phase. For tracing fluid bioturbation such as occurs when an infaunal organism irrigates its burrow, we need a tracer associated predominantly with the fluid phase (i.e. porewater).

For studying particle mixing, there are a number of nuclides associated with the sediment phase. These include $^{32}$Si, $^{14}$C, $^{230}$Th, $^{231}$Pa, and $^{210}$Pb. The half-lives of these nuclides span a range from tens to thousands of years, and hence, not all will be useful to determine mixing rates. If the rate of decay (as indicated by the half-life of a nuclide) is comparable to the rate at which it is mixed into the sediment pile, the nuclide should show a gradient near the sediment-water interface (e.g., $^{32}$Si, $^{210}$Pb). Longer-lived nuclides will behave more nearly as stable tracers, and their activities will be essentially constant over a continuously mixed depth in the sediment column. This property enables long-lived nuclides to be used to estimate the depth to which the sediments are mixed continuously.

The first attempt to use several of these nuclides to study mixing in deep-sea sediments was by Nozaki *et al.* (1977). They analyzed cores collected in the FAMOUS project (Fig. 3-4). In this study cores of calcareous ooze were recovered by the research submersible *Alvin* in the rift valley of the Mid-Atlantic Ridge (see Fig. 3-4). Coring by submersible insures that the sediment-water interface is recovered, a requirement for near-interface studies.

Fig. 3-4.    Recovery of deep-sea core by the *Alvin*. Cores ~25 cm long were taken during Project FAMOUS (French-American Mid-Ocean Undersea Study) by slowly inserting a plastic core barrel into the carbonate ooze. From Nozaki *et al.,* 1977.

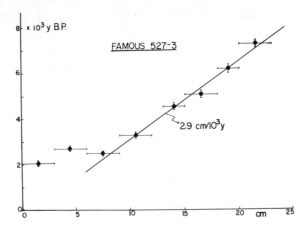

Fig. 3-5. $^{14}$C ages as a function of depth in the sediment column for Mid-Atlantic Ridge core FAMOUS 527-3. Note constancy of the age in the top 8 cm. The linear increase below 8 cm corresponds to a sediment accumulation rate of 2.9 cm/1000 y. The pattern is produced by continuous bioturbation to 8 cm with essentially no mixing below. Data from Nozaki *et al.*, 1977.

The patterns of $^{14}$C and $^{210}$Pb found in one of the cores studied by Nozaki *et al.* (1977) are shown in Figs. 3-5 and 3-6. $^{14}$C shows a nearly constant age with depth to ~8 cm (Fig. 3-5). Below 8 cm, age increases regularly, and the age-depth relationship gives a sediment accumulation rate of 2.9 cm/1000 y. $^{210}$Pb in the same core shows a gradient in the top 8 cm, decreasing until its activity is essentially equal to that of $^{226}$Ra (Fig. 3-6).

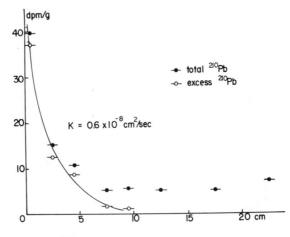

Fig. 3-6. $^{210}$Pb activity as a function of depth in core FAMOUS 527-3. Excess $^{210}$Pb (total $^{210}$Pb activity-$^{226}$Ra activity) is distributed to 8 cm. The solution to Eq. (3) applied to the profile yields a particle mixing coefficient of $6 \times 10^{-9}$ cm$^2$/sec. Data from Nozaki *et al.*, 1977.

THE USE OF NATURALLY OCCURRING RADIONUCLIDES AS TRACERS

The information yielded by these two tracers is that the sediments are mixed continuously to a depth of 8 cm. The rate of mixing is fast enough to allow the $^{14}C$ to be homogenized but sufficiently slow that there is a gradient in $^{210}Pb$. The $^{210}Pb$ profile allows a crude time scale to be attached to the reworking—viz., the 8 cm must be reworked completely in at least 100 years (i.e., about 5 half-lives of $^{210}Pb$) in order for there to be excess $^{210}Pb$ at all depths in the mixed zone.

A more exact way of expressing the rate at which sediment is bioturbated is by analogizing the mixing process to eddy diffusion [see, for example, Goldberg and Koide (1962) and Guinasso and Schink (1975)]. The equation commonly used to express all of the processes affecting the distribution of a radioactive tracer in the near-interface region is:

$$\frac{\partial A}{\partial t} = D_B \frac{\partial^2 A}{\partial z^2} - S \frac{\partial A}{\partial z} - \lambda A + P \tag{3}$$

where $A$ = radioactivity of nuclide

$D_B$ = particle-mixing coefficient

$S$ = sediment accumulation rate

$\lambda$ = decay constant of nuclide

$P$ = production rate of nuclide from decay of parent

$z$ = depth in the sediment column

$t$ = time

Equation (3) can be solved exactly by assuming the tracer distribution is in steady state ($\partial A/\partial t = 0$), and applying appropriate boundary conditions. The values of $\lambda$ and $P$ are known, and $S$ may be determined using the distribution of longer-lived tracers below the mixed zone. $D_B$ is determined by fitting the solution of Eq. (3) to the activity-depth data. It is important to note that although sediment mixing may not be occurring in a strict eddy-diffusionlike manner, the use of $D_B$ provides a parameter through which to compare mixing at different locations.

Detailed $^{14}C$ and $^{210}Pb$ analyses have been made for the near-interface regions of several other carefully collected carbonate-rich deep-sea cores (Peng et al., 1977, 1979). In general, these studies show results comparable to those discussed above (see Table 3-2).

Not all deep-sea sediments are carbonate rich. Much of the floor of the Pacific Ocean, for example, is characterized by slowly accumulating clays sometimes associated with the remains of siliceous plankton (e.g. Fig. 3-3). The results of a study of the distributions of $^{230}Th$ and $^{210}Pb$ in such sediment are shown in Figs. 3-7 and 3-8. Like the $^{14}C$ profiles in the carbonate core from the FAMOUS area, the excess $^{230}Th$ activity (measured $^{230}Th$ activity minus that of its parent, $^{234}U$) is constant over a depth interval of ~10 cm (Fig. 3-7). Below 10 cm, the decrease in $^{230}Th$ activity indicates a sediment accumulation rate of a few millimeters per thousand years (Cochran and Krishnaswami, 1980).

TABLE 3-2. Mixing Rates in Deep-Sea Sediments (Based on $^{210}$Pb Profiles)

| Location | Water Depth (m) | Sediment Type | Sediment Accumulation Rate (cm/1000 y) | Depth of Continuous Mixing (cm) | Mixing Coefficient, $D_B$ (cm²/s) | Reference[a] |
|---|---|---|---|---|---|---|
| *North Atlantic* | | | | | | |
| 36°49'N, 38°15'W | 2705 | calcareous ooze | 2.9 | 8 | $6 \times 10^{-9}$ | (1) |
| *South Atlantic* | | | | | | |
| 41°33'S, 20°12'E | 5000 | calcareous ooze | <0.4 | 9 | $1 \times 10^{-9}$ | (2) |
| *North equatorial Pacific* | | | | | | |
| 9°2'N, 151°11'W | 5040 | clay-siliceous ooze | 0.15 | 7 | $7 \times 10^{-9}$ | (3) |
| 11°15'N, 139°4'W | 4830 | clay-siliceous ooze | 0.31 | 8 | $8 \times 10^{-9}$ | (3) |
| 15°20'N, 125°54'W | 4640 | clay-siliceous ooze | 0.14 | 12 | $13 \times 10^{-9}$ | (3) |
| *South equatorial Pacific* | | | | | | |
| 2°14'S, 157°0'E | 1600 | calcareous ooze | ~2.0 | 8 | $4 \times 10^{-9}$ | (4) |
| *Antarctic* | | | | | | |
| 53°0'S, 35°38'E | 4340 | siliceous ooze | 1.5 | 15 | $1 \times 10^{-9}$ | (2) |
| 52°59'S, 35°43'E | 4730 | siliceous ooze | — | — | $5 \times 10^{-9}$ | (2) |
| 53°16'S, 35°54'E | 4540 | siliceous ooze | — | — | $3 \times 10^{-9}$ | (2) |
| 66°47'S, 30°08'E | 4080 | clay | <3.0 | 9 | $1 \times 10^{-9}$ | (2) |

[a] (1) Nozaki *et al.*, 1977.
(2) DeMaster, 1979.
(3) Cochran and Krishnaswami, 1979.
(4) Peng *et al.*, 1979.

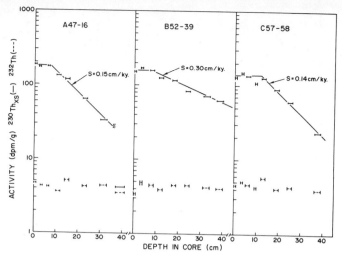

Fig. 3-7.   $^{230}$Th and $^{232}$Th activity as a function of depth in three north equatorial Pacific cores taken from the area shown in Fig. 3-3. $^{232}$Th is contained in the detrital fraction of the sediment, and its activity is essentially constant with depth. $^{230}$Th is in excess of its parent, $^{234}$U, and the decrease with depth (time) in the sediment column yields a sediment accumulation rate. Constancy of $^{230}$Th activity in the near-interface region is due to biological mixing which is rapid compared to the half-life of $^{230}$Th. Data from Cochran and Krishnaswami, 1980.

$^{210}$Pb in each of these cores is present to depths of 7 to 12 cm (Fig. 3-8). Values of the mixing coefficients $(D_B)$ range from 7 to 13 × 10$^{-9}$ cm$^2$/s, not substantially different from that found in the Mid-Atlantic Ridge core. An independent check is available on the $^{210}$Pb mixing rates determined for these cores. The sediments contain about 10% opaline silica (DeMaster, 1979), mostly the tests of diatoms and radiolaria.

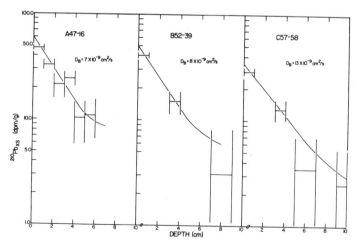

Fig. 3-8.   Excess $^{210}$Pb versus depth in three north equatorial Pacific cores. Gradients in excess $^{210}$Pb are due to bioturbation, and the mixing coefficients calculated using Eq. (3) are shown. See Table 3-2; data from Cochran and Krishnaswami, 1980.

J. KIRK COCHRAN

As these tests formed, they incorporated the cosmogenic nuclide $^{32}$Si. The half-life of $^{32}$Si is sufficiently short (~300 y) that its distribution in these sediments should be dominated by mixing rather than sediment accumulation.

DeMaster (1979) has determined the particle-mixing coefficient from the $^{32}$Si profile in core A47–16 to be $1.6 \times 10^{-9}$ cm$^2$/s, approximately a factor of five less than that obtained using $^{210}$Pb. The difference in $^{210}$Pb and $^{32}$Si mixing rates may reflect the possibility that the tracers are associated with particles of different sizes—$^{210}$Pb with the clay or fine-grained fraction and $^{32}$Si with larger siliceous tests and test fragments. Selectivity in particle sizes during ingestion by bottom-dwelling fauna may then cause the tracers to reflect two different mixing rates. In this case, the siliceous tests are mixed more slowly than is the fine-grained clay fraction.

$^{32}$Si and $^{210}$Pb also have been compared as tracers of particle mixing in highly siliceous Antarctic sediments (DeMaster, 1979). In such sediments the mixing rates calculated using the two tracers are in good agreement.

Table 3–2 summarizes particle-mixing coefficients and depths of mixing for a number of deep-sea cores from the world oceans. DeMaster and Cochran (1977) have made several observations from these results:

1. In sediments of different types with different accumulation rates, the depth of mixing varies by only a factor of two.
2. Mixing coefficients vary by about an order of magnitude, but show no clear relationship to sediment type or accumulation rate.

Turekian *et al.* (1978) have observed that the variation in particle-mixing rates for deep-sea sediments may be related to variability in benthic biomass (about a factor of 10 in deep-sea sediments) and in surface productivity (about a factor of three, excluding extremes), among other factors. Benthic biomass may vary locally and control variability in mixing rates within a small area (as in the Antarctic cores listed in Table 3–2), whereas variation in surface productivity may control variability from one oceanic area to another.

## GROWTH RATES OF DEEP-SEA ORGANISMS

The growth of organisms living at the deep-sea sediment-water interface is another process whose rate may be determined using naturally occurring radionuclides. In particular, it is possible to determine the growth rates of organisms (e.g. clams, corals) which secrete a skeleton of calcium carbonate. As the shell is formed, nuclides of the uranium and thorium decay series are incorporated from seawater. This is especially true of U, $^{226}$Ra, $^{228}$Ra, and $^{210}$Pb. The choice of a suitable nuclide for growth-rate determinations depends on the half-life of the nuclide relative to the growth rate, and both $^{228}$Ra and $^{210}$Pb have proved useful in this context.

There are two methods of using nuclides as chronometers for shell growth. One involves determining the activities in different size fractions of a given population. The other involves activity measurements in different growth increments of a single shell. Both methods require some knowledge of how the mass of the shell changes as a function of time, that is, of a "growth law" for the shell.

The first reported use of radionuclides to determine growth rates of deep-sea benthos was by Turekian *et al.* (1975). These workers measured $^{228}$Ra activities in four size fractions of *Tindaria callistiformis*, a deposit-feeding bivalve recovered from 3800 m in the North Atlantic (Fig. 3-9). As Fig. 3-9 shows, shells of *Tindaria* have regularly spaced increments which Turekian *et al.* (1975) interpreted as due to shell growth. A relationship between the length of a shell and the number of these increments, coupled with an observed cubic relationship between mass and length [i.e. mass = constant $\times$ (length)$^3$], gave a growth-rate function of mass versus time for *Tindaria*. If the ratio of $^{228}$Ra to Ca being added to the shell during growth is known, then the $^{228}$Ra activity in any size fraction is given by the integrated amount of mass added, weighted for the decay of $^{228}$Ra in each mass increment since the mass (and $^{228}$Ra) was added. Mathematically, this may be expressed as

$$A_{Ra}(T) = A_{Ra}^0 \int_{T_0}^{T} f(t)\, e^{-\lambda(T-t)} dt \tag{4}$$

where $A_{Ra}$ = activity of $^{228}$Ra at time, $T$, of recovery from the sea floor

$A_{Ra}^0$ = $^{228}$Ra/Ca ratio added during shell growth, assumed constant

$f(t)$ = growth-rate function expressed as change of mass per time

$\lambda$ = decay constant of $^{228}$Ra

$T_0$ = time of inception of an individual

and $T$-$T_0$ = age of an individual when collected

For *Tindaria*, Turekian *et al.* (1975) determined $f(t)$ from shell growth increments, as discussed above. To avoid having to know $A_{Ra}^0$, Eq. (4) was expressed as ratios of

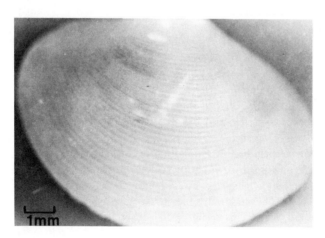

Fig. 3-9. Photograph of shell of *Tindaria callistiformis*. This small (~8 mm) deposit-feeding bivalve was recovered from a depth of 3800 m in the North Atlantic. Regularly spaced bands are inferred to be growth increments.

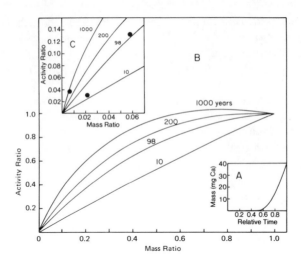

Fig. 3-10. $^{228}$Ra activity ratio ($^{288}$Ra activity in individual of a given size fraction normalized to that of the largest size fraction) versus mass ratio (mass of individual of a given size fraction normalized to that of the largest size fraction), for *Tindaria*. Curves are calculated from Eq. (4) for various ages of the oldest individual. The data suggest an age of 98 years. Data from Turekian *et al.*, 1975.

$^{228}$Ra activities in the smaller size fractions to that of the largest size fraction. By selecting the age of the largest individuals, the family of curves shown in Fig. 3-10 was calculated. Although there is considerable scatter, the data are not inconsistent with an age for the largest individuals of ~100 y. This result is in agreement with expectations of slow growth rates in deep-sea benthos based on low metabolic rates (Grassle and Sanders, 1973; Smith and Teal, 1973) and the general invariant pressure and temperature of the abyssal ocean floor.

The result obtained for *Tindaria* does not hold for all deep-sea bivalves, however. The hydrothermal springs of the Galápagos Rift support a considerable number of fauna, including large vesicomyid clams (Corliss *et al.*, 1979, see Fig. 3-11). Turekian *et al.* (1979) have determined the age of one such clam by using the $^{228}$Th/$^{228}$Ra activity ratio as a chronometer. As in *Tindaria*, $^{228}$Ra is incorporated into the shell as the clam grows. Once there, it decays to $^{228}$Th, and the $^{228}$Th/$^{228}$Ra ratio in a single specimen is an indicator of the age of the organism. Also, as with *Tindaria,* a growth law for mass as a function of time must be specified. For the case of a constant rate of mass addition with time ($m = ct$), Eq. (4) (plus a similar equation for the change of $^{228}$Th in the shell with time), applied to the vesicomyid analyzed by Turekian *et al.* (1979) yields the growth curve shown in Fig. 3-12. The measured $^{228}$Th/$^{228}$Ra activity ratio corresponds to an age for this 22-cm clam of ~7 y. This result, while different from that obtained for *Tindaria*, is in agreement with what is known about the East Pacific Rise hydrothermal springs (Corliss *et al.*, 1979). Unlike the ambient ocean floor, these springs are characterized by elevated temperatures. The circulation patterns of seawater through the mid-ocean ridge basalts can change with time, leading to extinction of a par-

THE USE OF NATURALLY OCCURRING RADIONUCLIDES AS TRACERS

Fig. 3-11. *Alvin* recovering vesicomyid clams from a Galápagos hydro-thermal spring. The clams are about 20 cm long. Photo courtesy R. Ballard, Woods Hole Oceanographic Institution.

ticular vent. Rapid growth rates are advantageous in such a relatively unpredictable and varying environment.

In addition to bivalves, growth rates for deep-sea solitary corals also have been determined using $^{228}$Ra activities in different size fractions of a population (Turekian *et al.*, 1981). As with clams, a growth law must be specified, and Fig. 3-13 (from Turekian *et al.*, 1981) shows two possibilities. In both cases, the coral is approximated as a cone, and the change in the total $^{228}$Ra in a coral of given size is plotted as a function of mass for cases of constant length increase with time and constant volume increase with time. For the best-fit activity versus mass curves shown in Fig. 3-13, the age of the largest specimen analyzed (~2.5 cm in length) is 6 y for the case of constant volume increase with time ($dv/dt$ = constant) and 60 y for constant length increase with time ($dL/dt$ = constant). One possible way of deciding between these two alternatives is to use another nuclide of different half-life. $^{210}$Pb determined in these corals is considerably in excess of $^{226}$Ra and shows little change between different size fractions. Although

Fig. 3-12. $^{228}$Th/$^{228}$Ra activity ratio as a function of age of vesicomyid clam recovered from the Galápagos hot spring area. The measured value of the ratio 0.68 ± 0.18, corresponds to an age of $6.5^{+5}_{-3}$ years for the 22-cm clam analyzed. Reprinted by permission from *Nature*, vol. 280, pp. 385–387. Copyright © 1979 Macmillan Journals Limited.

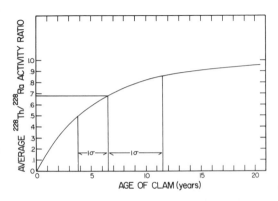

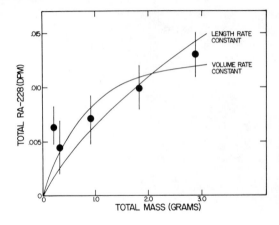

Fig. 3-13. Total $^{228}$Ra activity in a solitary coral as a function of the mass of the coral. The data were collected from different size fractions of a population of deep-sea corals from the South Atlantic. The curves are best fit model calculations of activity versus mass for growth functions of constant length rate (dL/dt = constant) and constant volume rate (dv/dt = constant). For the two models, the largest individual in the population has an age of 6 years for the constant volume rate case and 60 years for the constant length rate case. Data from Turekian *et al.*, 1981.

the $^{210}$Pb data must be interpreted with caution because $^{210}$Pb may not be incorporated into the coral skeleton (rather, it may be adsorbed on the surface), they tend to agree better with more rapid growth predicted by the constant volume rate model.

## SUMMARY

Naturally occurring radionuclides are useful time tracers for certain biologically related processes operating near the deep-sea sediment-water interface. This paper has considered two such processes: the rate and depth of mixing of bottom sediments by organisms, and the rate of growth of benthic fauna which secrete a calcium carbonate shell or skeleton. Selecting appropriate nuclides to use as chronometers depends on the chemical behavior of the nuclide and its half-life relative to the rate of the process in question.

Nuclides that are removed from seawater by particles are potential tracers for biological mixing of deep-sea sediments. Shorter-lived nuclides like $^{210}$Pb and $^{32}$Si can be used to determine rates of sediment mixing, whereas longer-lived nuclides (e.g. $^{230}$Th, $^{231}$Pa, and $^{14}$C) indicate the depth to which mixing occurs. $^{210}$Pb removed from the overlying water column is distributed in a mixed layer near the tops of carefully collected deep-sea sediment cores. The nature of the $^{210}$Pb decrease below the sediment-water interface is used to calculate particle-mixing coefficients, which range from 1 to 13 × $10^{-9}$ cm$^2$/s and correspond to turnover times of ~100 years for the mixed layer. Depths of continuous mixing of deep-sea sediments range from 7 to 15 cm.

Nuclides that are incorporated into the calcium carbonate hard parts of mollusks and corals are chronometers for growth rates of these organisms. $^{228}$Ra has proved especially useful in this context. $^{228}$Ra activities in different size fractions of a small deposit-feeding deep-sea bivalve, *Tindaria callistiformis*, suggest that this clam reaches a maximum size of ~8 mm in ~100 y. Other deep-sea bivalves grow more rapidly. The activity ratio of $^{228}$Th (a daughter of $^{228}$Ra) to $^{228}$Ra in a vesicomyid clam from the Galápagos hot spring area indicates that the clam grows to a size of ~20 cm in ~7 y.

$^{228}$Ra measurements on a population of deep-sea solitary corals show that the age of the largest coral examined ($\sim$2.5 cm) is 6 or 60 y, depending on whether constant volume or constant length increase with time is chosen as the growth function.

## ACKNOWLEDGMENTS

The research described in this paper was supported through National Science Foundation grant OCE76-02-39 and Department of Energy grant EY-76-5-02-3573, both to Karl K. Turekian, Principal Investigator. I am indebted to K. Turekian, R. Aller, L. Benninger, D. DeMaster, Y. Nozaki, and N. Landman for their many discussions with me of the ideas presented in this paper.

## REFERENCES

Broecker, W. S., 1965, An application of natural radon to problems in ocean circulation, *in* Ichiye, T., ed., *Symposium on Diffusion in Oceans and Fresh Waters,* Lamont-Doherty Geological Observatory, p. 116–141.

——, Goddard, J., and Sarmiento, J., 1976, The distribution of Ra–226 in the Atlantic Ocean: *Earth Planet. Sci. Lett.,* v. 32, p. 220–235.

Chan, L. H., Edmond, J. M., Stallard, R., F., Broecker, W. S., Chung, Y. C., Weiss, R. F., and Ku, T. L., 1976, Radium and barium at GEOSECS stations in the Atlantic and Pacific: *Earth Planet. Sci. Lett.,* v. 32, p. 258–267.

Chung, Y. C., and Craig, H., 1972, Excess-radon and temperature profiles from the eastern equatorial Pacific: *Earth Planet. Sci. Lett.,* v. 14, p. 55–64.

Cochran, J. K., 1979, The geochemistry of $^{226}$Ra and $^{228}$Ra in marine deposits: Ph.D. Thesis, Yale Univ., New Haven, Conn., 260 p.

——, and Krishnaswami, S., 1980, Radium, thorium, uranium and Pb–210 in deep-sea sediments and sediment pore waters from the north equatorial Pacific: *Amer. J. Sci.,* v. 280, p. 849–889.

Corliss, J. B., Dymond, J., Gordon, L. I., Edmond, J. M., von Herzen, R. P., Ballard, R. D., Green, K., Williams, D., Bainbridge, A., Crane, K., and van Andel, T. H., 1979, Submarine thermal springs on the Galápagos Rift: *Science,* v. 203, p. 1073–1083.

DeMaster, D., 1979, The marine silica and $^{32}$Si budgets: Ph.D. Thesis, Yale Univ., New Haven, Conn., 308 p.

——, and Cochran, J. K., 1977, Rates of particle mixing in deep-sea sediments using Pb–210 measurements: *EOS, Trans. Amer. Geophys. Un.,* v. 58, p. 1154.

Goldberg, E. D., and Koide, M., 1962, Geochronological studies of deep-sea sediments by the ionium/thorium method: *Geochim. Cosmochim. Acta,* v. 26, p. 417–450.

Grassle, J. F., and Sanders, H. L., 1973. Life histories and the role of disturbance: *Deep-Sea Res.,* v. 20, p. 643–659.

Guinasso, N. L. Jr., and Schink, D. R., 1975, Quantitative estimates of biological mixing rates in abyssal sediments: *J. Geophys. Res.,* v. 80, p. 3032–3043.

Jannasch, H. W., and Wirsen, C. O., 1973, Deep-sea microorganisms—*in situ* response to nutrient enrichment: *Science*, v. 180, p. 641–643.

Key, R. M., Guinasso, N. L. Jr., and Schink, D. R., 1979, Emanation of radon–222 from marine sediments: *Mar. Chem.*, v. 7., p. 221–250.

Ku, T. L., Li, Y. H., Mathieu, G. G., and Wong, H. K., 1970, Radium in the Indian-Antarctic Ocean south of Australia: *J. Geophys. Res.*, v. 75, p. 5286–5292.

Nozaki, Y., Cochran, J. K., Turekian, K. K., and Keller, G., 1977, Radiocarbon and Pb–210 distribution in submersible-taken deep-sea cores from Project FAMOUS: *Earth Planet. Sci. Lett.*, v. 34, p. 167–173.

Peng, T. H., Broecker, W. S., Kipphut, G., and Shackleton, N., 1977, Benthic mixing in deep-sea cores as determined by $^{14}$C dating and its implications regarding climate stratigraphy and the fate of fossil fuel $CO_2$, *in Fate of Fossil Fuel $CO_2$ in the Ocean:* Anderson, N. R., and Malahoff, A., eds., New York, Plenum, p. 355–373.

——, Broecker, W. S., and Berger, W. H., 1979, Rates of benthic mixing in deep-sea sediment as determined by radioactive tracers: *Quat. Res.*, v. 11, p. 141–149.

Scott, M. R., 1968, Thorium and uranium concentrations and isotope ratios in river sediments: *Earth Planet. Sci. Lett.*, v. 4, p. 245–252.

Smith, K. L. Jr., and Teal, J. M., 1973, Deep-sea benthic community respiration—an *in situ* study at 1850 meters: *Science*, v. 184, p. 72–73.

Turekian, K. K., Cochran, J. K., Kharkar, D. P., Cerrato, R. M., Vaisnys, J. R., Sanders, H. L., Grassle, J. F., and Allen, J. A., 1975, Slow growth rate of a deep-sea clam determined by Ra–228 chronology: *Proc. Nat. Acad. Sci. USA*, v. 72, p. 2829–2832.

——, Cochran, J. K., and DeMaster, D. J., 1978, Bioturbation in deep-sea deposits—rates and consequences: *Oceanus*, v. 21, p. 34–41.

——, Cochran, J. K., and Nozaki, Y., 1979, Growth rate of a clam from the Galápagos Rise hot spring field using natural radionuclide ratios: *Nature*, v. 280, p. 385–387.

——, Cochran, J. K., Vaisnys, J. R., and Delaney, M., 1981, Growth rate of a deep-sea solitary coral: in preparation.

Kon-Kee Liu* and I. R. Kaplan
Department of Earth and Space Sciences
Institute of Geophysics and Planetary Physics
University of California
Los Angeles, California 90024

# 4

# NITROUS OXIDE IN THE SEA OFF SOUTHERN CALIFORNIA[†]

*Present address:  Institute of Earth Sciences
Academia Sinica
P. O. Box 23–59
Taipei, Taiwan, R. O. C.

†Contribution No. 2077:  Institute of Geophysics and Planetary Physics
University of California
Los Angeles, California 90024

# ABSTRACT

The mean mole fraction of $N_2O$ in the ocean air off the coast of southern California was found to be 334 ± 5 ppb. The $N_2O$ measurements of the air collected at UCLA show possible influences from industrial $N_2O$ sources, but the three-month average is not significantly different from that of the ocean air. The $N_2O$ saturation in the surface water is 8% in the Santa Barbara Basin and 23-31% near the sewage outfall at Whites Point. The $N_2O$ concentrations in the subsurface waters near Whites Point are 2-6 n$M$ higher than those at a station 20 km away in the San Pedro Basin. The apparent $N_2O$ production is linearly correlated with $[NO_3^- + NO_2^-]$. The ratio of $\Delta N_2O/\Delta(NO_3^- + NO_2^-)$ is approximately 0.08% in the surface layer (0-100 m) and 0.3% in the deeper water. This probably reflects a change of $N_2O/(NO_3^- + NO_2^-)$ ratio in the products of nitrification. The fraction of organic nitrogen converted to $N_2O$ during oxidation of the organic matter is estimated to be 0.06-0.11%. In the bottom water of Santa Barbara Basin, $N_2O$ was preferentially consumed during denitrification under normal conditions, whereas as $N_2O$ production was temporarily enhanced following flushing of the basin water, but returned later to its normal steady state.

# INTRODUCTION

Nitrous oxide ($N_2O$) is a constituent of the atmosphere. The average mole fraction of $N_2O$ in the air is approximately 330 ppb (Pierotti and Rasmussen, 1977). Long-term measurements of $N_2O$ in the upper troposphere made between 1976 and 1979 suggest that the $N_2O$ content is increasing slowly at about 1% per year (Goldan et al., 1980). Growth of $N_2O$ content in the atmosphere is expected as a result of increasing use of fertilizer and fossil fuels (Hahn and Junge, 1977). Crutzen (1971) pointed out that the atmospheric $N_2O$ is an important controlling agent of stratospheric ozone. NO and $NO_2$, which are produced from $N_2O$, catalyze the destruction of ozone by undergoing the following reactions:

O ($^1$D) being activated monatomic oxygen,

$$N_2O + O(^1D) \rightarrow 2\,NO$$

$$NO + O_3 \rightarrow NO_2 + O_2$$

$$NO_2 + O \rightarrow NO + O_2$$

Therefore, increase of the $N_2O$ content in the atmosphere may lead to weakening of the ozone layer. It is estimated that a 20% increase in the global production of $N_2O$ may result in a 4% decrease of the stratospheric ozone (Crutzen, 1974). The ozone layer absorbs the ultraviolet light with wavelength 200-300 nm from the sun, and thus helps protect the biosphere on the surface of the earth from radiation damage. Also, the ozone controls the temperature profile in the stratosphere due to heat generated from absorption of ultraviolet light (Goody and Walker, 1972). Reduction of the ozone content in the stratosphere may result in damage to human health and injury to the growth of higher plants. That may also lead to a change in the global climate due to cooling of the stratosphere (Reck, 1976). However, the rate of $N_2O$ growth in the future is very difficult to predict because the sources and sinks of atmospheric $N_2O$ are poorly understood (Hahn and Junge, 1977; Liu *et al.*, 1977).

The first measurements of $N_2O$ in seawater were made by Craig and Gordon (1963) in the South Pacific. Their preliminary data suggested that a major part of the ocean is probably undersaturated with $N_2O$. Later Hahn (1974) did an extensive study of $N_2O$ in the North Atlantic. His measurements show that the North Atlantic Ocean is supersaturated with $N_2O$ and the degree of saturation in the surface water ranges from 104% to more than 200%, with an average of 180%. He concluded that the ocean is probably an important source of $N_2O$. Recently, Weiss (1978), using a more advanced technique, measured about 4200 samples of surface seawater in the North Atlantic and Indian Oceans. He found that the average degree of saturation is less than 104%. The concentrations of $N_2O$ were found nearly in equilibrium with the atmosphere in large areas, whereas significant supersaturation was discovered in areas of local upwelling or vertical mixing. This implies that emission of $N_2O$ from the major areas of the ocean is probably weak, but a significant contribution may be provided by the localized strong sources. Therefore, it is important to investigate areas where high supersaturation of $N_2O$ may occur, such as areas of upwelling, high productivity, or ocean dumping of nitrogen-containing wastes.

Nitrous oxide is generally considered to be an intermediate or a byproduct of the biological pathways of the nitrogen cycle (see Fig. 4-1). In soil, $N_2O$ is formed during denitrification (Alexander, 1971) as well as during nitrification (Breitenbeck *et al.*, 1980). In seawater, $N_2O$ was observed to accumulate during oxidation of organic matter, and was interpreted to result from $N_2O$ leakage during nitrification (Cohen and Gordon, 1978; Yoshinari, 1976). $N_2O$ can also enter the pathway of denitrification acting as an electron acceptor (Delwiche and Bryan, 1976). Consumption of $N_2O$ during denitrification in oxygen-deficient waters was observed in the eastern tropical North Pacific Ocean (Cohen and Gordon, 1978), and in the Saanich Inlet (Cohen, 1978).

Off the coast of southern California, upwelling occurs in the spring and summer (Sverdrup and Fleming, 1941). A high productivity is sustained by the nutrient-laden

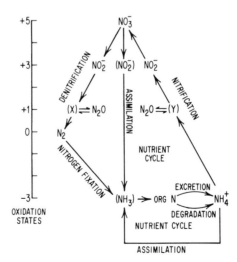

Fig. 4-1.    The biological pathways of the nitrogen cycle (after Liu, 1979). The parentheses indicate that the species are intermediates. $X$ and $Y$ are two unknown intermediates, possibly $N_2O$ or $HNO$ or $H_2N_2O_2$.

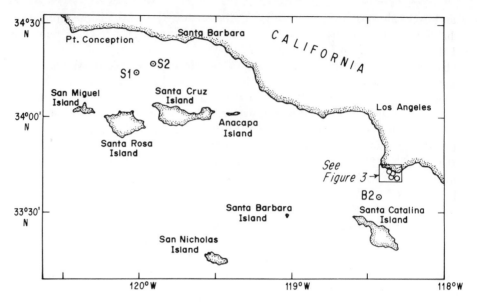

Fig. 4-2.    Sampling stations in the sea off southern California.

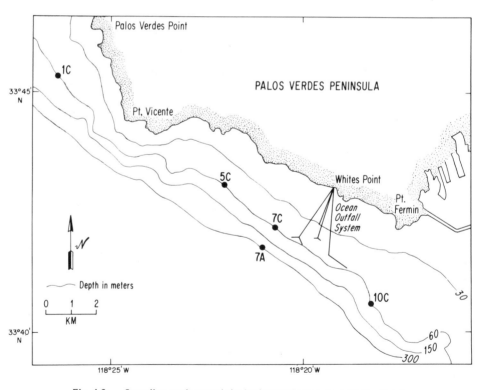

Fig. 4-3.    Sampling stations and the bathymetric map near Whites Point.

water in the upwelling areas (Emery, 1960). Santa Barbara Basin is one such area, and it is the shallowest basin on the southern California borderland. The sill depth is 475 m and the greatest bottom depth is 625 m. Rapid consumption of oxygen is observed in the basin, which implies intensive nitrification. The low oxygen content causes intensive denitrification to develop in its bottom water (Liu, 1979). Thus it provides an ideal environment for studying $N_2O$ in the marine nitrogen cycle.

Southern California is a very highly populated area. In 1977, an average of about one billion gallons of wastewater per day was discharged into southern California coastal waters from Oxnard to San Diego (Schaffer, 1979). The $N_2O$ distribution was studied in the San Pedro Basin off Los Angeles and near Whites Point, which is the sewage discharge site of the Joint Water Pollution Control Plant of the Los Angeles County Sanitation Districts. Thus, the significance of $N_2O$ emission to the atmosphere resulting from ocean dumping of the municipal wastes can be assessed.

A series of cruises were conducted on board *R/V Velero IV* of the University of Southern California between 1975 and 1978 to study the geochemistry of inorganic nitrogen compounds in the sea off southern California (Liu, 1979). On the *SNOW* cruise in August 1977, water samples from the Santa Barbara Basin were collected for $N_2O$ analysis. On the *GIN* cruise in August 1978, water samples and air samples were collected in the Santa Barbara Basin and the San Pedro Basin and near Whites Point. The sampling stations are shown in Figs. 4–2 and 4–3.

## METHOD

Seawater samples for $N_2O$ analysis were drawn into salinity bottles from Niskin bottles in a way similar to the collection of the oxygen samples. One milliliter of saturated $HgCl_2$ solution was added quickly to each sample as a preservative. The bottle was tightly stoppered immediately and stored in a reversed position. The bottles were covered to avoid light and kept refrigerated until analysis. Ocean air samples were collected in evacuated glass tubes. On the *SNOW* cruise, the samples were stored in 20 ml Vacutainers®, which were not used on the *GIN* cruise due to high background. These samples were returned to UCLA and analyzed by gas chromatography.

A gas stripping line similar to the apparatus described by Cohen (1977) was built and attached to a Varian® 3740 gas chromatograph equipped with a $^{63}Ni$ electron capture detector (Fig. 4–4). Twenty to fifty milliliters of water sample was injected into the stripping chamber. $N_2O$ was stripped out of the water sample by purging nitrogen gas (which was filtered by the HydroPurge® filter to remove water vapor and $N_2O$, and the DOW® Oxygen Stripper to remove oxygen) through the water sample at a flow rate of 200 ml/min. The stripping gas passed through a trap filled with drierite and ascarite to trap water vapor and $CO_2$. $N_2O$ was trapped in a U-shaped stainless steel tube ($\frac{1}{8}$ in. O.D. x 8 in.) packed with a 13X molecular sieve submerged in ice water. The molecular sieve that had trapped $N_2O$ was switched to the carrier gas line of the GC by means of a Valco® Zero Dead Volume gas sampling valve. $N_2O$ was then released by heating the molecular sieve with a furnace kept at 300°C. The GC column was a stainless steel column ($\frac{1}{8}$ in. O.D. x 10 ft) packed with 80-100 mesh Chromosorb

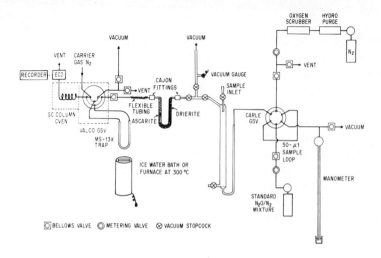

Fig. 4-4. The stripping line and analyzing system for $N_2O$.

101 kept at 60°C. $N_2O$ was detected by the electron capture detector kept at 390°C.

Standard gas mixtures of $N_2O/N_2$ with a mixing ratio of 239 ppmV prepared by Matheson® was run along with the samples for calibration purposes. A gas sample loop of 75 $\mu l$ mounted on a Carle® gas sampling valve coupled with a mercury manometer

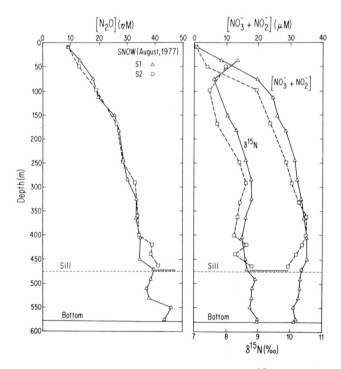

Fig. 4-5. Profiles of $N_2O$ and concentrations and $\delta^{15}N$ of ($NO_3^-$ + $NO_2^-$) observed in the Santa Barbara Basin on the *SNOW* cruise. The bottom of the shallower station (S2) is shown as the hatched bar.

was used to inject the standard gas mixture. An analyzed air sample of 335 ± 2 ppbV provided by Dr. Rasmussen of Oregon Graduate Center was used to check the calibration system (Rasmussen and Pierotti, 1978). The signal from the detector was plotted on a chart recorder and was also converted to a frequency modulated signal that was recorded on tape with a tape recorder. The recorded signal was later digitized and integrated with a system developed by D. Friedman at UCLA, which used a PDP-11/10 computer. The quantity of $N_2O$ was calculated from the integrated peak area. The precision of the procedure was ±2% for air samples, ±3% for seawater samples stored in glass bottles, and ±5% for samples stored in Vacutainers.

## RESULTS

The $N_2O$ profiles observed at two stations in the Santa Barbara Basin during two cruises are shown in Figs. 4-5 and 4-6. The deeper station (S1) is at the center of the basin. The shallower station (S2) is on the slope of the basin. Increase of $N_2O$ with depth is rapid in the surface layer, and becomes gradual and then levels off in the deeper water. Contrasting distributions near the bottom were observed on two different cruises. As observed on the *SNOW* cruise (1977), growth of $N_2O$ with depth was enhanced near the bottom on both stations. During the *GIN* cruise (1978), an abrupt decrease of $N_2O$

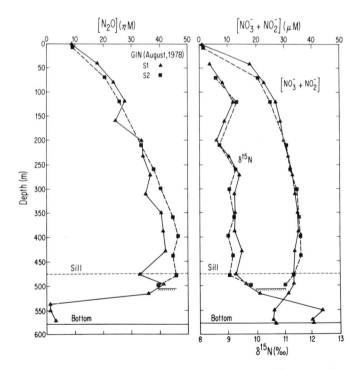

Fig. 4-6.   Profiles of $[N_2O]$ and concentration and $\delta^{15}N$ of ($NO_3^-$ + $NO_2^-$) observed in the Santa Barbara Basin on the *GIN* cruise. The bottom of the shallower station (S2) is shown as the hatched bar.

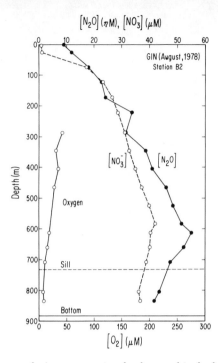

Fig. 4-7. Profiles of N₂O, nitrate, and oxygen observed in the San Pedro Basin on the *GIN* cruise.

was observed. Accompanying hydrographical observations indicated that flushing of the basin water occurred in spring 1977, prior to the *SNOW* cruise (Liu, 1979), whereas the hydrographic conditions were normal, as defined by Sholkovitz and Gieskes (1971), during the *GIN* cruise (Liu, 1979). Denitrification in the bottom water was operating as indicated by the depletion of $[NO_3^- + NO_2^-]$ and an abrupt increase of $\delta^{15}N^*$ (Cline and Kaplan, 1975) (see Fig. 4-6). The abrupt change of $N_2O$ and that of $\delta^{15}N$ are well matched, which suggests $N_2O$ distribution is probably sensitive to denitrification. During the *SNOW* cruise, a slight decrease of $[NO_3^- + NO_2^-]$ was observed near the bottom, but the $\delta^{15}N$ profiles indicate denitrification was not yet developed in the bottom water after the flushing disrupted the normal condition.

The $N_2O$ profile of Station B2, located in the San Pedro Basin, is shown in Fig. 4-7. The $N_2O$ concentration increases with depth reaching a maximum at 600 m, and then decreases with depth. As observed in the North Atlantic (Hahn, 1974; Yoshinari, 1973) and North Pacific (Cohen and Gordon, 1979), the $N_2O$ maximum usually corresponds to the oxygen minimum. However, the oxygen profile observed at Station B2 does not show a corresponding minimum but a monotonous decrease toward the bottom. Nevertheless, the nitrate profile shows features similar to the $N_2O$ profile. This similarity implies that the $N_2O$ content is probably controlled by the processes that control the nitrate content. Both $N_2O$ and nitrate are consumed in the deeper part of the basin, most likely by marine denitrifiers.

The $N_2O$ profiles of stations near Whites Point are shown in Fig. 4-8, and for comparison the $N_2O$ profile of Station B2 is also shown. The overall transport of the

$$*\delta^{15}N = \left( \frac{^{15}N/^{14}N_{sample}}{^{15}N/^{14}N_{air}} - 1 \right) \times 1000\%_0$$

NITROUS OXIDE IN THE SEA OFF SOUTHERN CALIFORNIA

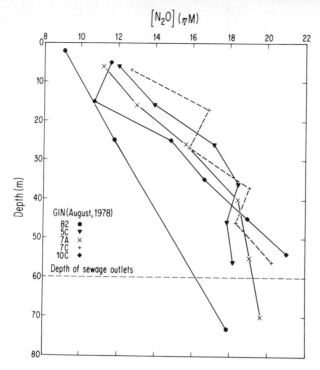

Fig. 4-8.    Profiles of $N_2O$ observed on the stations near the sewage outfall at Whites Point. The profile observed on Station B2 in the San Pedro Basin is also shown for comparison.

surface current in this area is in the SE–NW direction (Jones, 1971). Station 7C is very close to the outfall system on the downcurrent side. Station 10C is on the upcurrent side. Station 5C is further downcurrent from 7C. The terminals of the two major outfall pipes are at a depth of 60 m and the above three stations are all on the 60 m isobath. Station 7A is slightly more offshore than 7C, but it is much deeper.

The $N_2O$ contents in the waters from the Whites Point area were 2-6 n$M$ more enriched in $N_2O$ than the waters at the same depths on Station B2 about 20 km from Whites Point. The enrichment most probably can be attributed to sewage discharge. The distribution of $N_2O$ in the water above the depth of 45 m reflected the influence of surface current. At Station 10C, which is upstream from the outfall, the $N_2O$ profile shows the least enrichment, and at one depth (15 m) the data point aligns with the $N_2O$ profile of Station B2, indicating the least sewage contamination (see Fig. 4-8). The $N_2O$ profiles of Stations 7C and 5C, which are directly downstream from the outfall, show the highest enrichment of $N_2O$. The profiles converge at the depth of 45 m and the pattern of $N_2O$ distribution is almost reversed. This suggests that the water circulation below 45 m is quite different from the surface current. The profile of Station 7A approaches that of Station B2 below 60 m, where the influence of wastewater discharge diminishes.

The mole fractions of $N_2O$ in the ocean air samples are shown in Table 4-1. The average value is 334 ± 5 ppb. This value is consistent with the $N_2O$ mole fraction of 332 ± 9 ppb measured in the ocean air of the eastern Pacific Ocean by Pierotti and Rasmussen (1977). If corrected for humidity, the average mole fraction of $N_2O$ in dry air

TABLE 4-1. Mole Fractions (ppb) of N$_2$O in Ocean Air

| Date | Station | Latitude | Longitude | Mole Fraction (ppb) | | | Mean | Mole Fraction in Dry Air |
|------|---------|----------|-----------|------|------|------|------|--------------------------|
| Aug. 22, 1978 | 1C | 33°45.4'N | 118°26.4'W | 331 | 343 | 334 | 336 ± 6 | 342 ± 6 |
| Aug. 23, 1978 | S1 | 34°14.3'N | 120°1.7'W | 330 | 337 | 332 | 333 ± 4 | 339 ± 4 |
| Aug. 23, 1978 | S2 | 37°17.0'N | 119°55.4'W | 341 | 330 | 334 | 335 ± 6 | 340 ± 6 |
| Aug. 25, 1978 | B2 | 33°35.0'N | 118°25.9'W | 328 | 333 | — | 331 ± 4 | 337 ± 4 |
| | | | Average | | | | (334 ± 5) | (340 ± 5) |

TABLE 4-2. N$_2$O in Surface Seawater and Fluxes of N$_2$O Emission

| Date | Station | $T_{water}$ (°C) | $S$ (‰) | $P$ (mmHg) | $X_{N_2O}$ (ppb) | [N$_2$O]water (nM) | [N$_2$O]eq (nM) | Saturation (%) | Flux $10^{-15}$ M cm$^{-2}$ S$^{-1}$ |
|------|---------|------------------|---------|------------|------------------|--------------------|-----------------|----------------|--------------------------------------|
| Aug. 22, 1978 | 5C[a] | 17.7 | 33.452 | 758.4 | 336[b] | 10.9 | 8.30 | 131 | 10.1 |
| | 7A[a] | 19.2 | 33.485 | 758.4 | 336[b] | 10.1 | 7.93 | 123 | 8.8 |
| | 7C[a] | 16.3 | 33.469 | 758.4 | 336[b] | 10.7 | 8.67 | 127 | 7.6 |
| | 10C[a] | 17.4 | 33.400 | 758.4 | 336[b] | 10.8 | 8.39 | 129 | 9.3 |
| Aug. 23, 1978 | S1 | 18.22 | 33.523 | 760.0 | 333 | 8.8 ± 0.1[c] | 8.14 | 108 | 2.6 |
| Aug. 25, 1978 | B2 | 19.46 | 33.515 | 762.0 | 331 | 9.1 | 7.81 | 117 | 5.3 |

[a]The properties of the surface water on the stations near Whites Point are obtained by extrapolation. The top Nansen bottles of these hydrocasts were 5–7 m below surface.

[b]The measurement was done on an air sample collected at Station 1C, which is 6 miles west of Whites Point.

[c]The value represents the mean of triplicate samples.

is 340 ± 5 ppb. This value is 13% higher than the mean value of $N_2O$ mole fraction (300 ppb) in dry ocean airs of the North Atlantic and Indian Oceans reported by Weiss (1978). This discrepancy is mostly attributed to different primary standards used for calibration. The accuracy of the $N_2O$ measurements in this study is estimated to be ±5% (Rasmussen and Pierotti, 1978). Between July 13 and October 21, 1978, more than 60 air samples were collected on the balcony of Slichter Hall, UCLA and analyzed for $N_2O$. The mean value of $N_2O$ mole fractions was 337 ± 15 ppb. This mean value does not differ significantly from the $N_2O$ mole fraction in the ocean air off the coast of southern California. The variation of these measurements (15 ppb) is much larger than the standard deviation of the $N_2O$ analysis (6 ppb). Therefore this variation probably reflects real fluctuation of $N_2O$ content in the urban air of Los Angeles. A large variation of 20 ppb was also observed in New York City (Pierotti and Rasmussen, 1977). The relatively large variation of $N_2O$ in urban airs is probably attributed to strong point sources of $N_2O$ in the industrial area resulting from combustion of coal, oil, or natural gas (Pierotti and Rasmussen, 1976; Weiss and Craig, 1976).

## DISCUSSION

### Surface Water

The rate of $N_2O$ exchange between air and surface seawater can be estimated using the stagnant film model (Broecker and Peng, 1974), if the respective concentrations of $N_2O$ in the air and surface water as well as the thickness of the stagnant film are known. The flux of $N_2O$ can be computed using the following equation:

$$F = K_m \frac{[N_2O]_w - [N_2O]_{eq}}{Z}$$

where $K_m$ = the molecular diffusivity of $N_2O$ in water,

$[N_2O]_w$ = concentration of $N_2O$ in the surface water,

$[N_2O]_{eq}$ = equilibrium concentration of $N_2O$ in seawater with respect to $N_2O$ in the overlying air, and

$Z$ = thickness of the stagnant film.

The equilibrium concentration is calculated from the solubility, $S_V$ (mole liter$^{-1}$ atm$^{-1}$), of $N_2O$ and its partial pressure in air.

$$[N_2O]_{eq} = S_V \cdot P \cdot X_{N_2O}$$

where $P$ = atmospheric pressure and $X_{N_2O}$ = the mole fraction of $N_2O$ in the air. The solubility of $N_2O$ ($S_V$) in seawater is computed from temperature and salinity using the formula given by Weiss and Price (1979).

The $N_2O$ concentrations and fluxes of $N_2O$ escaping from the surface seawater are presented in Table 4-2. The hydrographic data for stations near Whites Point were obtained by extrapolation. The shallowest sampling depths of these hydrocasts were 5–7 m below surface. All of the surface waters were supersaturated with $N_2O$. The degree of saturation in the surface water is 108% in the Santa Barbara Basin. The values of saturation near Whites Point ($128 \mp 3\%$) are much higher, obviously resulting from wastewater discharge. The value for the San Pedro Basin is between the two afore-mentioned extremes. If the value for the Santa Barbara Basin is considered the natural background, the additional amount of $N_2O$ in the surface water of the San Pedro Basin can very probably be attributed to spreading of $N_2O$ from the sites of sewage outfalls on the coast of southern California.

Cohen and Gordon (1979) measured the $N_2O$ saturation in the surface water of the Santa Barbara Basin to be 105%, which is slightly lower than our measurement (108%). This difference may represent seasonal variation of $N_2O$ saturation in the Santa Barbara Basin. Their measurement was done in January 1977, before upwelling began in this area (Liu, 1979; Sverdrup and Fleming, 1941). Our measurement was conducted in August 1978, shortly after upwelling subsided (Liu, 1979). Therefore, our measure-ment probably reflects the remnant of the waters enriched in $N_2O$ brought to the sur-face during upwelling. On the other hand, the higher $N_2O$ saturation could be attributed to a higher production rate of $N_2O$ resulting from a higher rate of nitrification, which is suggested by the fact that the nitrate concentration in the euphotic zone was much higher in August 1978 than in January 1977 (Liu, 1979).

The fluxes of $N_2O$ emission from the sea surface are calculated using the average value of 50 $\mu m$ for the thickness of the stagnant film and the molecular diffusion co-efficient of $N_2O$, both given by Broecker and Peng (1974). The $N_2O$ flux at Station S1 in the Santa Barbara Basin is computed to be $2.6 \times 10^{-15}$ M cm$^{-2}$ sec$^{-1}$. Eppley *et al.* (1970) reported that the primary production rate at a station off La Jolla, Califor-

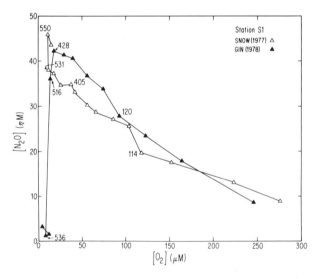

Fig. 4-9.   Relationship between [$N_2O$] and oxygen observed in the Santa Barbara Basin. The numbers indicate the depths.

nia, was 50 mg-atom C m$^{-2}$ day$^{-1}$ when upwelling subsides. If the same rate of $CO_2$ fixation is assumed for the Santa Barbara Basin, the flux of organic nitrogen fixed by phytoplankton is computed to be 7.1 mg-atom N m$^{-2}$ day$^{-1}$ ($8.3 \times 10^{-12}$ g-atom N cm$^{-2}$ sec$^{-1}$), using a C/N ratio of 7 for phytoplankton. Assuming that the $N_2O$ flux is sustained by oxidation of the planktonic organic matter, one can calculate the fraction of organic nitrogen converted to $N_2O$ to be 0.06%.

The mean fraction of organic nitrogen converted to $N_2O$ in the world ocean is estimated for comparison. The average mole fraction of $N_2O$ in air has been estimated to be 330 ppb (Pierotti and Rasmussen, 1977). The average $N_2O$ supersaturation is assumed to be 4% (Weiss, 1978). At an average surface temperature of 18°C (Bialek, 1966) and salinity of 35‰, the flux is computed to be $1.3 \times 10^{-15}$ M cm$^{-2}$ sec$^{-1}$. The total $N_2O$ production is estimated at 4 Tg N/y. The rate of organic nitrogen synthesized in the world ocean is estimated to be $3.7 \times 10^{-15}$ - $5.3 \times 10^{15}$ g N y$^{-1}$, based on the primary production of $2.2 \times 10^{16} \sim 3.2 \times 10^{16}$ g C y$^{-1}$ (Fogg, 1975; Koblentz-Mishke *et al.,* 1968). Therefore, the $N_2O$/organic N ratio is 0.08-0.11%, which is reasonably close to the value obtained for the Santa Barbara Basin.

## Subsurface Water and Basin Water

The distributions of $N_2O$ in the ocean were found to be negatively correlated with oxygen distribution and directly correlated with nitrate distribution (Cohen and Gordon, 1979; Yoshinari, 1976). These correlations were interpreted as suggesting generation of $N_2O$ during nitrification. The relationships between $N_2O$ and oxygen observed in the Santa Barbara Basin during two cruises are shown in Fig. 4-9. The $N_2O$ concentration increases with decreasing oxygen concentration in the water column above the sill depth. The slope of $N_2O$ increase is more gradual in the region of higher oxygen concentration, which corresponds to the zone above the depth of about 100 m. The slope becomes steeper in the water below ca. 100 m. The relationships in the bottom water observed on the two cruises are strikingly different. The $N_2O$ concentration increased abruptly by 8 n*M* between the depths of 531 and 550 m, as observed during the *SNOW* cruise, whereas it decreased drastically by 35 n*M* between the depths of 516 and 536 m as observed during the *GIN* cruise. The abrupt drop of $N_2O$ was obviously due to $N_2O$ consumption during denitrification in the bottom water. Similar features were observed in the eastern tropical North Pacific (Cohen and Gordon, 1978) and Saanich Inlet, British Columbia (Cohen, 1978).

As mentioned previously, the condition of the basin observed during the *GIN* cruise represented a normal condition, whereas the condition observed during the *SNOW* cruise represented a transient condition affected by flushing of the basin water, which later returned to the normal condition (Liu, 1979). The new water contained more oxygen, and therefore disrupted denitrification, and the $N_2O$ produced in the bottom water accumulated without being consumed by denitrifying bacteria. It is noteworthy that the increase of oxygen concentration in the basin water probably never exceeded 20 $\mu M$ during the course of flushing (Liu, 1979; Sholkovitz and Gieskes, 1971), but this was enough to surpass the critical oxygen level for suppressing denitrification. The measured oxygen level was reduced to 10–15 $\mu M$ in the bottom water at the time of the *SNOW* cruise. The abrupt increase of $N_2O$ in the bottom water probably suggests very efficient production of $N_2O$ at a low oxygen level. Extraordinarily high levels of $N_2O$ were also observed in a very thin layer of water near the upper boundary of the oxygen-deficient zone in the eastern tropical North Pacific[1]. This feature is probably related to a high rate of nitrification at a low oxygen level. Carlucci and McNally (1969) reported that the rate of nitrification is generally faster at lower oxygen levels (as low as 5 $\mu M$). Since $N_2O$ may be produced from the decomposition of an intermediate of nitrification (Hahn and Junge, 1977), at a fast rate of nitrification a higher concentration of accumulated intermediate is more probable, which may lead to a higher fraction decomposing into $N_2O$. However, prior to substantiation of this mechanism by microbial study, other possible mechanisms should not be excluded as the explanation. For example, $N_2O$ may be produced (instead of being consumed) by marine denitrifiers (rather than nitrifiers) in the transition zone.

It is notable that, in the bottom water during the *GIN* cruise, $N_2O$ was almost completely consumed while nitrate concentration was reduced by less than 30% (see Fig. 4-6). This suggests that $N_2O$ is preferentially consumed by marine denitrifyers in the bottom water. Thermodynamic calculations suggest that the use of $N_2O$ as an electron acceptor is thermodynamically more favorable than the use of nitrate (Liu, 1979), which may explain the preferential use of $N_2O$ by the denitrifying bacteria. However, exhaustion of $N_2O$ was not observed as corresponding to nitrate depletion in the bottom waters as shown in Figs. 4–5 and 4–7. This may be explained by assuming that the observed nitrate depletion did not result from *in situ* consumption but from diffusive loss to a sink most likely in the sediments or the sediment-water interface. Similarly, the $N_2O$ was also controlled by the diffusive process, which does not show a specific preference for $N_2O$. Also to be seen in Figs. 4–5 and 4–7, the corresponding $\delta^{15}N$ values of nitrate which show little change in the bottom water are consistent with the proposed diffusion-controlled distribution. Therefore, $N_2O$ distribution can be used to distinguish *in situ* denitrification in the water column from diffusive loss of $NO_3^-$ or $N_2O$ to the sediments: The sudden drop of $N_2O$ concentration suggests the former and the gradual decrease suggests the latter.

The rate of $N_2O$ consumption in the bottom water is estimated by calculating the flux of $N_2O$ diffusing toward the bottom along the sharp gradient observed on the *GIN* cruise. The gradient is estimated to be 1 n$M$/m (see Fig 4-6a). The eddy diffusion coefficient is assumed to be 3 cm$^2$/sec (Chung, 1973).[2] Thus the flux is computed to be 3 $\times$ $10^{-14}$ M cm$^{-2}$ sec$^{-1}$, which is about ten times the flux of $N_2O$ escaping the sea surface. This flux is about 1% of the flux of nitrate consumed in the zone of denitrification, whereas the concentration of $N_2O$ in the water above this zone is only 0.3% (on a $\mu$g-atom N basis) of the nitrate concentration (Liu, 1979). If the $N_2O$ distribution represents a steady state, the production of $N_2O$ above the consumption zone must be very efficient to sustain such a rapid consumption rate. If not, $N_2O$ would disappear with time.

[1] J. D. Cline, personal communication, 1979.

[2] Also from D. E. Hammond, personal communication, 1979.

## Apparent N₂O Production

Yoshinari (1976) defined the apparent $N_2O$ production as the quantity of $N_2O$ in seawater in excess of the concentration in equilibrium with the atmospheric $N_2O$ at the *in situ* temperature and salinity. It can be expressed as

$$[\Delta N_2 O] = [N_2 O]_{obs} - [N_2 O]_{eq}$$
$$= [N_2 O]_{obs} - S_V \cdot P \cdot X_{N_2 O}$$

where $[N_2 O]_{obs}$ is the $N_2O$ concentration observed in the seawater, and $[N_2 O]_{eq}$ is the equilibrium concentration which is calculated from the solubility at the *in situ* temperature and salinity (Weiss and Price, 1980), the average atmospheric pressure (1 atm) and the average mole fraction of $N_2O$ in the air (330 ppb). Cohen and Gordon (1979) considered apparent $N_2O$ production as a property of the seawater similar to the nutrient production during decomposition of marine organic matter.

The apparent $N_2O$ production is plotted against $[NO_3^- + NO_2^-]$ for the samples from Stations S1, S2, and B2 in Fig. 4-10. The samples in the bottom water of the Santa Barbara Basin are not included because of the unusual processes controlling $N_2O$. The two parameters are correlated linearly, but there is a distinct break of the relationship where the slope changes significantly. This break point corresponds to the depth of approximately 100 m. In the surface layer (0–100 m) the slope of $[\Delta N_2 O]$ versus $[NO_3^- + NO_2^-]$ is $0.85 \times 10^{-3}$ (on a $\mu$g-atom N basis) with a correlation coefficient of 0.96. In the deeper water, the slope is $3.0 \times 10^{-3}$ with a correlation coefficient of 0.91.

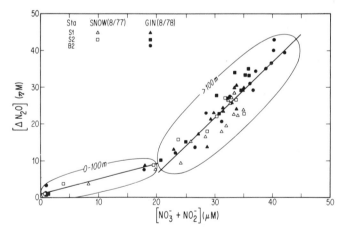

Fig. 4-10. Relationship between apparent $N_2O$ production ($\Delta N_2O$) and $[NO_3^- + NO_2^-]$ observed in the Santa Barbara Basin and the San Pedro Basin. The samples in the bottom water of SBB are not included. The least-squares fit for the samples in the surface layer (0–100 m) has a slope of $0.85 \times 10^{-3}$ (on a $\mu$g-atom N basis) and a correlation coefficient of 0.96. That for the samples in the deeper water has a slope of $3.0 \times 10^{-3}$ and a correlation coefficient of 0.91.

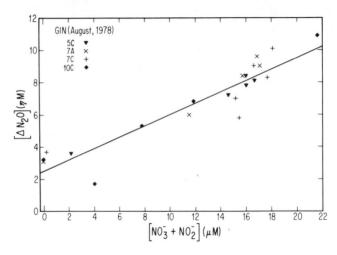

Fig. 4-11. Relationship between apparent $N_2O$ production and $[NO_3^- + NO_2^-]$ observed in the water near the sewage outfall at Whites Point. The least-squares fit has a slope of $0.70 \times 10^{-3}$ (on a $\mu$g-atom N basis) and a correlation coefficient of 0.93.

The relationship between apparent $N_2O$ production and $[NO_3^- + NO_2^-]$ in the water samples from the Whites Point area is similar to that in the surface layer of the offshore stations (see Fig. 4-11). The slope determined by linear regressions is 0.70 $\times 10^{-3}$ on a $\mu$g-atom N basis, with a correlation coefficient of 0.93. This $\Delta N_2O/\Delta(NO_3^- + NO_2^-)$ ratio is slightly lower than that $(0.85 \times 10^{-3})$ for the shallow water of the off-shore stations. If this value is taken as the yield of $N_2O$ during nitrification, the total production of $N_2O$ from the wastewater discharge can be computed. The discharge of ammonia nitrogen into the ocean at the Whites Point outfall is about $5 \times 10^7$ g/day (Schaffer, 1979). If all of the ammonia is nitrified, $4.2 \times 10^4$ g/day of ammonia nitrogen would be converted to $N_2O$. This quantity is equivalent to $1.5 \times 10^3$ M/day of $N_2O$. The average flux of $N_2O$ escaping the sea surface to the air in the Whites Point area is $9 \times 10^{-15}$ M cm$^{-2}$ sec$^{-1}$ (Table 4-2). If the flux at Station S1 is taken as the natural background, the $N_2O$ flux attributable to sewage outfall in the Whites Point area is $6.4 \times 10^{-15}$ M cm$^{-2}$ sec$^{-1}$ ($5.5 \times 10^{-6}$ M m$^{-2}$ day$^{-1}$). The amount of $N_2O$ produced from the sewage discharge can sustain such a flux in an area of $2.7 \times 10^8$ m$^2$ (270 km$^2$). Since the flux must decrease with increasing distance from the source, the area affected by the sewage outfall should be larger. If the flux at Station B2 is used for calculation, the area becomes 640 km$^2$, which corresponds to a half circle with a radius of 20 km. This is about the distance between Whites Point and Station B2. This calculation demonstrates that the estimated $N_2O$ production from sewage is adequate to sustain the excess $N_2O$ fluxes in the vicinity of Whites Point.

## The $\Delta N_2O/\Delta(NO_3^- + NO_2^-)$ Relationship

In the sea off southern California, the $\Delta N_2O/\Delta(NO_3^- + NO_2^-)$ ratio was found to be 0.07%–0.085% in the surface layer (0–100 m) and 0.3% in the deeper water. This change can be interpreted as change of the $N_2O/(NO_3^- + NO_2^-)$ ratio in the products of nitrification. As discussed previously, the yield of $N_2O$ is probably higher at lower oxygen levels. The oxygen content in the water below the thermocline (at roughly 100 m) is markedly lower than that in the surface layer. Therefore, a higher fraction of $N_2O$ is probably released during nitrification in the deeper water. Riley (1951) estimated that 90% of the particulate organic matter decomposes in the surface layer (0–100 m) and only 10% sinks to the deeper water. Assuming that 95% of this organic nitrogen is nitrified in the deeper water, one can calculate the fraction of organic nitrogen converted to $N_2O$ during organic decomposition to be 0.07% and 0.11%, respectively, for 100% and 50% nitrification achieved in the surface layer (i.e., 0–50% organic nitrogen recycled without being nitrified). The calculated range is consistent with the previous estimate of 0.06–0.11% based on the $N_2O$ fluxes evolving from the surface ocean to the atmosphere.

The foregoing discussion is based on an assumption that the relationship between $[\Delta N_2O]$ and $[NO_3^- + NO_2^-]$ is controlled only by the ratio of $N_2O/(NO_3^- + NO_2^-)$ in the products of nitrification. This assumption is certainly oversimplified. Several other processes may also significantly alter the relationship. In the surface layer, $N_2O$ constantly diffuses out of the seawater into the atmosphere, while nitrate and nitrite are recycled through photosynthesis and nutrient regeneration (see Fig. 4-1). The relationship may also be changed by mixing of water masses which are significantly different in compositions of $N_2O$ and $(NO_3^- + NO_2^-)$. The true relationship of $N_2O/(NO_3^- + NO_2^-)$ in the products of nitrification must be substantiated by direct observation of microbial cultures.

# CONCLUSION

The mean value of $N_2O$ mole fractions in the ocean air off the coast of southern California was determined to be $334 \pm 5$ ppb, which is consistent with the measurements in the eastern Pacific by Pierotti and Rasmussen (1977). The $N_2O$ mole fraction in air in the area of sewage outfall is not significantly different from the ocean air. Measurements of the air samples collected at UCLA show possible influences of industrial $N_2O$ emission, but the average over a three-month period is not significantly different from the value of the ocean air.

The $N_2O$ supersaturation in the surface water of the Santa Barbara Basin probably varies seasonally. In August 1978, we observed an 8% supersaturation which corresponds to a flux of $N_2O$ escaping the sea surface of approximately $2.6 \times 10^{-15}$ mol cm$^{-2}$ sec$^{-1}$. This flux represents about 0.06% of the nitrogen flux assimilated by the phytoplankton. The average $N_2O$ flux of the world oceans is calculated to be $1.3 \times 10^{-15}$ mol cm$^{-2}$ sec$^{-1}$ based on the average supersaturation of 4%. The total emission of $N_2O$ from the oceans is deduced to be 4 Tg N/y, which represents about 0.08–0.11% of the total production of organic nitrogen.

The $N_2O$ supersaturation in the surface waters near the sewage outfall at Whites Point was three to four times the value observed in the Santa Barbara Basin. The $N_2O$ concentrations in the subsurface water were 2–6 n$M$ higher than those observed at the station in San Pedro Basin, which is about 20 km from Whites Point. The excess $N_2O$ is obviously attributable to the large amount of ammonia and organic nitrogen discharged from the outfall.

Linear relationships between the apparent $N_2O$ production $[\Delta N_2O]$ and $[NO_3^- + NO_2^-]$ were observed for the subsurface water samples from the Santa Barbara Basin, the San Pedro Basin, and the area near Whites Point. There is a distinct break of the relationship in the surface layer (0-100 m), where the $\Delta N_2O/\Delta (NO_3^- + NO_2^-)$ ratio was $0.7 \times 10^{-3} - 0.85 \times 10^{-3}$, in contrast to the deeper water, where the ratio became $3 \times 10^{-3}$. This change probably indicates change of the $N_2O/(NO_3^- + NO_2^-)$ ratio in the products of nitrification, which may be affected by the oxygen level in the seawater. However, other processes, such as diffusive loss of $N_2O$ and mixing of different water masses, which may alter the relationship, cannot be excluded. If the observed ratios indeed represent the ratios in the products of nitrification, the fraction of organic nitrogen converted to $N_2O$ during oxidation of organic matter is computed to be 0.07% and 0.11%, respectively, for 100% and 50% nitrification achieved in the surface layer. This range is consistent with the estimate (0.06-0.11%) based on $N_2O$ fluxes evolving from the sea surface, but only about half of the value (0.2%) reported by Cohen and Gordon (1979), whose calculation was based on the data representing mostly waters below the surface layer. According to Riley (1951), 90% of the marine organic matter is decomposed in the surface layer.

The bottom water of the Santa Barbara Basin is normally very low in oxygen, which induces development of denitrification. The normal condition was observed during the *GIN* cruise in 1978. The $N_2O$ profile showed an abrupt drop from 36.1 n$M$ to 1.2 n$M$ near the bottom, whereas the nitrate decrease was less than 30%. This suggests that the $N_2O$ distribution is probably a sensitive indicator of *in situ* denitrification. Strong upwelling in the spring and summer sometimes causes flushing of the basin water, which results in a slight rise of oxygen level and disrupts denitrification. Prior to the *SNOW* cruise in 1977, the basin water was flushed. The $N_2O$ profile showed, instead of a drop, an abrupt increase of 8 n$M$ near the bottom. This feature and the aforementioned increase of $\Delta N_2O/\Delta(NO_3^- + NO_2^-)$ in the water below the surface layer imply that the yield of $N_2O$ during nitrification is probably higher at a lower oxygen level, provided that the oxygen level is above the critical concentration required to suppress the onset of denitrification.


## ACKNOWLEDGMENTS

We thank Mr. Ed Ruth and Mr. David Friedman for technical support on instrumentation, and Captain Andy Valois and the crew of *R/V Velero IV* of the University of Southern California for assistance on the sample collecting cruises. Special thanks are due to the DCPG group of the Scripps Institution of Oceanography for loaning us the Niskin bottles and other oceanographic equipment. The manuscript was typed by Mrs. Susan Yamada and the figures were drafted by Ms. Janice Mayne, both at Caltech. This work was supported by National Science Foundation grant OCE 76-14481.

Alexander, M., 1971, *Microbial Ecology*: New York, Wiley, 425 p.

Bialek, E. L., 1966, *Handbook of Oceanographic Tables*: Special Publ. SP-68, U.S. Naval Oceanogr. Office, Washington, D.C.

Breitenbeck, G. A., Blackmer, A. M., and Bremner, J. M., 1980, Effects of different nitrogen fertilizers on emission of nitrous oxide from soil: *Geophys. Res. Lett.*, v. 7, p. 85–88.

Broecker, W. S., and Peng, T. -H., 1974, Gas exchange rates between air and sea: *Tellus*, v. 26, p. 21–35.

Carlucci, A. F., and McNally, P. M., 1969, Nitrification by marine bacteria in low concentrations of substrate and oxygen: *Limnol. Oceanog.*, v. 14, p. 736–739.

Chung, Y., 1973, Excess radon in the Santa Barbara Basin: *Earth Planet. Sci. Lett.*, v. 17, p. 319–323.

Cline, J. D., and Kaplan, I. R., 1975, Isotopic fractionation of dissolved nitrate during denitrification in the eastern tropical North Pacific Ocean: *Mar. Chem.*, v. 3, p. 271–299.

Cohen, Y., 1977, Shipboard measurement of dissolved nitrous oxide in seawater by electron capture gas chromatography: *Anal. Chem.*, v. 49, p. 1238–1240.

——, 1978, Consumption of dissolved nitrous oxide in an anoxic basin, Saanich Inlet, British Columbia: *Nature*, v. 272, no. 5650, p. 235–237.

——, and Gordon, L. I., 1978, Nitrous oxide in the oxygen minimum of the eastern tropical North Pacific—evidence for its consumption during denitrification and possible mechanism for its production: *Deep-Sea Res.*, v. 25, p. 509–524.

——, and Gordon, L. I., 1979, Nitrous oxide production in the ocean: *J. Geophys. Res.*, v. 84 (Cl), p. 347–353.

Craig, H., and Gordon, L. I., 1963, Nitrous oxide in the ocean and the marine atmosphere: *Geochim. Cosmochim. Acta*, v. 27, p. 949–955.

Crutzen, P. J., 1970, The influence of nitrogen oxides on the atmospheric ozone content: *Quart. J. Royal Met. Soc.*, v. 96, p. 320–325.

——, 1971, Ozone production rates in an oxygen, hydrogen, nitrogen oxide atmosphere: *J. Geophys. Res.*, v. 76, p. 7311–7327.

——, 1974, Estimates of possible variations in total ozone due to natural causes and human activities: *Ambio*, v. 3, p. 201–210.

Delwiche, C. C., and Bryan, B. A., 1976, Denitrification: *Ann. Rev. Microbiol.*, v. 30, p. 241–262.

Emery, K. O., 1960, *The Sea Off Southern California*: New York, Wiley, 366 p.

Eppley, R. W., Reid, F. M. H., and Strickland, J. D. H., 1970, Estimates of phytoplankton crop size, growth rate, and primary production, *in* Strickland, J. D. H., ed., The ecology of the plankton off La Jolla, California, in the period April through September, 1967: *Bull. Scripps Inst. Oceanogr.*, v. 17, p. 33–42.

Fogg, G. E., 1975, Primary productivity, *in* Riley, J. P., and Skirrow, G., eds., *Chemical Oceanography*, 2nd Ed., Vol. 2: London, Academic Press, p. 385–453.

Goldan, P. D., Kuster, W. C., Albritton, D. L., and Schmeltekopf, A. L., 1980, Stratospheric $CFCl_3$, $CF_2Cl_2$, and $N_2O$ height profile measurements at several latitudes: *J. Geophys. Res.*, v. 85(Cl), p. 413–423.

Goody, R. M., and Walker, J. C. G., 1972, *Atmospheres*: Englewood Cliffs, N.J., Prentice-Hall, 150 p.

Hahn, J., 1974, The North Atlantic Ocean as a source of atmospheric $N_2O$: *Tellus*, v. 26, p. 160–168.

———, and Junge, C., 1977, Atmospheric nitrous oxide—a critical review: *Zeitschrift für Naturforschung*, v. 32a, p. 190–214.

Jones, J. H., 1971, General circulation and water characteristics in the Southern California Bight: *Southern California Coastal Water Research Project Rep., Tech. Rep. 101*, SCCWRP, El Segundo, California.

Koblentz-Mishke, O. J., Volkovinsky, V. V., and Kabanova, J. G., 1968, Plankton primary production of the world ocean, *in* Wooster, W. S., ed., *Scientific Exploration of the South Pacific*: N.A.S., Washington, D.C., p. 183–193.

Liu, Kon-Kee, 1979, Geochemistry of inorganic nitrogen compounds in two marine environments—the Santa Barbara Basin and the ocean off Peru: Ph.D. dissertation, Univ. California, Los Angeles, Calif., 354 p.

Liu, S., Cicerone, R., Donahue, T., and Chameides, W., 1977, Sources and sinks of atmospheric $N_2O$ and the possible ozone reduction due to industrial fixed nitrogen fertilizers: *Tellus*, v. 29, p. 251–263.

Pierotti, D., and Rasmussen, R. A., 1976, Combustion as a source of nitrous oxide in the atmosphere: *Geophys. Res. Lett.*, v. 3, p. 265–267.

———, and Rasmussen, R. A., 1977, The atmospheric distribution of nitrous oxide: *J. Geophys. Res.*, v. 82, p. 5823–5832.

Rasmussen, R. A., and Pierotti, D., 1978, Interlaboratory calibration of atmospheric nitrous oxide measurements: *Geophys. Res. Lett.*, v. 5, no. 5, p. 353–355.

Reck, R. A., 1976, Atmospheric temperature calculated for ozone depletions: *Nature*, v. 263, p. 116–117.

Riley, G. A., 1951, Oxygen, phosphate and nitrate in the Atlantic Ocean: *Bingham Oceanog. Bull.*, v. 13, p. 1–126.

Schaffer, H. A., 1979, Characteristics of municipal wastewater discharges, 1977, *in* Bascom, W., ed., *Coastal Water Research Project, Annual Report for the Year 1978*: Southern California Coastal Water Research Project, El Segundo, California, 253 p.

Sholkovitz, E. R., and Gieskes, J. M., 1971, A physical-chemical study of the flushing of the Santa Barbara Basin: *Limnol. Oceanogr.*, v. 16, p. 479–489.

Sverdrup, H. U., and Fleming. R. H., 1941, The waters off the coast of southern California, March to July, 1937: *Bull. Scripps Inst. Oceanogr.*, v. 4, p. 261–378.

Weiss, R. F., 1978, Nitrous oxide in the surface water and marine atmosphere of the North Atlantic and Indian Oceans (abstract): *Transactions, Amer. Geophys. Union*, v. 59, no. 12, p. 1101–1102.

———, and Craig, H., 1976, Production of atmospheric $N_2O$ by combustion: *Geophys. Res. Lett.*, v. 3, p. 751–753.

———, and Price, B. A., 1979, Nitrous oxide solubility in water and seawater: *Mar. Chem.*, v. 8, p. 347–359.

Yoshinari, T., 1976, Nitrous oxide in the sea: *Mar. Chem.*, v. 4, p. 189–202.

M. I. Venkatesan and I. R. Kaplan
Department of Earth and Space Sciences
Institute of Geophysics and Planetary Physics
University of California
Los Angeles, California 90024

P. Mankiewicz
Science Applications, Incorporated
1200 Prospect Avenue
La Jolla, California 92037

W. K. Ho and R. E. Sweeney
Global Geochemistry Corporation
6919 Eton Avenue
Canoga Park, California 91303

# 5

# DETERMINATION OF PETROLEUM CONTAMINATION IN MARINE SEDIMENTS BY ORGANIC GEOCHEMICAL AND STABLE SULFUR ISOTOPE ANALYSES*

*Contribution No. 1976:   Institute of Geophysics and Planetary Physics
University of California at Los Angeles
Los Angeles, California 90024

# ABSTRACT

Sediment samples from pristine and polluted marine environments were analyzed for their hydrocarbon distributions. Stable isotope ratios of sulfur in the extractable organic matter were also determined for the same samples. $\delta^{34}S$ values could be correlated with petroleum contamination indicated by the hydrocarbon profiles.

In the pristine marine environment, extractable sulfur is derived mainly from bacterial reduction of seawater sulfate. In both reducing and oxidizing sediment, the $\delta^{34}S$ value of extractable sulfur is $\leq -15\permil$. Petroleum-derived sulfur, concentrated in the refractory compounds resistant to weathering and transported to marine sediment, is isotopically heavier than bacterial sulfur: $\sim 0\permil$ for Prudhoe Bay, Alaska; $-5$ to $+5\permil$ for the Gulf Coast oils; and $+8$ to $+15\permil$ for the southern California natural seeps. In marine environments adjacent to each of these areas, abnormally positive $\delta^{34}S$ values were measured in certain sediments. These sediments are the same as those shown by organic geochemical analyses to be contaminated by crude oil.

Extractable sulfur from sediment-trap particulates does not appear to originate from *in situ* bacterial processes and the $\delta^{34}S$ may therefore be directly related to the $\delta^{34}S$ value of petroleum contaminants.

# INTRODUCTION

Normal alkanes frequently are useful indicators of hydrocarbon sources; marine organisms, land plants, and petroleum each have characteristic distributions of *n*-alkanes which can be distinguished for correlation purposes (Clark and Blumer, 1967; Eglinton and Hamilton, 1963; Kollatukudy and Walton, 1972; Speers and Whitehead, 1969). However, weathering processes may extensively alter the composition of petroleum hydrocarbons introduced into the marine sedimentary environment, often resulting in an "unresolved complex mixture" (Farrington *et al.*, 1977). Therefore, identifying the predominant source of hydrocarbons requires other correlations. The branched and especially the cyclic hydrocarbons, more resistant than *n*-alkanes, can yield information correlating sediment chemistry with sources (Dastillung and Albrecht, 1976; Simoneit and Kaplan, 1980). Stable isotopic fractionation studies of carbon, nitrogen, and sulfur, which may be less affected by weathering, could also be useful as source indicators. Particularly, petroliferous sulfur, concentrated in the refractory compounds, can yield information independent of the compositional complexity of the sample. The range of $\delta^{34}S$ values (stable isotope ratio of sulfur) for a series of oils studied by Sweeney and Kaplan (1978) were significantly different from those of the various sulfur species present in offshore sediments (Kaplan *et al.*, 1963). In fact, the isotopic ratio was so characteristic for some oils that they could be distinguished from one another. However, $\delta^{13}C$ and $\delta^{15}N$ were not helpful for tracing petroleum pollution in the marine environment due to the similar ranges of values measured for the local oils. Based on these observations, an attempt was made to correlate the hydrocarbon data of sediments from different environments with the stable isotope composition of sulfur. The southern California Bight region, known to be contaminated by crude oil (Reed *et al.*, 1977; Crisp *et al.*, 1979; Simoneit and Kaplan, 1980; Venkatesan *et al.*, 1980), was chosen as the primary site of study. Sediments and trap particulates were analyzed from the area. The main alternative source of organic sulfur in marine sediments is bacterially produced hydrogen sulfide from *in situ* sulfate reduction. Therefore, sediments from pristine environments such as the Gulf of Alaska, the eastern Bering Sea, and Walvis Bay, were collected to document the distribution of $\delta^{34}S$ values in uncontaminated samples.

Pertinent data were also obtained on the potential sources of oil pollution such as beach tars and imported oils to the Los Angeles Harbor in the southern California Bight to compare with the natural seeps and sediments of this area. The locations of the natural seeps previously measured for $\delta^{34}S$ (Sweeney and Kaplan, 1978) from southern California Bight are shown in Fig. 5-1. Coal Oil Point (+14 to +15‰) and Carpenteria (+10‰) seeps were shown to have different isotopic ratios and several beach tars collected along the shore from Jalama to Venice Beach were shown by isotope comparison to have been derived from Coal Oil Point (Sweeney and Kaplan, 1978). The beach tars in the Santa Monica Bay region were also reported to come mainly from the Coal Oil Point seepage or from a submarine seepage in the Redondo Canyon ($\delta^{34}S$, +8 to +12‰) by Hartman (1978) after his detailed study of the $\delta^{34}S$ values.

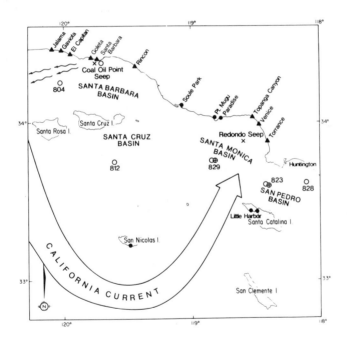

Fig. 5-1. Index map of southern California Bight. The locations of the Redondo and Coal Oil Point seeps are shown along with the sites of tar and sediment collections.

| Seep | × |
| Sediment | ○ |
| Trap | ⊕ |
| Beach Tar Locations | ▲ |
| Ancient Tar | ● |

The transport of the Coal Oil Point material was considered to occur via the counterclockwise California current that travels around offshore islands before returning to the coast near Santa Monica Bay. Hartman (1978) compared the rate of seep discharge at Coal Oil Point to the rate of accumulation of beach tars (with this isotopic label) along the coast. Using suitable corrections for evaporation and weathering, he estimated that most of the refractory portion of the seep-discharged material was ultimately transported to the beaches. Tars on the beaches are soon agglomerated with sand grains and transported back to the ocean by tidal action and bottom currents. The ultimate destination of the natural marine seep material off southern California, therefore, appears to be the marine sediments of the continental shelf.

A study of hydrocarbon distributions and sulfur isotope ratio measurements of sulfur in the extractable organic matter from these sediments were carried out to test if crude oil contamination can be determined by using $\delta^{34}$S measurements.

## EXPERIMENTAL

Freeze-dried sediments were extracted using a toluene:methanol (3:7) azeotropic mixture in a Soxhlet apparatus for 100 hours with one solvent change after 24 hours (Venkatesan *et al.*, 1981). The total extracts were washed down a copper column to purify the samples of elemental sulfur, saponified and then chromatographed on a column packed with 1:2 alumina over silica gel, successively eluting with two column volumes each of hexane, benzene, and methanol. The three solvents eluted the *n*-alkanes and olefins, aromatic compounds and olefins, and polar compounds, respectively.

Gas chromatographic analyses of the hexane and benzene fractions were conducted using a Hewlett Packard Model 5840A instrument, equipped with a linear temperature programmer, flame ionization detector, and an electronic integrator. A 30 m $\times$ 0.25 mm glass capillary column wall-coated with OV–101 (J&W, Inc.) was used and programmed at $4°$C/min from $35°$ to $260°$, then held isothermal for about 2 hours, with a flow rate of 4 ml/min of helium.

The GC/MS analyses of the hexane and benzene fractions were carried out on a Finnigan Model 4000 Quadrupole Mass Spectrometer interfaced directly with a Finnigan Model 9610 Gas Chromatograph. The GC was equipped with a 30m $\times$ 0.25 mm glass capillary column wall-coated with OV–101 (J&W, Inc.). The mass spectrometric data was acquired and processed using a Finnigan Incos Model 2300 data system with 3.1 software.

The methanol fraction was combusted in a Parr bomb with 30 atm oxygen pressure and the produced sulfate was precipitated with barium chloride and converted to sulfur dioxide by the method of Bailey and Smith (1972). A minimum of 0.1 mg of sulfur was necessary for the isotope preparation technique. Therefore, 20 mg of sample had to be used for a typical sulfur content of 0.5%. Though organic sulfur compounds are present to a lesser extent in the hexane and benzene fractions, the methanol fraction, for simplicity, will be termed extractable sulfur in this chapter. The methanol fraction is seldom used for gas chromatographic identification and so is readily available from sediments analyzed in organic geochemical studies. Stable isotope ratios were determined following the method of Kaplan *et al.* (1970), and are presented in the standard "δ" notation relative to Canyon Diablo troilite (FeS) with a precision of 0.1‰.

$$\delta = \left[ \frac{R_{\text{sample}} - R_{\text{standard}}}{R_{\text{standard}}} - 1 \right] \times 1000$$

where $R = {}^{34}S/{}^{32}S$ for sulfur.

## RESULTS AND DISCUSSION

Table 5-1 lists the contents and $\delta^{34}S$ values of extractable sulfur of various sediment and trap particulate samples. All sediment samples were collected at the sediment surface except for two sections of core from the Santa Barbara Basin.

The sediment from Walvis Bay, off the west coast of Africa, consists predominantly of diatomaceous ooze. The bottom waters have low oxygen concentrations and aid in the preservation of this algal organic detritus. The *n*-alkanes and *n*-fatty acids of the samples exhibit a predominantly marine autochthonous origin (Simoneit *et al.*, 1978). No petroleum residues were detected and only trace amounts of terrigenous lipids are present. The sediments from this area represent a typical example of a marine end member environment. Hydrogen sulfide exists in the surface pore water of this reducing basin. The $\delta^{34}S$ value of the organic sulfur (-21‰) is typical of near-shore reducing marine sediments (Goldhaber and Kaplan, 1974).

The Gulf of Alaska and eastern Bering Sea samples are from oxidizing sedimentary environments and contain 1-2% organic carbon. Hydrocarbons in the Gulf of Alaska sediments appear to be derived in approximately equal abundance from marine and terrigenous sources, while in Bering Sea, a terrigenous source predominates. The sediments show no indication of petroleum input. Sulfate reduction may occur in microreducing environments of oxidizing sediment (Kaplan *et al.*, 1963) and the sulfur in the extractable organic matter, as denoted by the $\delta^{34}S$ value of -17‰, appears to be of this origin.

Of the 40 samples analyzed from the eastern Bering Sea and Gulf of Alaska areas, only one sample at Station 35 (EBBS 35, 56°12.4'N, 168°20.4'W) shows anomalous data on hydrocarbons. The total hydrocarbon contents range from 2 to 27 $\mu g/g$ in all of the stations typical of unpolluted recent marine sediments, whereas at Station 35, the total extractable hydrocarbons were measured to be 241 $\mu g/g$ dry sediment. The gas chromatographic profile of *n*-alkanes also substantiates that this station has been exposed to weathered petroleum hydrocarbons similar to southern California Bight sediments while the other stations are not (Fig. 5-2). As a comparison, a chromatogram of a clean sediment (EBBS 37) is also included. The extended triterpanes ($\geq C_{31}$), analyzed by GC/MS exist as two diastereomers at position 22 in EBBS 35, also indicating petroleum input (Dastillung and Albrecht, 1976). The preponderance of the 17α(H)-hopanes in this station further corroborates a petroleum source for these compounds

TABLE 5-1. Content and $\delta^{34}$S of Sulfur in the Extractable Organic Matter from Recent Sediments and Sediment Traps

| Location | Sulfur (%) | $\delta^{34}$S(‰) | Comments |
|---|---|---|---|
| Walvis Bay[a] | | | |
| Core 38 | 0.8 | −21.2 | Reducing sediment, no petroleum |
| Box Core 2 | 0.5 | −21.1 | contamination |
| Gulf of Alaska[b] | | | |
| GASS composite | 0.4 | −17.3 | Oxidizing sediment, no petroleum contamination |
| Eastern Bering Sea[c] | | | |
| EBBS−composite | 0.2 | −17.9 | Oxidizing sediment, no petroleum contamination |
| EBBS−35 | 0.5 | −7.1 | Petroleum contamination |
| Gulf of Mexico[d] | | | |
| Big Hill | − | −15.7 | (Increasing |
| West Hackberry I | − | −12.7 | petroleum |
| West Hackberry II | − | −6.1 | contamination) |
| Southern California Bight[e] | | | Petroleum contamination |
| Coal Oil Point | 1.6 | +13.5 | 500 meters from seep |
| Santa Barbara Basin | | | |
| (0-131 cm) | 1.1 | +5.6 | Composite sections of core |
| (131–400 cm) | | +2.1 | with the break (131 cm) representing 1000 years |
| (Surface) | − | +2.3 | BLM Station 804 |
| San Pedro Basin | 0.7 | −2.5 | BLM Station 823 |
| San Pedro Basin | 1.1 | −6.8 | BLM Station 823 |
| Offshore Huntington Bch. | − | +2.2 | BLM Station 828 |
| Santa Cruz Basin | − | −3.3 | BLM Station 812 |
| Sediment Traps[f] | | | |
| Santa Monica Basin | − | +7.3 | BLM Station 829 |
| San Pedro Basin | − | +3.0 | BLM Station 823 |

For station locations, see references:
[a]Simoneit et al., 1978.

[b]Kaplan et al., 1977.

[c]Venkatesan et al., 1981.

[d]Shokes and Mankiewicz, 1978.

[e]Figure 5-1.

[f]Crisp et al., 1979.

(Dastillung and Albrecht, 1976; Simoneit and Kaplan, 1980). The $\delta^{34}$S value of this sample is 10‰ more positive than other samples from the Alaska offshore. Alaskan oil at Prudhoe Bay has a $\delta^{34}$S value of about 0‰ (Grizzle et al., 1979). Petroleum from this source or from an unidentified oil seep on the eastern Bering Sea shelf could be the contaminant in EBBS 35. GC/MS analyses of other samples in this area such as at Stations 8, 59, and 65 (for exact locations, see Venkatesan et al., 1981) indicate that the triterpenoids consist of predominantly the 17$\beta$(H), 21$\beta$(H)–hopanes and hopenes

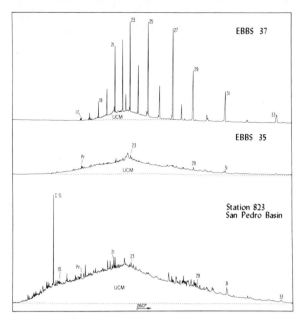

Fig. 5-2. Representative gas chromatograms of hexane fractions of lipids from sediments in eastern Bering Sea (EBBS), Alaska, and San Pedro Basin (southern California Bight). Numbers 15–33 refer to carbon-chain length of n-alkanes; Pr: pristane; UCM: unresolved complex mixture; I.S.: Internal standard (Hexamethylbenzene).

with only traces of the $17\alpha(H)$, $21\beta(H)$-hopanes. The extended hopanes ($\geqslant C_{31}$) are present as single C-22 diastereomers. The presence of $17\beta(H)$ stereomers and of the triterpenes suggests a recent biogenic source for these compounds, originating directly from the regional biota.

Three near-shore sediment samples from the Gulf of Mexico were selected at sites which were considered to have different degrees of petroleum contamination (Shokes and Mankiewicz, 1978). A direct correlation exists between the $\delta^{34}S$ value of organic sulfur and the methylated/condensed ratio of the polynuclear aromatic series. The low methylated/condensed ratio (Big Hill), indicative of low-level petroleum contamination, corresponds to a $\delta^{34}S$ of -16‰, similar to that of many other uncontaminated marine environments. The $\delta^{34}S$ value becomes progressively more positive as the petroleum hydrocarbon contamination increases (Table 5-1). The set of data comparing methyl/condensed phenanthrenes and $\delta^{34}S$: 0.4, -15.7‰, 0.7, -12.7‰, and 11.6, -6.1‰, for the three stations illustrates this argument. Several Gulf Coast oils have $\delta^{34}S$ values[1] ranging from -5 to +5‰ and could be the source of contamination. However, the Big Hill sample has a higher concentration of extractable hydrocarbons (35 ppm), compared to the two West Hackberry samples (9 and 13 ppm). The hydrocarbon contamination in the former sample is indicated by the unresolved complex mixture on the gas chromatograms, which is 80% of the total hydrocarbons. This

[1] UCLA, unpublished data.

is attributed to a combination of finer sediment texture and higher input of pyrolytic hydrocarbons transported via atmospheric fallout and urban runoff rather than to crude oil hydrocarbons from seawater. This is also supported by our observation that benzopyrene, a combustion product, is more predominant in the Big Hill than in the West Hackberry samples (Mankiewicz, 1980). The possible important implication is that the sulfur isotope data here may indicate that crude petroleum is not the direct source of hydrocarbons in the Big Hill sample, unlike the West Hackberry samples, which exhibit sulfur isotopic data characteristic of direct petroleum pollution.

The above results indicate that extractable sulfur in marine sediments which have not been contaminated by petroleum hydrocarbons has a $\delta^{34}S$ value similar to that of bacterially produced hydrogen sulfide, or generally, in the range of –15 to –22‰. Although only a few samples were analyzed, the sediments chosen in uncontaminated areas represented both oxidizing and reducing conditions and seem to show little differences in $\delta^{34}S$ values.

The extractable sulfur in petroleum-contaminated sediments has $\delta^{34}S$ values intermediate between that of the uncontaminated sediment (–15 to –22‰) and values of sulfur in petroleum-rich source areas. In areas off the western coast of Alaska and in the Gulf of Mexico, the $\delta^{34}S$ value of the sulfur in the petroleum contaminant is more positive than the bacterially produced sulfur. Protein-bound sulfur in marine organisms has been reported to have $\delta^{34}S$ values around +20‰ (Kaplan et al., 1963). However, under the experimental conditions employed here, only lipid-related sulfur is extracted into the hexane fraction that is used in the column chromatography and subsequent sulfur isotopic analysis. Any protein-bound sulfur extracted by the initial toluene:methanol Soxhlet extraction would be hydrolyzed and remain in the aqueous phase in the saponification step. Therefore, the relatively positive $\delta^{34}S$ values would indicate only petroleum input in these areas.

Sediment samples from the southern California Bight have $\delta^{34}S$ values in the methanol extracted sulfur between –6.8 and +13.5‰ (Table 5-1). The most positive value (+13.5‰) is for Coal Oil Point sediment, about 500 meters from the seep. Trap particulates from the Santa Monica Basin and the San Pedro Basin have $\delta^{34}S$ values more positive than those for the underlying sediment: +7.3, compared to +0.3‰; and +3.0, compared to –6.8‰, respectively. Although isotopically light sulfur (pyrite) existed in these trap particulates (Crisp et al., 1979), the humic acid fraction contained no measurable sulfur. Humic acids in marine sediment commonly contain about 1% bacterial sulfur (Nissenbaum and Kaplan, 1972) and it could be argued that extractable organic sulfur in the trap particulates is not of a bacterial origin. The more negative values observed in the sediments are probably due to the incorporation of bacterial sulfur. For the San Pedro Basin, the hydrocarbon distribution pattern indicates a mixed petroleum input from natural seeps as well as from heavy anthropogenic activities such as shipping, sewage outfalls, and industrial effluents (Venkatesan et al., 1980). But the sources for the Santa Monica Basin are either sedimented tar that had been washed off the local beaches or anthropogenic inputs. The $\delta^{34}S$ value for the Santa Monica Basin sediment-trap particulate sample (+7.3‰) is similar to that for the nearby Redondo seep, which could be a dominant source of petroleum input into this area. The triterpanes in the Redondo seep consist predominantly of the $17\alpha(H),21\beta(H)$ hopane series containing the two major analogs $C_{28}H_{48}$ and $C_{30}H_{52}$. The mass spectrum of the $C_{28}$ triterpane has been correlated with that of $17\alpha(H),18\alpha(H),21\beta(H)$-28,30 bisnorhopane. This compound is abundant in the Monterey Shale and in offshore seeps (Coal Oil Point) at Santa Barbara, and is either absent or present only in trace amounts in most other crude oils (Seifert et al., 1978) and has therefore been suggested to be a potentially characteristic molecular marker for petroleum from the southern California area (Simoneit and Kaplan, 1980). The presence of bisnorhopane in the trap particulates and the sediments thus indicates that seeps in the southern California Outer Continental Shelf may be the source of petroleum contamination in the Santa Monica and San Pedro area. The $\delta^{34}S$ values could be a potential indicator in differentiating between seeps, especially in the trap particulates. The ability to distinguish the source seep by $\delta^{34}S$ will be enhanced by the absence of anthropogenic input that dilutes the petroleum contribution from the seeps (Crisp et al., 1979).

Listed in Table 5-2 are the ranges of $\delta^{34}S$ values for southern California seeps and beach tars. Different chemically separated fractions (i.e., total, methanol fraction, and asphaltenes) have relatively the same $\delta^{34}S$ composition for the Coal Oil Point samples. The methanol fraction has generally the same sulfur concentration as the total

TABLE 5-2. Summary of Sulfur Content and Isotope Ratio for Seep and Beach Tars Collected in the Southern California Borderland

| Location | No. of Samples | Sulfur Content (%) | $\delta^{34}S$ (‰) | References and Comments |
|---|---|---|---|---|
| Natural seeps | | | | |
| Coal Oil Point | | | | |
| Total | 5 | 4.6–5.2 | +14.1–+15.1 | Sweeney & Kaplan (1978) |
| Methanol Eluate | 3 | 4.4–4.5 | +12.2–+14.0 | |
| Asphaltenes | 13 | 7.5–8.3 | +13.7–+15.0 | Hartman (1978) |
| Redondo Seep | | | | |
| Asphaltenes | 4 | 4.5–6.7 | +7.9–+12.3 | Hartman (1978) |
| Manhattan Seep | | | | |
| Asphaltenes | 1 | 5.0 | +7.7 | Hartman (1978) |
| Recent beach tars (asphaltenes) | | | | |
| Jalama to Venice Beach | 9 | — | +13.2–+14.7 | Sweeney & Kaplan (1978) |
| Topanga Canyon | 2 | — | +13.0–+14.8 | Hartman (1978) |
| Venice Beach | 20 | — | [a](+1.9) +6.8–+15.9 | Hartman (1978) |
| Torrance Beach | 16 | — | [a](+4.9) +9.7–+16.4 | Hartman (1978) |
| Ancient tars (total sample) | | | | |
| San Nicolas Island | | 3.6 | +15.7 | Age unknown |
| Catalina Island, west side | | 2.4 | +14.4 | ~2000 years old[b] |
| Little Harbor | | 3.9 | +12.9 | |
| Paradise Cove | | 2.2 | +10.8 | |
| Point Mugu (Simomo) | | 0.3 | +7.2 | ~1000 years old[b] |
| Ventura Co. (Soule Park) | | 1.2 | +6.5 | |

[a] Anomalous results.

[b] Kaufman, personal communication, 1980.

M. I. VENKATESAN, I. R. KAPLAN, P. MANKIEWICZ, W. K. HO, R. E. SWEENEY crude oil, but the asphaltene fraction is significantly enriched in sulfur. Of the 47 recent beach tars listed in Table 5-2, only two have $\delta^{34}$S values significantly different from the seep sources. Six ancient tars scraped from archaeological specimens from both the offshore islands and coastal sites are also listed in Table 5-2. The San Nicolas and Catalina Island samples appear to come from Coal Oil Point, substantiating the transport system suggested by Hartman (1978). The coastal material appears to be derived from another seep (possibly an onshore seep).

The sulfur content and $\delta^{34}$S value of some typical oils imported into Los Angeles Harbor from various countries have been determined.[2] $\delta^{34}$S of these oils range from -0.2 to +10.5‰. Only oils from Venezuela, Indonesia, and Algeria have $\delta^{34}$S values greater than 0‰. In summary, it is interesting to note that the trap particulates from Santa Monica Basin have $\delta^{34}$S values more positive (+7.3‰) than all but the Venezuela crude oil imported into Los Angeles Harbor. Although contribution from the Redondo seepage may be important, anthropogenic contamination has been demonstrated to be predominant in the samples studied. A mixture of sulfur from this source and sulfur (+15‰) from Coal Oil Point could therefore lead to the intermediate isotope ratio measured in the trap particulates.

## CONCLUSIONS

1. The sulfur isotope ratio of extractable organic sulfur in marine sediment appears to be altered by the contribution of petroliferous material. Sulfur compounds are concentrated in the refractory (polar and asphaltene) fraction of oils and exhibit little isotopic fractionation during transport and weathering. The isotopic ratio of petroleum sulfur is often significantly different from that added to the sediment by bacterial reduction of seawater sulfate. The difference between Coal Oil Point seepage (+15‰) and bacterial sulfur (-20‰) is over 30‰. Therefore, petroliferous material can be characterized and its presence in marine sediments measured.

2. The interpretation of sulfur isotopic distribution is consistent with the distributions of hydrocarbons analyzed in the sediments and aids in delineating the sources of pollution. Sediments known to be clean by organic geochemical measurements show $\delta^{34}$S values of < -16‰. Extractable sulfur in petroleum-contaminated sediments has been found to have $\delta^{34}$S values intermediate between those of uncontaminated sediment and the value of the sulfur in the crude oil source.

3. Sediment traps may collect material that do not contain a bacterial contribution to extractable organic sulfur, hence measured $\delta^{34}$S values should be similar to those of the source(s) of deposition.

## ACKNOWLEDGMENTS

We thank Mr. E. Ruth for GC-MS data acquisition, Dr. B. R. T. Simoneit for the identifications of the triterpenoids, D. Winter for stable isotope measurements, and T. S. Kaufman for collection of archaeologic samples. This work was supported by: a sub-

[2] UCLA, unpublished data.

contract from Science Applications, Inc. (Contract No. AA550-CT6-40), a prime contractor to the Bureau of Land Management; National Oceanic and Atmospheric Administration (Contract No. 03-6-022-35250); Department of Energy and Bureau of Land Management (Contract No. EY-76-S-03-0034, P.A. 134); and Department of Energy contract (Strategic Petroleum Reserve Office); (Contract No. EL-77-C-01-8780 to Science Applications, Inc.)

## REFERENCES

Bailey, S. A., and Smith, J. W., 1972, Improved method for the preparation of sulfur dioxide from barium sulfate for isotope ratio studies: *Anal. Chem.*, v. 44, p. 1542–1543.

Clark, R. C., and Blumer, M., 1967, Distribution of paraffins in marine organisms and sediments: *Limnol. Oceanogr.*, v. 12, p. 79–87.

Crisp, P. T., Brenner, S., Venkatesan, M. I., Ruth, E., and Kaplan, I. R., 1979, Organic chemical characterization of sediment-trap particulates from San Nicolas, Santa Barbara, Santa Monica, and San Pedro Basins, California: *Geochim. Cosmochim. Acta*, v. 43, p. 1791–1801.

Dastillung, M., and Albrecht, P., 1976, Molecular test for oil pollution in surface sediments: *Mar. Pollut. Bull.*, v. 7, p. 13–15.

Eglinton, G., and Hamilton, R. J., 1963, The distribution of alkanes, *in* Swain, T., ed., *Chemical Plant Taxonomy*: New York, Academic Press, p. 187–217.

Farrington, J. W., Frew, N. M., Gischwend, P. M., and Tripp, B. W., 1977, Hydrocarbons in cores of northwestern Atlantic coastal and continental margin sediments: *Estuarine Coastal Mar. Sci.*, v. 5, p. 793–808.

Goldhaber, M. B., and Kaplan, I. R., 1974, The sulfur cycle, *in* Goldberg, E. D., ed., *The Sea, Vol. 5—Marine Chemistry*: New York, Wiley, p. 569–655.

Grizzle, P. L., Coleman, H. J., Sweeney, R. E., and Kaplan, I. R., 1979, Correlation of crude oil source with nitrogen, sulfur, and carbon stable isotope ratios: Symp. Preprints, *ACS/CSJ, Div. of Petroleum Chem.*, Apr. 1–6, 1979, Honolulu, p. 39–57.

Hartman, B. A., 1978, The use of carbon and sulfur isotopic ratios and total sulfur content for identifying the origin of beach tars in Santa Monica Bay, California: Ph.D. Thesis, Univ. Southern California.

Kaplan, I. R., Emery, K. O., and Rittenberg, S. C., 1963, The distribution and abundance of sulfur in Recent marine sediments off southern California: *Geochim. Cosmochim. Acta*, v. 27, p. 297–331.

——, Reed, W. E., Sandstrom, M., and Venkatesan, M. I., 1977, Characterization of organic matter in sediments from Gulf of Alaska, Bering and Beaufort Seas: *Annual Report to U.S. Dept. of Commerce, NOAA, Contract No. 03-6-022-352. R.U. No. 480*, p. 16.

——, Smith, J. W., and Ruth, E., 1970, Carbon and sulfur concentration and isotopic composition in Apollo 11 lunar samples: *Proc. Apollo 11 Lun. Sci. Conf. (Geochim. Cosmochim. Acta Suppl. 2)*, p. 1317–1329.

Kollatukudy, P. E., and Walton, T. J., 1972, The biochemistry of plant cuticular lipids, *in* Holman, R. T., ed., *Progress in the Chemistry of Fats and Lipids, Vol. 13, Pt. 3*: New York, Pergamon, p. 121–175.

Mankiewicz, P. J., 1980, Hydrocarbon composition of sediments, water, and fauna in selected areas of the Gulf of Mexico and southern California marine environments: Ph.D. Thesis, Univ. California at Los Angeles.

Nissenbaum, A., and Kaplan, I. R., 1972, Chemical and isotopic evidence for the *in situ* origin of marine humic substances: *Limnol. Oceanogr.*, v. 17, p. 570–582.

Reed, W. E., Kaplan, I. R., Sandstrom, M., and Mankiewicz, P., 1977, Petroleum and anthropogenic influence on the composition of sediments from the southern California Bight: *Proc. 1977 Oil Spill Conference*, New Orleans, Louisiana, *API*, *EPA*, and *USCG*, p. 183–188.

Seifert, W. K., Moldowan, J. M., Smith, G. W., and Whitehead, E. V., 1978, First proof of $C_{28}$-pentacyclic triterpane in petroleum: *Nature*, v. 271, p. 436–437.

Shokes, R., and Mankiewicz, P., 1978, Geochemical baseline study TEXOMA offshore brine disposal sites: *Annual Report to Strategic Petroleum Reserve Office*, DOE Contract No. *EL-77-C-01-8788* to SAI, Inc., La Jolla, Calif.

Simoneit, B. R. T., and Kaplan, I. R., 1980, Triterpenoids as molecular indicators of paleoseepage in Recent sediments of the southern California Bight: *Mar. Environ. Res.*, v. 3, p. 113–128.

——, Mazurek, M. A., Stuermer, D. H., Toth, D., Kalil, E., and Kaplan, I. R., 1978, Organic geochemistry of recent sediments from Walvis Bay, Southwest Africa, *Distribution and Fate of Biogenic and Petroleum-Derived Substances in Marine Sediments: Annual Report to DOE/BLM, Contract No. EY-76-S-03-0034*, UCLA, Los Angeles, Calif., p. 8.

Spears, G. C., and Whitehead, E. V., 1969, Crude petroleum, *in* Eglinton, G., and Murphy, M. T. J., eds., *Organic Geochemistry*: New York, Springer-Verlag, p. 638–675.

Sweeney, R. E., and Kaplan, I. R., 1978, Characterization of oils and seeps by stable isotope ratios: *Proc. Energy/Environment California*, 1978, *SPIB*, p. 281–293.

Venkatesan, M. I., Brenner, S., Ruth, E., Bonilla, J., and Kaplan, I. R., 1980, Hydrocarbons in age-dated sediment cores from two basins in the southern California Bight: *Geochim. Cosmochim. Acta*, v. 44, p. 789–802.

——, Sandstrom, M., Brenner, S., Ruth, E., Bonilla, J., Kaplan, I. R., and Reed, W. E., 1981, Organic geochemistry of surficial sediments from Eastern Bering Sea, *in* Hood, D. W., and Calder J. A., eds., *The Bering Sea Shelf—Oceanography and Resources*, OMPA, Univ. of Washington Press, Seattle, Washington, v. 1, p. 389–409.

G. Ross Heath
School of Oceanography
Oregon State University
Corvallis, Oregon 97331

6

# DEEP-SEA
# FERROMANGANESE
# NODULES

# ABSTRACT

Ferromanganese nodules are ubiquitous over regions of the deep-sea floor where sediments accumulate more slowly than about 7 m per million years. Variations in nodule composition are thought to reflect variations in the vertical flux of transition metals carried by settling detritus (largely of biogenic origin), and in the oxic and suboxic diagenetic reactions that break down biogenic debris at the sea floor. Over the next several years, *in situ* experiments should help resolve the roles of bacteria, inorganic redox reactions, and nodule mineralogy in partitioning trace metals such as copper and nickel between sediments, nodules, and bottom waters of the oceans.

# INTRODUCTION

Deep-sea ferromanganese (or manganese) nodules have fascinated marine geologists since the initial recovery of large quantities during the first great oceanographic expedition by HMS *Challenger*, from 1872–1876 (Murray and Renard, 1891). This fascination derives from their appearance (particularly when contrasted with the featureless pelagic clay that surrounds them), their high contents of many transition metals, and their abundance over large areas of the floor of the abyss (Fig. 6-1).

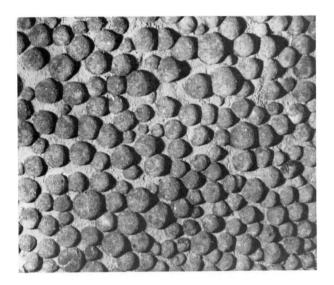

Fig. 6-1.  Deep-sea ferromanganese nodules on the floor of the South Pacific at 36°37′S, 149°00′W, 5487 m. Nodules range up to about 10 cm in diameter. Lamont-Doherty Geological Observatory photo C9-67, #13, courtesy L. Sullivan.

DEEP-SEA FERROMANGANESE NODULES

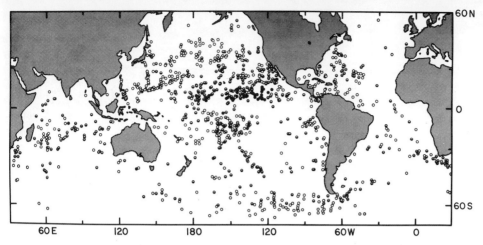

Fig. 6-2. Global distribution of nodules recovered by scientific expeditions. Filled circles are analyzed nodules containing more than 1% copper; half-filled contain less than 1% copper; open circles are nodules for which analyses are not available. Heath, 1978.

## OCCURRENCE

### Regional Distribution

Manganese nodules occure worldwide (Fig. 6-2). At first glance, the distribution appears to be fairly uniform, at least within the limitations of our samples. Closer examination, however, reveals that nodules are restricted to areas where pelagic sediments are accumulating at rates of less than about 7 m per 1 million years. Thus, they are rare or absent from the hemipelagic deposits of the continental slope and rise, from turbidite abyssal plains, from the great deep-sea "drift" deposits laid down by contour-following bottom currents such as those of the western Atlantic, and from calcareous and siliceous biogenic oozes laid down beneath biologically productive surface waters along the equator and at high latitudes.

Most of the sediments associated with nodules are very fine grained, highly oxidized pelagic clays. In some areas siliceous tests of radiolarians and diatoms are abundant (Horn *et al.*, 1973). The content of organic carbon in the sediments generally is much less than 1% (values of the order of 0.01% have been reported).

The absence of nodules from the more rapidly accumulating pelagic sediments is not well understood. Burial soon after nucleation, or the existence of near-surface, mildly reducing conditions inimical to manganese precipitation, are possible explanations, but the crucial comparative observations have yet to be made.

### Local Distribution

At scales of hundreds of meters to kilometers, the surface distribution of nodules is very patchy. This patchiness can take the form either of marked variations in nodule

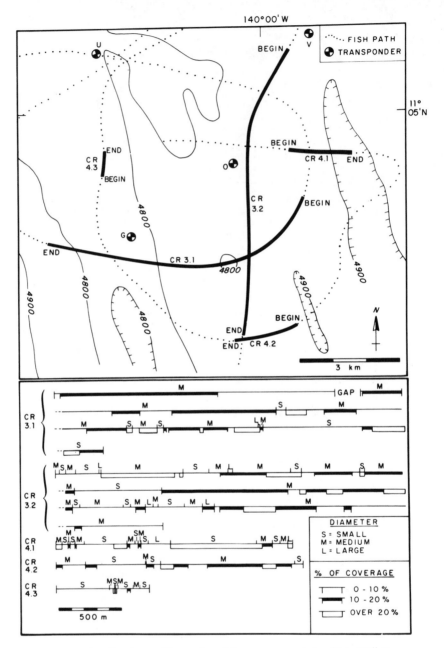

Fig. 6-3.    Small-scale patchiness of nodule coverage and size in a small region of the northern equatorial Pacific at 11°N, 140°W. Properties of nodules within the units recognized on each camera traverse (C.R.) are uniform. From Karas, 1978.

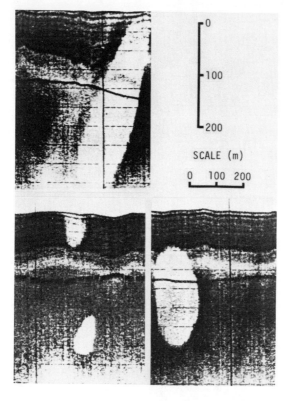

Fig. 6-4. Nodule-free areas (white) in a dense nodule field at 6°30′N, 92°45′W (eastern Pacific) as revealed by side-looking sonar records from Scripps Institution of Oceanography's Deep-Tow vehicle. After Spiess *et al.*, 1977.

size and coverage, often with zones of fairly uniform characteristics separated by sharp boundaries (Karas, 1978; Fig. 6–3), or of "bare" patches in areas that otherwise have a ubiquitous nodule cover (Spiess *et al.*, 1977; Fig. 6–4).

Such small-scale patchiness is difficult to define in an areal sense because our bottom imaging and sampling techniques are ill suited for investigating this length scale. Consequently, the patchiness is difficult to relate to local environmental variables. Moore and Heath (1966) showed a positive correlation between nodule occurrence and thickness of Quaternary sediments in a small area of the equatorial North Pacific. Such a pattern is surprising, inasmuch as slow sedimentation would be expected to reduce the probability of burial of the slowly growing nodules. In contrast, M. Lyle[1] found that thin Neogene sediments have a dense cover of nodules whereas thick sections are capped by widely spaced nodules. Recent near-bottom acoustic profiling combined with photographic surveys in the equatorial Pacific (Karas, 1978; Spiess and Greenslate, 1976) lend indirect support to this latter relationship. Additional field studies to back up such surveys have yet to be carried out, however.

[1] Oral communication, 1979.

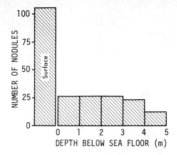

Fig. 6-5. Surficial and buried nodules in piston cores collected by Scripps Institution of Oceanography and Lamont-Doherty Geological Observatory. From Heath, 1978.

## Buried Nodules

The abundance of nodules at the sea floor, relative to their abundance in layers deeper in the sediments, has long puzzled marine geologists (Cronan and Tooms, 1967a; Glasby, 1977; Menard, 1964; Mero, 1965; Moore and Heath, 1967). As Bender *et al.* (1966) and Heath (1979) have pointed out, this pattern is *consistent* with the accumulation rates of nodules relative to sediments (discussed in a subsequent section), but this does not *explain* the phenomenon. The extent of the surface concentration is revealed by examination of deep-sea sediment cores in the collections of Scripps Institution of Oceanography (Cronan and Tooms, 1967a) and Lamont-Doherty Geological Observa-

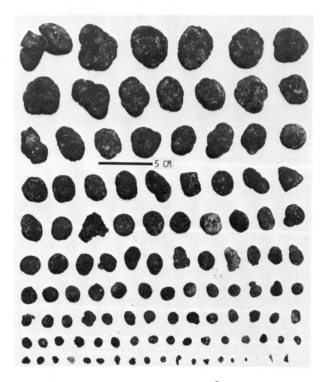

Fig. 6-6. Nodules collected by a 0.25 m$^2$ box core from the eastern North Pacific at 14°15'N, 124°58'W. After Macdougall, 1979.

DEEP-SEA FERROMANGANESE NODULES

tory (Horn *et al.*, 1972). These data (Fig. 6-5; Heath, 1978) show that there are as many nodules at the sea floor as in the underlying 4 m of sediment. For an average accumulation rate of 4 m/my for the sediments associated with deep-sea nodules, this implies that the average nodule spends about 1 million years at the sea floor prior to burial.

The process or processes that keep nodules from being buried are essentially un-known. However, where the size distributions and growth and burial rates of populations of nodules are well characterized, many seem to fit a simple model. Figure 6-6, which is typical of many box-core and grab samples, shows that nodules at a single location are quite variable in size. If the abundance of nodules as a function of size is plotted on a semilog scale (Fig. 6-7), the data for each sample are roughly linear (Heath, 1979). This implies a relation of the form

$$\ln N = \ln N_0 - M (D - D_0) \tag{1}$$

where $N$ is the number of nodules of diameter $D$ per unit area, $M$ is the slope of the lines in Fig. 6-7, and $N_0$ is the number of the smallest nodules (diameter $D_0$) in the sample. Macdougall's (1979) data show that the growth rate of nodules at a single lo-cation is independent of size (Fig. 6-8), so that

$$D = D_0 + 2Gt \tag{2}$$

where $G$ is the growth rate of a nodule and $t$ is its age. Substituting from (2) into (1) gives

$$\ln N - \ln N_0 - 2 MGt \tag{3}$$

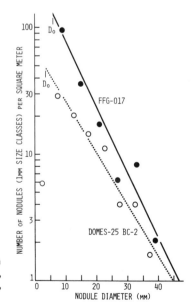

Fig. 6-7. Size distributions of nodules from 9°3'N, 146°29'W (FFG-017) and 14°15'N, 124°59'W (DOMES-25 BC-2). From Heath, 1979.

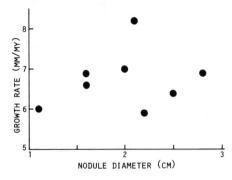

Fig. 6-8. Growth rates of nodules from a box core at 14°15'N, 124°58'W, showing the lack of dependence of growth rates on nodule size. Data from Macdougall, 1979.

or, differentiating

$$\frac{dN}{dt} = -2M \cdot G \cdot N \qquad (4)$$

In other words, the burial rate of surface nodules ($dN/dt$) depends only on the number of nodules present, not on their size, mass, age, or similar variables. This relation is equivalent to

$$\frac{dN}{dt} = -B \cdot N \qquad (5)$$

where $B$ is a "burial" constant (the reciprocal of the average residence time of a nodule at the sea floor). From (4) and (5)

$$M = -\frac{B}{2G} \qquad (6)$$

so that the slopes of curves like those in Fig. 6-7 are a direct measure of the ratio of burial to growth rates.

Table 6-1 summarizes values for $M$, $B$, and $G$. The similarity of the $B$ values derived from equation (6) and Fig. 6-7 to the "global" burial rate inferred from Fig. 6-5 lends credence to the model. These results suggest that small nodules are just as likely to be buried as large ones. Thus, the mechanism that keeps nodules at the sea floor is independent of size.

A number of theories have been put forward over the years to explain the surface abundance. Most of them (such as dissolution at depth and diffusion of the dissolved metals back to the surface, or flotation of the nodules when seismic events liquefy the pelagic clays) have been rejected as our knowledge of the chemical and physical properties of the deep seabed have improved. Two remain, however. Cronan and Tooms (1967a) suggest that downslope rolling and displacement of nodules in hilly areas could keep them at the sea floor, particularly if bottom currents were strong enough to exhume buried nodules. Such a mechanism should concentrate nodules in depressions. Alter-

DEEP-SEA FERROMANGANESE NODULES

TABLE 6-1. Burial and Growth Rates of Nodules from Two Pacific Sites Derived from Heath's (1979) Model

| $B$ (my$^{-1}$) | $G$ (mm/my) | $M$ (mm$^{-1}$) | $D_0$ (mm) | $N_0$ (m$^{-2}$) | Sample No. |
|---|---|---|---|---|---|
| 1.20 | 5.0[a] | -0.12 | 6.0 | 130 | FFG-017[d] |
| 1.26 | 6.7[a] | -0.094 | 3.5 | 49 | BC-02[e] |
| 1.63 | 6.8[b] | -0.12 | 6.0 | 130 | FFG-017[d] |
| 1.28 | 6.8[b] | -0.094 | 3.5 | 49 | BC-02[e] |
| 0.96[c] | 4.0 | -0.12 | 6.0 | 130 | FFG-017[d] |
| 0.96[c] | 5.1 | -0.094 | 3.5 | 49 | BC-02[e] |

[a]Growth rates for nodules from BC-02 and near FFG-017 from Macdougall (1979).
[b]Mean growth rate of Pacific nodules from Andersen and Macdougall (1977).
[c]Global burial rate from Heath (1978) assuming a sedimentation rate of 4 m/my.
[d]FFG-017 from Andrews *et al.* (1974), 9°31'N, 14°29'W, 5180 m
[e]BC-02 from DOMES Station 25 (Macdougall, 1979), 14°15'N, 124°59'W, 4560 m.

natively, a number of authors (Glasby, 1977; Menard, 1964; Moore and Heath, 1967) have suggested that large benthic organisms foraging for food intermittently displace nodules and keep them at the surface. No direct observations of this process have yet been made. However, indirect support comes from Paul's (1976) photographs of small animals cleaning off nodule tops by feeding on the newly deposited (and presumably nutritious) sediment (Fig. 6-9), and from the baited camera pictures of Sessions *et al.* (1968), Hessler and Jumars (1977), and others that prove the existence of a large population of mobile epibenthic scavengers in even the most barren red-clay areas (Fig. 6-10). Inasmuch as a nodule would have to be displaced only every few thousand years to avoid being buried, our failure to observe the phenomenon is not too surprising. Neither of the theories explains how highly asymmetrical nodules with well-differentiated

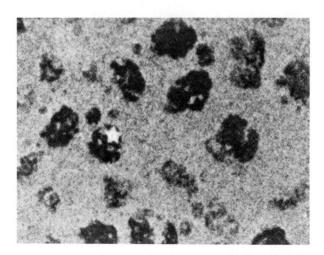

Fig. 6-9. Small starfish feeding on sediment on a North Pacific manganese nodule. Nearby sediment-free nodules probably were cleaned by such activity. After Paul, 1976.

Fig. 6-10.  Large benthic scavengers (rattail fish and amphipods) that are ubiquitous in nodule-bearing regions, and could conceivably move nodules. Hessler and Jumars, 1977.

tops and bottoms (see subsequent section) can remain at the sea floor for hundreds of thousands, if not millions, of years without losing their orientations.

Relatively little is known of deeply buried nodules because so few have been sampled. The Deep Sea Drilling Project (DSDP) has recovered a number of nodules, but their recorded depths of occurrence must be treated with caution because surface nodules readily fall down the hole when the circulation of drilling fluid is interrupted to add a length of drill string or to recover a sediment core.

Careful examination of associated microfossils has proven that some nodules are not displaced (Cronan, 1973). These older nodules differ in composition from the surface deposits at the same location, particularly when the associated lithologies differ. No systematic study of such differences has yet been made.

Menard (1976) has summarized the distribution of DSDP nodules in Pacific sediment of Cenozoic age. He suggests that the nodules are not uniformly distributed in space and time but rather are concentrated in the older deposits. In contrast, Glasby's (1978) review of DSDP data suggests that nodules are enriched in the late Cenozoic sediments. The number of nodules recovered from sediments of a given age, lithology, and sediment type is so small that statistical uncertainties are very large.

## DESCRIPTION

### Shape

The most attractive nodules are spherical and range from a centimeter or so to several tens of centimeters in diameter (Fig. 6-11). Such nodules tolerate handling reasonably well and tend to appear in display collections. In reality, though, spherical nodules are quite rare. Discoidal or irregular shapes are much more common (Fig. 6-12). In some cases the nucleus controls the shape, but generally the shape is governed by ir-

DEEP-SEA FERROMANGANESE NODULES

Fig. 6-11. Simple, subspherical nodules, typical of areas of strong bottom currents, from 32°43'N, 158°16'W, 1300 m. After Sorem and Fewkes, 1979.

regular growth of the layers of ferromanganese oxyhydroxides, leading to botryoidal masses often modified by the fusion of two or more nodules. Where the *in situ* orientation of a nodule is known, it is often found to have a top that is relatively simple, whereas the bottom consists of a mass of irregular gritty protuberances (Fig. 6–13).

Very few generalizations can be made relating nodule shapes to depositional conditions. However, the great uniformity of nodules collected at a single station [first pointed out by Murray and Renard (1891) who stated that "it is usually possible for us to state

Fig. 6-12. Asymmetric nodules typical of more quiescent depositional areas (smooth surface is exposed to seawater), from 7°25'S, 152°15'W (right) and 33°31'S, 74°43'W. After Murray and Renard, 1891.

Fig. 6-13.    Contrast between textures of tops (smooth) and bottoms (rough) of typical North Pacific nodules from 10°N, 140°W. After Sorem and Fewkes, 1979.

at sight from which *Challenger* station any particular nodule had been produced"] suggests that local conditions, rather than the unique history of each nodule, governs its shape. The author's impression is that simple forms, especially spheres, typify non-depositional or erosional areas (Payne and Conolly, 1972; Watkins and Kennett, 1972) whereas complex forms are more common in areas of relatively rapid deposition. This question has not been investigated in a systematic way, however.

### Size

Most nodules are a few centimeters in diameter. Where protected from burial (such as on the seamounts of the northern Emperor chain, for example), they can grow to sizes approaching 30 cm. In depositional areas, samplers often recover a range of sizes (Fig. 6-6), suggesting a steady nucleation of new nodules, presumably balanced by burial of an equivalent number of older ones. At many locations, however, all the nodules are of roughly equal size, implying a single nucleation event and subsequent uniform growth. Such a size distribution would be predicted for nondepositional or erosional areas, or for areas lacking nuclei (in which case the number of nodules should decrease exponentially with time if the model discussed earlier is correct).

Most sediments associated with nodules contain micronodules 1 mm or less in diameter. There is a distinct size gap between these micronodules and the smallest nodules, which usually are 3-6 mm in diameter. The origin of this gap is unclear, although it may be related to the process that keeps nodules at the surface. For instance, if benthic organisms are responsible for intermittently displacing nodules, they may not be able to distinguish micronodules from the associated sediment, and so allow them to be buried as soon as they form. If this explanation is correct, it suggests that the presence of nuclei a few millimeters in diameter is essential for nodule formation. The absence of appropriate nuclei could explain the bare areas of Fig. 6-4, for example. Horn *et al.* (1973) have postulated that the availability of nuclei (particularly volcanic debris) is *the* factor governing nodule density at any deep-sea location. This hypothesis has yet to be adequately tested, particularly in terms of separating the influence of nuclei from other environmental variables such as sediment type, sedimentation rate, etc.

## Surface Textures

The surfaces of nodules range from smooth to gritty to botryoidal to irregular. As mentioned previously, the most striking pattern is that of smooth, relatively featureless surfaces on portions of the nodules exposed to seawater, versus gritty, dendritic surfaces on buried portions (Fig. 6-14). Such a difference could exist if the smooth surfaces result from dissolution whereas the gritty areas are growing actively. Radiometric analy-

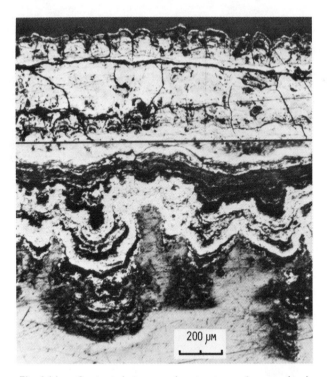

200 μM

Fig. 6-14.    Contrast between microstructure of upper (top) and lower (bottom) surfaces of a North Pacific nodule from 15°46′N, 126°W, 4514 m. After Sorem and Fewkes, 1979.

ses show that this interpretation is not correct, at least where sediments are highly oxidized, but such analyses do not establish whether the smooth surfaces result directly from the depositional process, or whether initial, small irregularities are removed by the gentle abrasion of bottom currents or by the feeding activities of the small benthic organisms photographed by Paul (1976).

## Attached Organisms

Greenslate (1974) and Dugolinsky *et al.* (1977) have pointed out that nodule surfaces often support a diverse assemblage of attached microorganisms (Fig. 6-15). Such assemblages are dominated by benthic foraminiferans that construct shelters or tests of calcite or detrital particles. Such habitats may play an important role in providing a mechanically strong framework that allows the accumulation of the chem-

Fig. 6-15.    Structures of encrusting protozoa on the surfaces of nodules. The unidentified forms in both the upper and lower right panels aggregate opal tests (mainly radiolarian), sponge spicules, and detrital siliceous grains into habitats. The tubular agglutinated foraminifera *Sacorrhiza ramosa* (lower left) is the most abundant encrusting species identified thus far. After Dugolinksy *et al.*, 1977.

DEEP-SEA FERROMANGANESE NODULES

ically precipitated ferromanganese oxyhydroxides that make up the bulk of the nodules.

Dugolinsky *et al.* (1977) have suggested that the epifauna is densest and most diverse on nodules from the equatorial North Pacific that are rich in copper and nickel. It is unclear, however, whether the surface organisms are in any way responsible for the metal enrichment, or whether both phenomena result from an additional factor (such as the steady rain of biogenic debris resulting from the high surface biological productivity). The relation of epifauna to nodule genesis has been likened to the relation of birds to trees; trees provide an excellent habitat for birds, but the presence of birds has very little to do with the origin and characteristics of trees!

## Internal Structure

A slice through the center of even the simplest-looking nodule reveals a complexly layered interior (Fig. 6-16). In many, if not most nodules, the internal layers surround well-defined nuclei (Fig. 6-17). These nuclei, which figure prominently in the descriptions by Murray and Renard (1891) as well as in many subsequent studies, can be as exotic as sharks' teeth or cetacean earbones, or as mundane as mud lumps or fragments of older nodules. Other common nuclei include volcanic debris and large detrital particles (particularly in areas of ice-rafting). In extreme cases, the shape of a nucleus controls the external form of a nodule. More commonly, however, nuclei seem to have little influence on the external characteristics of nodules. No systematic geographic pattern

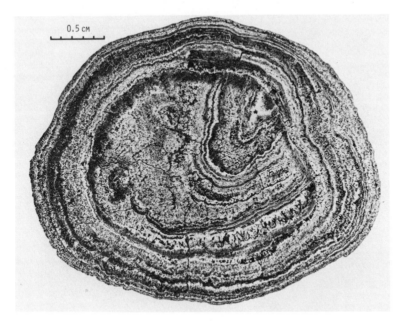

0.5 CM

Fig. 6-16.   Typical section through a North Pacific pelagic nodule from 15°8'N, 136°22'W showing the complex internal layering. Bright layers are well crystallized manganese oxyhydroxides (often todorokite-rich). Dull layers are rich in detritus and iron oxyhydroxide. After Sorem and Fewkes, 1979.

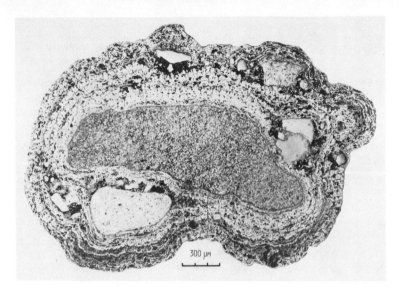

Fig. 6-17.   Section through a complex Drake Passage nodule from 58°S, 60°W showing ferromanganese deposition around multiple clastic nuclei. After Sorem and Fewkes, 1979.

of distinctive nuclei has been recognized, and the compositions of nodules appear to be independent of the nature of their nuclei. As discussed previously, nuclei may be essential for the growth of nodules in areas of active sedimentation, but further coordinated sediment-nodule studies are required to test this possibility.

For many years, the genesis of the internal layering of nodules has been a sub-

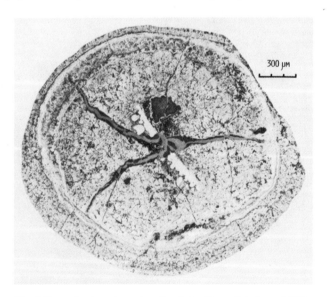

Fig. 6-18.   Section through a North Pacific nodule from 30°N, 160°W showing internal shrinkage cracks resulting from recrystallization of initially loosely packed and highly hydrated ferromanganese oxyhydroxides. After Sorem and Fewkes, 1979.

ject of vigorous debate, with strong proponents of both authigenic and diagenetic origins. Despite the efforts of Foster (1970) and Sorem (1973), who attempted to correlate individual layers between nodules from single locations as well as much larger areas, recent detailed radiochemical and mineralogical studies have tipped the scale in favor of diagenesis.

The most obvious (although far from universal) evidence of diagenesis is internal shrinkage (Fig. 6-18). Such shrinkage results from the conversion of highly hydrated oxyhydroxides deposited at nodule surfaces to more massive layers and bodies of better defined minerals within the heart of the nodule. The external layer, which is fairly homogeneous, readily segregates into manganese oxyhydroxide (highly reflective), and iron oxyhydroxide and silicate-rich (dull) layers. The factors that control the dimensions of the layers and their diversity of forms (simple laminae to cusps to mottled masses) are not understood.

At the microscopic scale, Burns and Burns (1978a) have recorded the appearance of small but well-developed crystals of barite, phillipsite, and todorokite in electron micrograph traverses from the exterior to the interior of a nodule (Fig. 6-19). The ap-

Fig. 6-19. High resolution scanning electron micrographs of todorokite (upper right) and phillipsite (lower right) crystals typical of the outer layers or cavities in the interiors of many nodules. The panels on the left show massive laminae of manganese oxyhydroxides (todorokite-bearing) typical of recrystallized nodule interiors. After Burns and Burns, 1978a.

pearance of these new phases is accompanied by the disappearance of siliceous tests and other unstable components.

Radiochemical data supporting the importance of diagenesis have been described by Krishnaswami *et al.* (1980) and more recently by MANOP (the NSF/IDOE-supported Manganese Nodule Program) workers (Moore *et al.*, 1980). These results, summarized in Table 6-2, suggest that the outer half-millimeter or so of a nodule is chemically "open," with ready loss or gain of radioactive daughters (and presumably other soluble elements). At greater depths, however, activity ratios begin to approach unity, suggesting that most elements are bound in well-crystallized secondary minerals. In the case of Ra, persistent disequilibria and the similarity of the concentration profiles inward from both sides of the MANOP nodule (Moore *et al.*, 1980) imply continued molecular diffusion. It is still unclear whether diagenesis at depth (which converts platy todorokite crystals to dense masses, for example) are isochemical, or whether the nodule interior is open to dissolved metals. The latter possibility is supported by decreases in Th concentrations (presumably by dilution) away from the surfaces of some Pacific nodules (Krishnaswami and Cochran, 1978; Moore *et al.*, 1980).

TABLE 6-2. Activities (dpm/g) of Uranium-Series Radionuclides in the Tops and Bottoms of North Pacific Nodules[a]

| *Depth* (mm) | $^{230}$Th | $^{226}$Ra | $^{210}$Pb | $\dfrac{^{226}\text{Ra}}{^{230}\text{Th}}$ | $\dfrac{^{210}\text{Pb}}{^{226}\text{Ra}}$ |
|---|---|---|---|---|---|
| *Top* | Core MN7601–20–BX2 | | (11°04.3′N, 140°02.8′W, 4722 m) | | |
| 0–0.3 | 719 | 208 | 138 | 0.29 | 0.66 |
| 0.3–0.7 | 241 | 145 | 36 | 0.60 | 0.25 |
| 0.7–1.2 | 116 | 57 | 47 | 0.49 | 0.82 |
| 1.2–1.3 | 154 | – | – | – | – |
| 1.3–1.5 | 52 | – | – | – | – |
| 1.5–1.7 | 30 | – | – | – | – |
| 2.2–2.8 | 6 | – | – | – | – |
| *Bottom* | | | | | |
| 0–0.4 | 30.9 | 175 | 30 | 5.7 | 0.17 |
| 0.4–0.8 | 9.7 | 83 | 23 | 8.6 | 0.28 |
| 0.8–1.3 | 6.6 | 27 | 33 | 4.1 | 1.20 |
| 1.3–2.1 | 4.5 | 11 | 15 | 2.4 | 1.37 |
| 2.1–2.8 | 2.7 | 15 | 15 | 5.7 | 1.00 |
| 2.8–4.0 | 1.3 | 3 | – | 2.2 | – |
| *Top* | Core A47–16–1 | | (9°2.3′N, 151°11.4′W, 5049 m) | | |
| 0–0.2 | 478 | 173 | 377 | 0.4 | 2.18 |
| 0.2–0.5 | 222 | 154 | 86 | 0.7 | 0.56 |
| 0.5–0.9 | 134 | 134 | 76 | 1.0 | 0.57 |
| 0.9–1.2 | 71 | 77 | 53 | 1.1 | 0.69 |
| 1.2–1.5 | 42 | 46 | 48 | 1.1 | 1.04 |
| 1.5–1.8 | 14 | 23 | 24 | 1.6 | 1.03 |
| 1.8–2.3 | 14 | 18 | 20 | 1.3 | 1.10 |
| 15.0–15.2 | 14 | 14 | 16 | 1.0 | 1.11 |

DEEP-SEA FERROMANGANESE NODULES

TABLE 6-2. (Continued)

| Depth (mm) | $^{230}Th$ | $^{226}Ra$ | $^{210}Pb$ | $\dfrac{^{226}Ra}{^{230}Th}$ | $\dfrac{^{210}Pb}{^{226}Ra}$ |
|---|---|---|---|---|---|
| *Bottom* | | | | | |
| 0–0.2 | 96 | 888 | 352 | 9.2 | 0.40 |
| 0.2–0.4 | 136 | 685 | 269 | 5.0 | 0.39 |
| 0.4–0.7 | 109 | 455 | 152 | 4.2 | 0.33 |
| 0.7–0.85 | 65 | 303 | 158 | 4.7 | 0.52 |
| 0.85–1.0 | 37 | 225 | 127 | 6.1 | 0.56 |
| *Top* | Core C57–58–1 | | $(15°19.5'N, 125°54.4'W, 4638 m)$ | | |
| 0–0.03 | 725 | 145 | 2505 | 0.2 | 17.28 |
| 0.03–0.1 | 570 | 191 | 246 | 0.4 | 1.29 |
| 0.1–0.2 | 390 | 180 | 62 | 0.5 | 0.34 |
| 0.2–0.3 | 200 | 155 | 59 | 0.8 | 0.38 |
| 0.3–0.4 | 95 | 118 | 63 | 1.2 | 0.53 |
| 0.4–0.6 | 89 | 105 | 51 | 1.2 | 0.49 |
| 0.7–0.8 | 25 | 63 | 41 | 2.6 | 0.64 |
| *Bottom* | | | | | |
| 0–0.4 | 40 | 340 | 183 | 8.6 | 0.54 |
| 0.4–0.9 | – | 316 | – | – | – |
| 0.9–1.1 | 5 | 128 | 68 | 26.9 | 0.53 |
| 1.1–1.4 | 5 | 62 | 54 | 11.7 | 0.87 |
| 1.4–1.6 | 5 | – | – | – | – |
| 1.6–1.8 | – | 35 | 31 | – | 0.90 |

[a] After Krishnaswami and Cochran (1978) and Moore *et al.* (1980).

An internal feature of some nodules, the origin of which is still unclear, is sequences of very fine laminae. Margolis and Glasby (1973) attribute such laminae to paleoceanographic changes with periodicities in the 25 to 10,000 year range. In view of the recent radiochemical data, however, it appears that Margolis and Glasby's interpretation should be tested by detailed microprobe traverses across the laminae to determine whether the range of chemical variation is compatible with such a primary origin.

## Growth Rates

Over the past decade or so, a large number of nodules have been dated by a variety of techniques. The results through 1975 have been thoroughly reviewed and summarized in a comprehensive paper by Ku (1977). The dating techniques include K-Ar and fission track analyses of volcanic nuclei of nodules, amino acid racemization analyses of cetacean earbone nuclei, and $^{10}Be$, $^{230}Th$, $^{231}Pa$, biostratigraphic (Harada and Nishida, 1976), paleomagnetic, and $\alpha$-track analyses of the outer oxyhydroxide layers. The results show remarkable agreement, given the diverse assumptions of the various methods and the differing chemical behaviors of the elements in question. In all cases, the estimated accretion rates are in the range of one to a few millimeters per million years. This rate, which is equivalent to a few unit cells (tens of angstrom units) per year, is one of the slowest, if not the slowest, ongoing depositional processes in nature. As discussed previously, this rate is consistent with observed size distributions and burial rates of nodules (Bender *et al.*, 1966; Heath, 1979).

Despite the diversity of dating techniques and the geological reasonableness of the results, these very slow accretion rates have been challenged (Arrhenius, 1967; Lalou and Brichet, 1972; Lalou *et al.*, 1979). These workers claim that nodules grow very rapidly (thereby eliminating the need to keep them at the surface for times equivalent to the deposition of several meters of sediment). As evidence, they cite the presence of young carbon (containing $^{14}$C) in nodule nuclei, and the presence of unsupported $^{230}$Th deep within some nodules. They suggest that the exponentially inward-decreasing concentrations of radionuclides used by other workers to infer slow growth rates result from diffusion of the dissolved nuclides into the porous outer layers of the nodules, or contamination of deeper layers by surface material.

The validity of accumulation rates based on decay profiles of elements in the U–Th decay series has been confirmed by detailed radiochemical studies by Krishnaswami and Cochran (1978), Krishnaswami *et al.* (1980), and Moore *et al.* (1980), as well as by Ku and Broecker's (1967) earlier work. These workers have shown that $^{230}$Th and $^{231}$Pa decay profiles yield consistent growth rates, despite the differences in their half-lives. Furthermore their data show that $^{210}$Pb is abundant only in the surface layer of nodules, rather than mimicking the distributions of $^{230}$Th or $^{231}$Pa as would be expected if downward diffusion or contamination were responsible for the profiles of U and Th daughters. Thus, the suggestion that nodule accumulation rates based on radiometric analyses are invalid seems untenable, not only because of the recent radiochemical experiments, but also because it is completely inconsistent with the size and burial-rate data cited earlier.

A new technique that was hoped would allow very high resolution dating of the outer layers of nodules, and which is rapid enough to allow many nodules (and several points around a single nodule) to be dated has been examined by Heye and Beiersdorf (1973), and Andersen and Macdougall (1978). This technique uses a thin polyester film to record alpha particles released from a polished nodule section over a period of several months. The $\alpha$-tracks are then counted over very narrow depth intervals with the aid of a petrographic microscope. For the exponential decrease in track numbers with depth to be converted to accretion rates, it is necessary that none of the decay products of $^{230}$Th migrate from their point of formation. This demanding requirement is not met for the outermost half millimeter or so of most nodules (Fig. 6-20), as might be suspected from radiochemical data by Krishnaswami and Cochran (1978) and Moore *et al.* (1980), and it is questionable at greater depths because of the diffusion of Ra. Thus, the exponential decrease in $\alpha$-tracks with depth contains information about both diffusion and radioactive decay. Perhaps fortuitously, however, the $\alpha$-track decay curves yield accretion rates within about 30% of those from other dating methods (Anderson and Macdougall, 1977; Krishnaswami and Cochran, 1978).

Figure 6-21 shows a histogram of accretion rates of Pacific nodules determined from $\alpha$-track profiles (Andersen and Macdougall, 1977). The distribution is almost

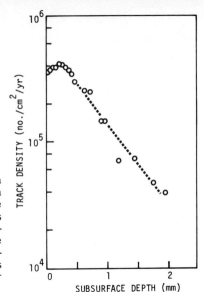

Fig. 6-20. Abundance of alpha tracks as a function of depth below the top surface of a MANOP nodule from 11° 4′N, 140° 3′W. The near-surface decrease in abundance reflects diffusive loss of $^{222}$Rn and $^{226}$Ra. The decrease in abundance deeper in the nodule reflects the exponential decay of unsupported $^{230}$Th and its daughter elements in secular equilibrium to $^{206}$Pb. After Moore et al., 1980.

normal for the 49 nodules accumulating at <16 mm/my, with a mean of 6.8 mm/my and a standard deviation of 3.1 mm/my.

Because of the evidence for migration of Ra and perhaps even Th and Pa in ferromanganese nodules (Ku et al., 1979b), efforts are underway to refine $^{10}$Be dating so that it can be applied to average-size nodules. This isotope, with a half-life of 1.5 million years, is formed by the interaction of cosmic rays with nitrogen in the upper atmosphere. Because of its very low concentration in nodules, it can be used only to date large ferromanganese slabs or nodules analyzed by conventional radiochemical (β-counting) techniques. But efforts are underway to measure $^{10}$Be directly using high-energy tandem van der Graaf accelerators as mass spectrometers (Nelson, 1977; Turekian et al., 1979). Progress to date suggests that analyses will be possible on about 100 mg of the nodule material, an amount equivalent to that now used for $^{230}$Th or

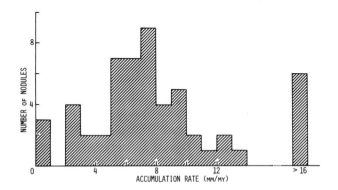

Fig. 6-21. Histogram of growth rates of Pacific nodules determined from alpha-track profiles. After Andersen and Macdougall, 1977.

[231]Pa analyses. [10]Be determinations will not only serve to check nodule chronologies based on uranium-series analyses, but also will allow the reconstruction of much longer growth histories because the half-life of this nuclide is an order of magnitude greater than either [230]Th (75,200 years) or [231]Pa (34,300 years).

## Elemental Composition

Over the past century an enormous number of nodules have been analyzed chemically. Several thousand analyses have been published in the open literature (see compilation by Monget *et al.*, 1976) and a comparable number exist in the files of the major nodule-mining consortia. Table 6–3, after Cronan (1976) shows that the bulk chemical composition of nodules varies significantly from ocean basin to ocean basin. The possible explanations for such differences are explored in a subsequent section of the paper. It is sufficient here to point out that relative to the global average, Atlantic Ocean nodules are enriched in Fe, Pacific Ocean nodules are enriched in Mn, Cu, and Ni, and Indian Ocean nodules are intermediate in character.

TABLE 6–3.  Average Elemental Compositions of Abyssal Nodules from the Major Oceans[a]

| Element | Pacific | Atlantic | Indian |
|---------|---------|----------|--------|
| Mn | 19.3 | 15.5 | 15.3 |
| Fe | 11.8 | 23.0 | 13.4 |
| Ni | 0.85 | 0.31 | 0.53 |
| Cu | 0.71 | 0.14 | 0.30 |
| Co | 0.29 | 0.23 | 0.25 |

[a]After Cronan, 1976. Values in weight percent.

Within a single ocean basin such as the Pacific, Calvert and Price (1977), Cronan and Tooms (1969), Mero (1965), Piper and Williamson (1977), and Skornyakova (1979 and references therein) have all outlined regions enriched in Fe and Co, Mn (with and without Cu and Ni), as well as other elements (Fig. 6–22). These provinces correspond in a gross way to the major geologic provinces of the Pacific Ocean basin (e.g., Heezen and Fornari, 1979; Horn *et al.*, 1972). The compositional differences between populations of nodules are greatly emphasized if the nodules are grouped by geomorphic (e.g., seamount) or depositional (e.g., hemipelagic) setting. Table 6–4, from Cronan (1977), summarizes a homogeneous data set accumulated over the past 10 to 15 years. The differences from group to group are striking. Of particular note are the enrichment of Co and Pb in seamount nodules, and the enrichment of Mn in hemipelagic nodules.

At smaller scales (nodules from areas of a few square kilometers, for example), the variability is reduced, but even in an area as small as 230 km$^2$, Calvert *et al.* (1978) found factor-of-two variations in Mn/Fe ratios (Table 6–5).

Statistical analyses of elemental compositions show that certain groups of elements vary coherently from nodule to nodule (Calvert and Price, 1977; Calvert *et al.*, 1978;

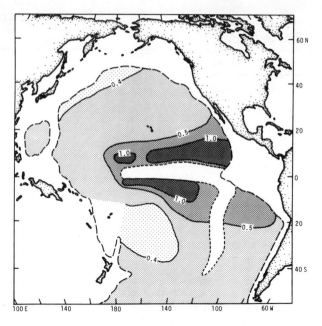

Fig. 6-22. Copper abundance in Pacific deep-sea nodules (after Skornyakova, 1979), showing strong concentrations of economically attractive nodules in the sub-equatorial region. Isopleths in weight percent copper. Hemipelagic deposits between long-dashed lines and the continents are characterized by anoxic diagenesis. Short-dashed line outlines nodule-free areas beneath the equatorial zone of high biological productivity, and along the crest of the East Pacific Rise. Blank area around Fiji — no data.

Cronan and Tooms, 1969). Thus, for abyssal nodules the covariance of Mn, Ni and Cu is well established (Cronan, 1977). Calvert and Price (1977) add to this group Mg, Ba, Mo, and Zn. A second group of elements that covary in these nodules includes Fe, Ti, P, Co, and Pb. These groupings reflect the abilities of elements to react with or substitute into oxyhydroxides of Mn and Fe (see the following section). If the Mn content of Pacific nodules is plotted against the contents of other correlated transition metals,

TABLE 6-4. Average Elemental Compositions of Nodules from Different Geologic Settings[a]

| Element | A | B | C | D | E | F | G |
|---------|------|------|------|------|------|------|------|
| Mn | 14.6 | 17.2 | 15.5 | 19.7 | 15.7 | 38.7 | 16.8 |
| Fe | 15.8 | 11.8 | 19.2 | 20.1 | 19.3 | 1.34 | 17.3 |
| Ni | 0.35 | 0.64 | 0.31 | 0.34 | 0.30 | 0.12 | 0.54 |
| Cu | 0.06 | 0.09 | 0.08 | 0.05 | 0.08 | 0.08 | 0.37 |
| Co | 1.15 | 0.35 | 0.40 | 0.57 | 0.42 | 0.01 | 0.26 |

[a]After Cronan, 1977. $A$ = seamounts; $B$ = oceanic plateaus; $C$ = active mid-ocean ridges; $D$ = inactive ridges; $E$ = marginal banks and seamounts; $F$ = continental borderlands; $G$ = abyssal sea floor. Values in weight percent.

TABLE 6-5. Elemental Composition of Pacific Nodules from a 230 km² Area at
8°20′N, 153°Wᵃ

| Element | Min. | Max. | Ave. |
|---------|------|------|------|
| Mn | 16.9 | 26.7 | 24.6 |
| Fe | 4.97 | 9.24 | 6.77 |
| Ni | 0.82 | 1.32 | 1.08 |
| Cu | 0.82 | 1.76 | 1.14 |
| Co | 0.11 | 0.21 | 0.16 |
| Zn | 0.07 | 0.11 | 0.09 |

[a]After Calvert *et al.*, 1978. Values in weight percent. Based on 19 analyses.

it is clear that the correlation is different depending on whether the samples contain more or less than about 17% Mn (Fig. 6–23, for example). As Calvert and Price (1977) point out, the nodules with high metal/Mn ratios contain the mineral todorokite, whereas the low-ratio nodules do not.

Although nodules can contain major proportions of nonmetallic elements, such elements can be explained by the accidental inclusion of sedimentary particles in the nodules. Calvert and Price's (1977) data for Si and Al in associated nodules and sediments (Fig. 6–24) support this explanation; scatter in the Si/Al ratio probably reflects the diagenetic loss of biogenic silica revealed by scanning electron micrographs. Diagenesis of the detrital particles in nodules is also suggested by the large frac-

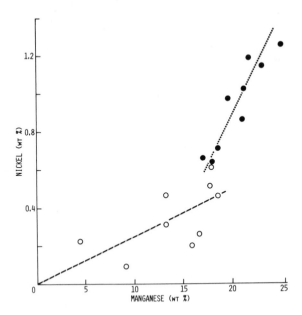

Fig. 6-23. Differing Ni-Mn relations for Pacific nodules containing only δ-MnO₂ (open circles) and those containing todorokite as well as δ-MnO₂ (solid circles). After Calvert and Price, 1977.

DEEP-SEA FERROMANGANESE NODULES

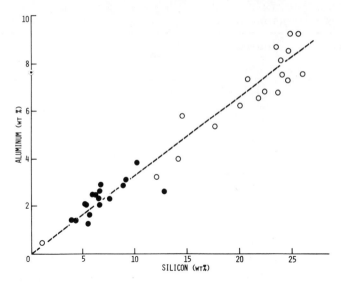

Fig. 6-24. Similarity of Si-Al correlations in Pacific nodules (solid circles) and associated sediments (open circles) suggests that these elements are present in the nodules as trapped detritus. After Calvert and Price, 1977.

tion of "detrital" elements that can be removed by mild chemical leaches (Moore *et al.*, 1980).

Examination of chemical data for single nodules reveals a considerable degree of heterogeneity, as would be anticipated from the textural and radiochemical data. The major systematic pattern, first documented by Raab (1972), is a consistent enrichment of nodule bottoms in Mn, Ni, and Cu relative to nodule tops, which are enriched in Fe, Co, and Pb (Fig. 6-25). Such a pattern draws attention to the importance of the substrate in determining nodule chemistry, a topic to which the author will return later in this chapter. The tight correlations between Mn and Ni or Cu that characterize the bulk nodule data begin to break down in data on parts of nodules (Calvert and Price, 1977).

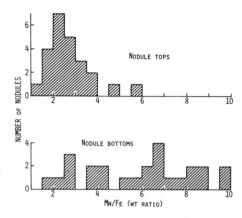

Fig. 6-25. Histograms of Mn/Fe in the tops and bottoms of Pacific nodules, showing relative enrichment of Mn in the nodule bottoms. Data from Raab, 1972; after Calvert and Price, 1977.

| Element | Reflective ("Crystalline") | | | Dull ("Amorphous") | | |
|---|---|---|---|---|---|---|
| | Min. | Max. | Ave. | Min. | Max. | Ave. |
| Mn | 23.5 | 47.5 | 35.8 | 12.3 | 27.6 | 19.4 |
| Fe | 0.12 | 4.36 | 1.28 | 10.7 | 27.1 | 16.9 |
| Ni | 0.05 | 3.67 | 1.74 | 0.02 | 0.83 | 0.27 |
| Cu | 0.13 | 2.95 | 1.25 | 0.01 | 0.41 | 0.16 |
| Co | 0.00 | 0.43 | 0.07 | 0.08 | 0.74 | 0.31 |
| Zn | 0.01 | 0.77 | 0.23 | 0.00 | 0.12 | 0.01 |

[a]From Banning, *in* Sorem and Fewkes, 1979. Values in weight percent.

For example, Raab's (1972) data show about a factor-of-two variation in the Ni content of nodule bottoms containing 25% to 30% Mn.

The simple interelemental correlations break down even further if individual laminae are analyzed by electron microprobe. Banning (in Sorem and Fewkes, 1979) has examined a large number of strongly reflective ("crystalline") and weakly reflective ("amorphous") layers in Pacific nodules. Although the average elemental abundances (Table 6-6) are reminiscent of bulk nodules, the correlation of Cu + Ni with Mn in the crystalline phases, for example, is negative (Fig. 6-26)—the exact opposite of Calvert and Price's (1977) finding for 21 bulk Pacific nodules [but consistent with Cronan's (1977) analyses of 63 abyssal Pacific nodules!]. Microprobe data are helping to relate nodule chemistry and mineralogy and to throw light on the diagenetic reactions within nodules. Thus far, however, no one has succeeded in creating a chemical stratigraphy for nodules from a single collection, or in relating the complex internal chemistry of nodules to their genesis or depositional environment.

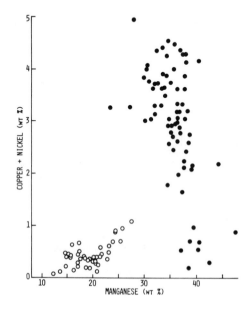

Fig. 6-26.   Relations between Cu + Ni and Mn in reflective layers (solid circles; manganese oxyhydroxides, including todorokite) and dull layers (open circles; iron oxyhydroxides and detritus) of manganese nodules. Electron microprobe data from Banning, *in* Sorem and Fewkes, 1979.

## Mineralogy

Because of the large number of known oxides and oxyhydroxides of iron and manganese, many of which are poorly crystalline and difficult to identify by standard X-ray diffraction techniques, considerable confusion surrounds the mineralogy of ferromanganese nodules. This confusion is compounded by controversy as to the equivalence of naturally occurring and artificially prepared minerals, and by the ease with which cations other than Fe and Mn can substitute into the crystal structures to yield different minerals.

The recent use of very high resolution transmission electron microscopy (Turner and Buseck, 1979) and selected area electron diffraction techniques have done much to clarify the situation. The subsequent discussion draws heavily on Burns and Burns' (1977a,b; 1980) recent critical reviews of the status of nodule mineralogy.

The iron and manganese minerals reported (with varying degrees of reliability) from ferromanganese nodules include $\delta$-$MnO_2$ ( = vernadite), birnessite [(Ca, Na) ($Mn^{2+}$, $Mn^{4+}$)$_7$ $O_{14}$ $\cdot$ $3H_2O$ = 7Å manganite = manganese manganate], todorokite [(Ca, Na, K, Ba, $Mn^{2+}$)$_2$ $Mn_5O_{10}$ $\cdot$ $H_2O$ = buserite? = 10Å manganite), psilomelane [(Ba, K, Mn, Co)$_2$ $Mn_5O_{10}$ $\cdot$ $xH_2O$], nsutite ($\gamma$-$MnO_2$), pyrolusite ($\beta$-$MnO_2$), cryptomelane [(K, Ba)$_{1-2}$ $Mn_8O_{16}$ $\cdot$ $xH_2O$], rancieite [(Ca, Mn) $Mn_4O_9$ $\cdot$ $3H_2O$], ferrihydrite ($5Fe_2O_3$ $\cdot$ $9H_2O$), goethite ($\alpha$-FeOOH), hydrogoethite (mixed oxyhydroxides), akaganeite ($\beta$-FeOOH), lepidocrocite ($\gamma$-FeOOH), feroxyhyte ($\delta$-FeOOH), hematite ($\alpha$-$Fe_2O_3$), maghemite ($\gamma$-$Fe_2O_3$), and magnetite ($Fe_3O_4$), according to Burns and Burns (1980). Deep-sea nodules, however, are dominated by $\delta$-$MnO_2$, todorokite, birnessite, and X-ray amorphous ferric oxyhydroxide.

$\delta$-$MnO_2$ is a very fine grained (reflected in broad diffraction peaks at 2.43 and 1.41Å) component of virtually all deep-sea nodules. It may be a finely crystalline form of birnessite (Bricker, 1965), but most workers now treat it as a separate phase. Its structure is unknown, but it appears to readily incorporate elements like Co and Ce. Todorokite, with characteristic diffraction lines at 9.6-9.8 and 4.8Å, is also abundant in deep-sea nodules, particularly those enriched in Cu and Ni (Fig. 6-23). Its structure has not been determined directly, but its fibruous crystal morphology (Burns and Burns, 1977a), electron diffraction pattern (Chukrov et al., 1979), and ultrastructure revealed by transmission electron microscopy (Turner and Buseck, 1979) point to a framework structure containing long tunnels surrounded by [$MnO_6$] octahedra (Fig. 6-27). This structure, analogous to hollandite or psilomelane, can accommodate large

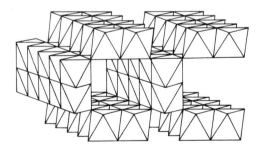

Fig. 6-27. Arrangement of [$MnO_6$] octahedra in the hollandite structure, a presumed analog of todorokite. The large tunnels can accommodate large cations and water. After Burns and Burns, 1977.

cations or water in the tunnels, as well as smaller cations substituting in the $[MnO_6]$ lattice.

Birnessite, the third common manganese phase found in deep-sea nodules, is characterized by diffraction lines at 7.0–7.3Å and 3.5–3.6Å. It is most abundant in rapidly deposited ferromanganese crusts (near hydrothermal vents) or in nodules from hemipelagic sediments. Because many deep-sea nodules contain the zeolite, phillipsite, which has an intense X-ray diffraction line at 7Å, some of the published birnessite occurrences (e.g., Barnes, 1967) should not be accepted uncritically. The structure of natural birnessite has not been determined, but electron diffraction studies of synthetic $MN_7O_{13} \cdot 5H_2O$ point to a chalcophanite-type structure with sheets of edge-shared $[MnO_6]$ octahedra separated by layers of water and hydroxyl ions which can accommodate large cations. The platy hait of birnessite crystals (Sorem and Fewkes, 1977) is consistent with this structure. Birnessite, like todorokite, appears readily to incorporate Ni, Cu, and Zn, as predicted by radius ratio and crystal field stabilization energy data (Burns and Burns, 1977b).


# GENESIS


The origin of nodules in general and nodules enriched in minor metals in particular involves the consideration of two major questions: Where do the metals come from? and, How are they concentrated into nodules at the sea floor?


## Sources of Metals


Four sources for the transition metals in nodules have been proposed:

**1. Continental weathering followed by transport of dissolved metals in rivers to the sea** This is the classic source for all the dissolved elements in seawater. Recent work by K. Turekian and his associates, however, suggests that the abundance of particles, change in salinity and pH, and widespread reducing conditions at the river-ocean confluence extract virtually all the dissolved metals in fluvial systems, leaving but a small residue to escape to the open ocean.[2] This suggestion has yet to be tested for a broad range of rivers. Even if it is correct, it may be offset by subsequent diagenesis [see (3) and (4), below].

**2. Leaching of newly formed oceanic crust by seawater hydrothermal systems at mid-ocean spreading centers** This possibility, in its modern form (Fig. 6-28) as proposed by Corliss (1971), has been confirmed by recent observations of the Galapagos Rift Zone and East Pacific Rise at 21°N by the submersible *Alvin* (Corliss *et al.*, 1978, 1979a,b; Edmond *et al.*, 1979a,b,c).

[2] K. Turekian, oral communication, 1979.

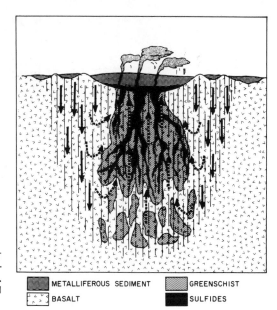

Fig. 6-28. Idealized view of the seawater hydrothermal system that develops in newly-formed oceanic crust (after Corliss, 1973). The debouching solutions add Mn and perhaps other transition metals to seawater.

| ▦ METALLIFEROUS SEDIMENT | ▨ GREENSCHIST |
| ⛰ BASALT | ■ SULFIDES |

The compositions of the debouching seawater sampled to date at the two hydrothermal sites are quite different. The Galapagos solutions are enriched in Mn but depleted in other transition metals, whereas the 21°N vents, particularly the "black smokers" (Fig. 6-29), are spewing out solutions markedly enriched in Cu and Zn, as well as Mn.

Laboratory experiments (Bischoff and Dickson, 1975; Hajash, 1975; Menzies and Seyfried, 1979; Mottl and Holland, 1978; Seyfried and Bischoff, 1977, 1979; Seyfried and Mottl, 1977) suggest that the water/rock ratio in the hydrothermal system may be responsible for the difference, but differences between the natural open systems and closed laboratory experiments, as well as the difficulty in modeling or duplicating multiple dissolution-precipitation events in the natural systems have thus far precluded a quantitative explanation of the difference between the two sites.

Based on an assumed constant ratio of hydrothermal leachates to $^3$He and to released heat in the hydrothermal solutions, Corliss *et al.* (1979) and Edmond *et al.* (1979b,c) have calculated global fluxes from this source to the oceans (Table 6-7). Clearly, hydrothermal Mn and Si are of major importance, in contrast to Ni and Cu. These values will almost certainly change as data from 21°N are fed into the calculations, but the initial realization that mid-ocean ridge hydrothermal systems play a major role in the chemistry of the oceans will continue to hold true.

**3. Anoxic diagenesis of hemipelagic sediments** In addition to the dissolved metals described under (1) above, rivers carry large loads of metals in particulate form. The fraction of these metals that is sorbed to clay particles or present as oxyhydroxides is very labile, particularly if deposited in a hemipelagic environment.

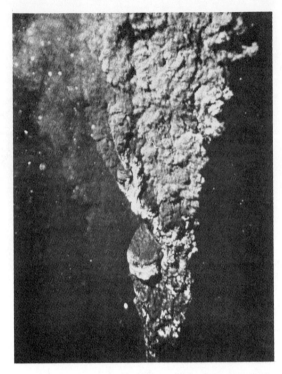

Fig. 6-29. Hot seawater hydrothermal solution erupting from a vent on the axis of the East Pacific Rise at 21°N. The turbidity is due to the precipitation of sulfides as the hydrothermal solution reacts with cold bottom water. The vent is about 10 cm across. Photo taken from the research submersible *Alvin*.

Recent studies of porewaters from hemipelagic deposits (Bender *et al.*, 1977; Froelich *et al.*, 1979) have suggested the following sequence of redox reactions:

$$O_2 + 4e^- \rightarrow 2O^=$$

$$2NO_3^- + 10e^- \rightarrow N_2 + 6O^=$$

$$Mn^{4+} + 2e^- \rightarrow Mn^{2+}$$

$$Fe^{3+} = e^- \rightarrow Fe^{2+}$$

$$SO_4^= + H_2O + 8e^- \rightarrow H_2S + 5O^=$$

Observations at MANOP site C (1°N, 140°W) in the equatorial Pacific suggest that the third reaction, at least, is bacterially mediated (Rosson *et al.*, 1979). The zone of $Fe^{3+}$ and $Mn^{4+}$ reduction is a region where a number of trace metals associated with ferromanganese oxyhydroxides are also solubilized (Klinkhammer, 1980a).

Redox profiles suggest that anoxic diagenesis should transport Mn, Fe, and associated metals to a thin, oxidized surface layer (Lyle, 1978; Lynn and Bonatti, 1965),

TABLE 6-7. Comparison of Elemental Fluxes to the Oceans from Rivers and
Mid-Ocean Ridge Hydrothermal Systems[a]

| Element | River Flux (M/y) | Hydrothermal Flux (M/y) |
|---|---|---|
| Mn | $50 \times 10^{9}$[b] | $60–180 \times 10^{9}$ |
| Si | $6.4 \times 10^{12}$ | $3.1 \times 10^{12}$ |
| Ba | $10 \times 10^{9}$ | $2.4–6.1 \times 10^{9}$ |
| Cu | $1.7 \times 10^{9}$ | [c] |
| Ni | $3 \times 10^{8}$ | [c] |

[a]Edmond et al., 1979a,b.
[b]Accumulation of Mn in deep-sea sediments (Bender et al., 1977).
[c]Cu and Ni are removed from seawater in the Galapagos hydrothermal systems (Edmond et al., 1979b).

but should release these metals to the ocean only where bottom oxygen values are very low (between about 200 and 800 m in the eastern Pacific, for example; Love, 1975). The quantitative importance of such areas, and the "leakage" of reduced metals from surficial sediments because of disequilibria due to the slow kinetics of the oxidation reactions has yet to be determined.

**4. Oxic diagenesis of pelagic sediments** The porewaters of oxidized sediments are significantly enriched in trace metals (Callender and Bowser, 1980; Klinkhammer, 1980a) even though the presence of free oxygen strongly favors the oxidation of these metals. It is unclear whether the porewater values reflect equilibration with solid phases or sorbed metals, or reflect redox disequilibria due to kinetic factors, but in any case oxic diagenesis does appear to constitute a source of dissolved transition metals to the sea floor. The existing data are inadequate to evaluate the global significance of this source relative to the other three.

## Concentration Mechanisms

Eight basic processes have been invoked to account for the availability of transition metals for incorporation into ferromanganese nodules at the sea floor. They are:

1. Vertical transport by settling particles;
2. Precipitation from bottom waters;
3. Upward diffusion of metals dissolved in porewaters;
4. Authigenic reactions in surficial sediments;
5. Bacterial activity in surface sediments and on nodules;
6. Activity of epibenthic microfauna on nodules;
7. Preferential incorporation of dissolved metals in specific iron/manganese oxy-hydroxide minerals; and
8. Local sea-floor volcanism.

The actual mechanism likely will involve a combination of these processes, and may well be geographically variable. The following discussion highlights our present understanding of these processes and summarizes some of the ongoing research designed to better understand their significance.

**1. Vertical transport by settling particles** Recent studies of particle-size distributions of material suspended in seawater (McCave, 1975), and of debris caught by sediment traps (Cobler and Dymond, 1980; Honjo, 1978; Spencer *et al.*, 1978) have shown that most of the particulate material reaching the sea floor is carried in large (hundreds of micrometers) fecal pellets, rather than by a slowly settling "lutite veil" (Arrhenius, 1954) made up of individual clay particles. Because the mass of a particle is proportional to the cube of its diameter and the settling velocity proportional to the square of its diameter, a mixture of equal numbers of 10 $\mu$ m and 100 $\mu$ m particles at the sea surface will result in a mass flux ratio of 1 to 10,000 at the sea floor.

Analyses of plankton (Bostrom *et al.*, 1978; Martin and Knauer, 1973) and of the vertical distribution of dissolved transition metals in the water column (Bender and Gagner, 1976; Bender *et al.*, 1977a,b; Boyle *et al.*, 1977; Bruland *et al.*, 1978; Sclater *et al.*, 1976) show that surface plankton extract metals in the same way that they extract major nutrients like phosphate, nitrate, and silica. The physiological pathways involving these metals are poorly understood, but the comparison of metal distributions with other nutrients suggests a complex picture, with some metals (Ni, for example) being incorporated in opal or calcite test material, and others (such as Cu) lodging in protoplasm. In all cases, the metals are carried in large particles, with some loss or gain en route, to the sea floor, where subsequent dissolution and decomposition reactions can release the metals to react with sediments or nodules, or to escape to the bottom waters of the ocean.

The coincidence of the northern equatorial Pacific zone of Cu-Ni-enriched nodules with the northern fringe of the equatorial zone of high biologic productivity has been pointed out by a number of workers (Greenslate *et al.*, 1973; Lyle, 1978; Lyle *et al.*, 1977). This coincidence is strong circumstantial evidence of the importance of biogenic debris from the surface in determining manganese nodule compositions. Current efforts by MANOP include the placement of sediment traps at five sites (Fig. 6–30) with a broad range of surface biologic productivities to assess the vertical flux of metals relative to nodule compositions and sediment properties. Published flux data (Martin and Knauer, 1980; Spencer *et al.*, 1978) yield values (g/cm$^2$/10$^6$ y) of 10–50 for Fe, 1 or more for Mn, and 0.35 for Cu. For Fe and Cu, these are more than adequate to account for observed accumulation rates in pelagic sediments (5.5, 1.6, and 0.03, respectively, for example; Heath *et al.*, 1970) or in three northern equatorial Pacific nodules (0.7, 2.6, and 0.15, respectively; Heath *et al.*, 1970). Clearly, coordinated sediment-trap and nodule-sediment studies are necessary to adequately assess flux balances, but the role of particles settling from the sea surface is unlikely to be insignificant.

**2. Precipitation from bottom waters** Many of the detailed vertical profiles of dissolved trace metals in seawater, made possible by the recent, improved analytical techniques cited earlier, show well-defined near-bottom maxima (Fig. 6–31). Such

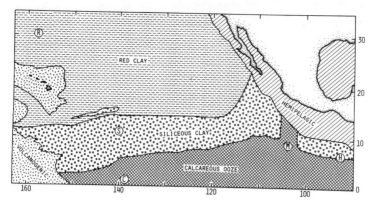

Fig. 6-30.    Locations of the MANOP detailed study areas in relation to the major sediment types of the eastern North Pacific. H = hemipelagic clay; M = metalliferous (rise crest) clay; C = calcareous ooze; S = siliceous (opal-bearing) clay; R = red (pelagic) clay.

maxima imply bottom sources (see the following sections), but they also form an important reservoir to supply transition metals to oxyhydroxides precipitating at the sea floor. Although bottom waters are assumed to influence the compositions of nodule tops and ferromanganese crusts on rock outcrops, it has not yet been determined whether geographic variations in the composition of bottom waters is reflected in the compositions of exposed oxyhydroxides.

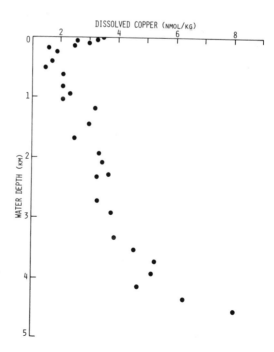

Fig. 6-31.    Concentrations of dissolved copper in a North Pacific seawater profile at 10°28'N, 123°38'W (after Boyle *et al.*, 1977). Surface enrichment is due to eolian input, bottom enrichment is due to the oxidative breakdown of biogenic debris at the sea floor.

**3. Upward diffusion of metals dissolved in porewaters** As discussed previously, both anoxic and oxic diagenetic reactions lead to increases in the transition-metal content of porewaters (Fig. 6-32). The resulting concentration gradient must lead to diffusion of these metals towards the sea floor. Within the zone of $Mn^{2+}$ migration in equatorial Atlantic sediments, for example (Froelich *et al.*, 1979), the upward flux is about 5 gm $Mn/cm^2/10^6y$, a value again comparable to the uptake rate by pelagic nodules [(1) above]. Where the zone of Mn reduction is separated from the sea floor by an oxidized layer, upward transport may still occur by bioturation of reactive oxyhydroxides (Froelich *et al.*, 1979), by oxic diagenetic reactions in the surface layer, or by the persistence of reduced metal ions due to the slow kinetics of the oxidation reactions.

Very little is known of the oxic diagenesis of different types of pelagic sediments. Both Volkov (1977) and Calvert *et al.* (1978) have shown, however, that nodule compositions are correlated with the contents of reactive metals in the subjacent sediments. Calvert *et al.* (1978) found a crude positive correlation between the oxalate-soluble Fe (Schwertmann, 1964) in the sediments and the total Fe in the nodules (Fig. 6-33) for a small area in the eastern equatorial Pacific. Volkov (1977) found correlations of 0.70, 0.43, 0.65, and 0.20 between Fe/Mn, Ni/Mn, Co/Mn, and Cu/Mn in nodules and in Chester and Hughes' (1967) solution-leachable sediment for 25 samples covering most of the North Pacific. These relations confirm the important role that near-surface sediments play in determining nodule chemistries (Raab, 1972), even though they throw little light on the reactions and pathways that link sediments and nodules.

**4. Authigenic reactions in surficial sediments** Two lines of evidence point to the occurrence of significant alteration of detrital particles at the sea floor prior to burial. First, the fluxes of a number of metals and nutrients measured by sediment traps exceed the accumulation rates of the same components in the sediments. Second, careful microscopic examination of the very top layer of sediment collected by box cores (Adelseck and Berger, 1975) reveals the presence of fragile and easily dissolved tests of microorganisms that are absent from deeper sedimentary layers.

The occurrence of such reactions is not surprising, since the time spent by a fecal pellet at the sea floor can be one or two orders of magnitude longer than the year or less that it spends settling through the water column. Yet we know that significant degradation does occur in the water column (Honjo, 1976).

Very little is known of the rate or nature of these sea-floor reactions. They are

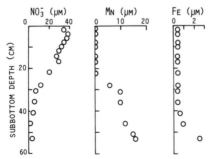

Fig. 6-32. Profiles of dissolved nitrate, Mn, and Fe in porewaters of an eastern equatorial Atlantic core from 1°5'N, 8°12'W, 4956 m (after Froelich *et al.*, 1979), indicating successive use of $NO_3^-$, $Mn^{4+}$ and $Fe^{3+}$ as oxidants of organic matter, leading to the mobilization of soluble $Mn^{2+}$ and $Fe^{2+}$.

DEEP-SEA FERROMANGANESE NODULES

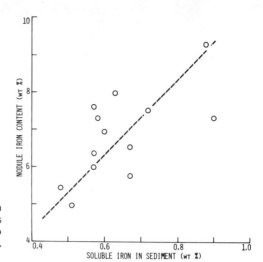

Fig. 6-33.  Relation of oxalate-soluble iron (Schwertmann, 1964) in pelagic sediments from the *Wahine* area (8°20′N, 153°W) to the total iron content of associated nodules. After Calvert *et al.,* 1978.

virtually impossible to measure on core samples both because of the difficulty of realistically duplicating the sea-floor conditions, and because the coring operation itself commonly results in loss or modification of the surfacemost layer. Recently K. Smith and co-workers (Smith, 1978; Smith *et al.,* 1978, 1979) and Hinga *et al.* (1979) have made *in situ* measurements of benthic respiration. These "bell jar" experiments confirm that virtually all the organic matter that rains on pelagic sediments is returned to the ocean prior to burial. No comparable data are available for transition metals. As part of the MANOP program, however, a "Lander" (Fig. 6-34) capable of collecting a time sequence of water samples from three chambers of "bell jars" enclosing 30 × 30 cm sections of the sea floor, is under construction (Weiss, 1979). This free-vehicle device, which will be able to transmit data to a surface ship and receive instructions during the course of an experiment, will also be able to measure the pH and oxygen content of the chambers as a function of time and to add trace quantities of various radioactive elements to the chambers during a deployment. The Lander will allow us finally to complete testing of the balance of metal fluxes into and out of the sea floor, thereby establishing the importance of metal uptake by nodules and sediments relative to escape back into the bottom water of the ocean.

## 5.  Bacterial activity in surface sediments and on nodules
Manganese-oxidizing bacteria from shallow-water environments have long been known and have been cultured from recovered deep-sea nodules (Ehrlich, 1963, 1971; Trimble and Ehrlich, 1968). It is still unclear, however, whether such bacteria actually play a role in the deposition of manganese on the nodules *in situ*, let alone whether they influence the accumulation of other transition metals. Recent shipboard experiments by K. Nealson[3] show minimal bacterial activity on freshly recovered nodules, but show steadily increasing activity with time, suggesting the activation of dormant forms or contamination by shallow-living bacteria during or after recovery. Such experiments were not designed to test for barophilic manganese oxidizers, however. The low bacterial activity on freshly col-

[3] Oral communication, 1979.

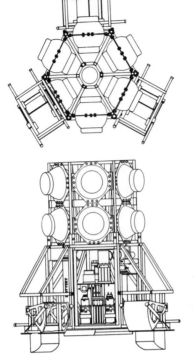

Fig. 6-34. MANOP's autonomous Bottom Lander, which is designed to carry out *in situ* experiments on nodule formation. Addition and removal of solutions from enclosed box-core chambers is controlled by microprocessors that can be monitored and even reprogrammed through an acoustic link to a surface ship. Courtesy of O. Kirsten and R. Weiss (see Weiss, 1979).

lected nodules contrasts strikingly with the high concentrations of manganese-oxidizing bacteria in the zone of manganese oxidation-reduction in equatorial Pacific sediments from 1°N, 140°W (Rosson *et al.*, 1979). The definitive answer on the role of bacteria in the formation of manganese nodules is most likely to come from experiments with the MANOP Lander, in which $^{54}$Mn, antibiotics, and resistant bacteria can be added in various sequences and combinations to assess bacterial activities and Mn-uptake rates.

**6. Activity of epibenthic microfauna on nodules** Greenslate (1974) and Dugolinski *et al.* (1977) have shown that the habitats of foraminifera and other microscopic animals cover the surfaces of manganese nodules. Such organisms may contribute to nodule growth by producing a durable framework within which and upon which more friable oxyhydroxides can be deposited. There is no direct evidence that the surface dwellers affect the rate of metal deposition or the composition of the oxyhydroxide phases. Experiments with the MANOP Lander may help to answer this question, although the present rudimentary knowledge of the physiology of these organisms tends to inhibit the development of testable hypotheses.

**7. Preferential incorporation of dissolved metals in specific iron/manganese oxyhydroxide minerals** Since the initial work of Barnes (1967) and Cronan and Tooms (1969) it has been clear that there is a statistical correlation between the abundance of todorokite relative to $\delta$-$MnO_2$, and the Cu and Ni content of ferromanganese nodules. The mineralogic and electron microprobe work of Burns and Burns (1977a,b, 1978a,b;

see also Sorem and Fewkes, 1979) has shown the strong preference of Cu and Ni for todorokite layers within nodules. Copper values up to 4% and nickel values up to 5% have been determined by spot analyses of individual todorokite patches.

Although the correlation between high trace-metal abundances and the presence of todorokite is clear, the cause-effect relation is not. Does the growth of todorokite (due to environmental conditions as yet unknown) result in the extraction of Cu and Ni from seawater and porewater solutions that are no different from those surrounding trace-metal-poor $\delta$-$MnO_2$ nodules? Or high concentrations of Cu and Ni in porewaters surrounding nodules favor the formation or preservation of todorokite by stabilizing its crystal structure? Mineral exposure experiments, in which suites of natural and artificial manganese and iron oxyhydroxides are being exposed for periods of years to ambient seawater in the five MANOP areas (Burns et al., 1978), should help to resolve this question, although additional experiments involving the exposure of the same minerals to porewaters may also be necessary.

**8. Local sea-floor volcanism** Ever since the discovery of deep-sea manganese nodules by HMS *Challenger* (Murray and Renard, 1891) there has been speculation that submarine volcanism supplies the transition metals to these nodules. The recent discovery of mid-ocean ridge hydrothermal systems (discussed previously) and of associated ferromanganese crusts (Corliss et al., 1978) have renewed interest in this source of metals to the oceans and to nodules. Because the areal distribution of trace-metal-enriched nodules shows no resemblance to the distribution of current oceanic volcanism, however, hydrothermal activity is likely to influence the globally integrated composition and growth rate of nodules more than the composition of a specific patch of nodules. Two observations require further study, however, before the local influence of hydrothermal activity can be discounted. Lupton and Craig (1978) have shown that midwater $^3$He of hydrothermal origin is being carried west from the East Pacific Rise between about 15°N and 30°S. The $^3$He plume roughly parallels the band of Cu-Ni-enriched nodules southeast of Hawaii (Fig. 6-2). Klinkhammer (1980b) has reported Mn enrichment in the plume, but an east-west transect of dissolved-metal profiles is needed to assess the importance of this source to the nodule field. In addition, during 1976 the research vessel *Valdivia* recovered warm water of presumed hydrothermal origin at 14°N, 153°W (Gundlach et al., 1976). Unfortunately, attempts to re-sample this material have been unsuccessful. Because of questions surrounding the shipboard sampling technique, as well as the unlikely variations in composition reported for the warm water and associated "normal" seawater (Gundlach et al., 1976), however, even the reality of this "hydrothermal" activity, let alone its possible influence on nearby manganese nodules, must be considered doubtful at the present time.

## Relation between Composition and Abundance

Until the development of box cores and free-fall grabs (reviewed in Moore and Heath, 1978) little was known of variations in the abundance of nodules on the sea floor beyond the qualitative estimates derived from deep-sea photographs. As coordinated measurements of nodule density (mass per unit area) and abundance have become available, however, it has become apparent that the two do not vary independently. Menard and

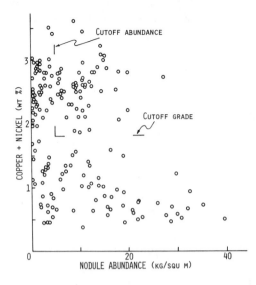

Fig. 6-35. Relation between areal coverage and grade (shown as Cu + Ni) of pelagic nodules (after Menard and Frazer, 1978). Figure shows crude negative correlation for entire data set, but not for potentially exploitable nodules.

Frazer (1978) first drew attention to the negative correlation between the Cu and Ni content of nodules and their areal abundance (Fig. 6-35). This correlation appears to apply on a local scale (e.g., Moritani *et al*., 1979; Fig. 6-36) as well as on the global scale presented by Menard and Frazer (1978). An exception is the case of Cu-Ni-enriched nodules from the northeastern equatorial Pacific; here, the correlation appears to break down, with no statistically significant relationship between coverage in excess of 5 kg/m² and Cu + Ni values in excess of 1.8% (McKelvey *et al*., 1979; Menard and Frazer, 1978).

In addition to their clear economic implications, these results suggest that only a limited supply of Ni and Cu are available to most areas of nodule deposition. Where the nodule cover is dense, the available Ni and Cu are dispersed, whereas scattered nodules are able to accumulate higher concentrations. Such an explanation is simple, but the existing data are inadequate to test it. Certainly the vertical flux of Cu and Ni with particles does not suggest that these metals are in limited supply. Further fieldwork is necessary to determine whether the negative correlations result from simple mass-

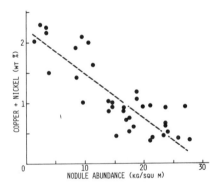

Fig. 6-36. Marked negative correlation of nodule coverage and grade (Cu + Ni) for the equatorial North Pacific around 10°N, 175°W. After Moritani *et al*., 1979.

DEEP-SEA FERROMANGANESE NODULES

balance constraints or from more complex conditions (involving variations in nodule growth rates, for example), which result in correlated Cu + Ni and nodule coverage variations without reflecting any cause-effect relationship between these two parameters.

**1. Why are nodules concentrated at the sea floor?** As discussed previously, the abundance of surface versus buried nodules and the consistency of this ratio with a simple model of random burial seems well established. Still unresolved is the mechanism or mechanisms that prevent the nodules from being buried. Downslope movement of sediment (Cronan and Tooms, 1967a) and nondeposition (Watkins and Kennett, 1971) are plausible explanations for limited areas of the deep sea, but many nodules are lying on actively accumulating sediments yet are not buried. Purely chemical processes (repeated dissolution at depth and reprecipitation at the sea floor) or physical processes (flotation when underlying sediments are liquefied by seismic events, for example) are ruled out, particularly for oxic sediments, by porewater analyses, the presence of buried nodules, and the observation that the wet bulk density of nodules (Greenslate, 1977) exceeds the wet bulk density of surface sediments.

The remaining explanation of repeated jostling and displacement by benthic organisms is not supported by direct observations, but is consistent with recent photos of small animals clearing off nodule tops by consuming the recently fallen sediment (Paul, 1976), of large mobile opportunistic benthic feeders that unquestionably have the physical capacity to move nodules (Hessler and Jumars, 1977; Sessions *et al.*, 1968), and of large, slowly moving benthic animals that have been photographed making their way across a small section of the sea floor over a period of $6\frac{1}{2}$ months (Paul *et al.*, 1978). Inasmuch as a nodule on a pelagic clay need be moved only once in a few thousand years to prevent it from being buried, time lapse photography of a field of several thousand nodules for a period of several years will be necessary before the probability of observing the biological movement of one or more nodules becomes high enough to test this hypothesis.

**2. Why is the composition of nodules geographically variable?** The preceding sections have outlined current theories on concentration mechanisms. From a consideration of these mechanisms and their influence on metal accumulation rates in nodules, Lyle (1978) has constructed a plausible model of nodule chemistry in the eastern South Pacific. He noticed that the growth rate of nodules is inversely related to their Fe contents. This relation, which implies a relatively constant rate of Fe deposition, led Lyle (1978) to a simple empirical formula that predicts a nodule's growth rate from its Fe content. Data from Heye and Marchig (1977) suggest that Co behaves much like Fe. Their data also show that the ratios of Cu, Mn, and Ni to Fe or Co increase with increasing growth rates (Fig. 6-37). Lyle (1978) suggests that the growth rate of nodules is constant in areas of very low biologic productivity, being controlled by precipitation from seawater. In areas of oxic diagenesis, the rate of growth increases with increasing biologic productivity in the overlying surface waters due to the influx of biogenic par-

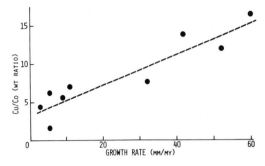

Fig. 6-37.    Relation of Cu/Co ratio to the growth rate of manganese nodules from the central equatorial North Pacific (data from Heye and Marchig, 1977). If Co accumulates at a constant rate, this figure implies an increase in Cu accumulation with increasing growth rate.

ticulates bearing transition metals (e.g., Spencer *et al.*, 1978). Because biogenic particles are richer in Cu and Ni relative to Mn than are nodules, the percentage of Cu and Ni increases as the surface productivity increases.

The situation changes dramatically, however, as the rate of input of organic debris increases to the level at which anoxic diagenesis begins in the sediment. The mobilization of Mn under anoxic conditions results in rapid nodule growth that virtually swamps the input of other metals, leading to lower Cu and Ni contents.

Analytical data from a range of depositional environments supports Lyle's (1978) hypothesis. Nodules show increases in Cu, Ni, and Mn, and decreases in Fe and Co from sediment-free rock areas to red clay to siliceous clay, but decreases in all elements except Mn from siliceous to hemipelagic clay. However, if Co is assumed to precipitate at a constant rate, the ratios of the other elements to Co (Fig. 6-38) show steady increases in Cu and Ni from the regions of lowest to highest productivity, with Mn increasing dramatically between siliceous and hemipelagic sites. The Fe/Co ratio varies irregularly on such a plot, suggesting that biogenically precipitated iron does not dominate other sources of this element.

An examination of Figs. 6-38 and 6-39 explains the occurrence of Cu-Ni-enriched nodules at the margins of high productivity areas in the equatorial Pacific. At lower

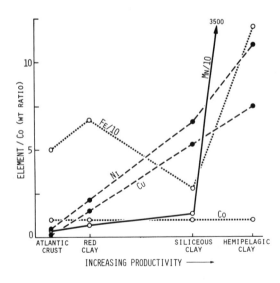

Fig. 6-38.    Ratios of elemental abundances to Co in ferromanganese deposits collected below surface waters of low to high biological productivity. If Co accumulates at a constant rate, the ordinate is proportional to elemental accumulation rates in the nodules. Data for Atlantic seamount crusts, and red clay and hemipelagic clay nodules from Cronan (1977), and for siliceous clay nodules from Lyle *et al.* (1977). Data are spaced along abscissa so as to produce a linear trend in Cu/Co ratios.

DEEP-SEA FERROMANGANESE NODULES

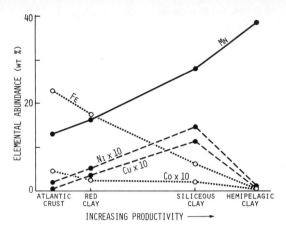

Fig. 6-39. Concentrations of transition metals in ferromanganese deposits collected below surface waters of low
to high biological productivity. Scale on abscissa and data
sources as Fig. 6-38.

productivities, these elements are diluted by background iron, whereas at higher productivities they are diluted by diagenetically mobilized manganese.

Although the lack of zones of Cu and Ni enrichment at the fringes of all pelagic
areas of high biological productivity is still to be explained, Lyle's (1978) hypothesis best fits our current knowledge of compositional variations in ferromanganese
nodules.

**3. Why is the proportion of the sea floor covered by nodules so variable
at virtually all scales?** The answer to this question requires reliable maps of nodule
coverage, combined with geochemical studies of nodules and sediments to assess variations in growth and burial rates of nodules relative to variations in accumulation rates
and lithology of the sediments. At present, only Scripps' deep-tow vehicle has
been used to map nodule coverage in sufficient detail to identify well-defined and
fairly homogeneous "patches" (Karas, 1978). The patches have not been sampled
sufficiently to relate the density of coverage to variations in nodule and sediment
properties.

CONCLUSIONS

Scientific and commercial endeavors directed at ferromanganese nodules during the past
decade have dispelled many of the myths that grew up during the preceding century.
It is clear that nodules are chemical precipitates from seawater and porewaters that most
likely are kept at the sea floor by benthic activity. Minor metal constituents, such as Cu
and Ni, are extracted from seawater by plankton at shallow depths, and settle to the
sea floor as tests and organic matter in fecal pellets. The transformation of particle-
associated metals to nodules currently is the subject of intense research activity focusing
on dissolved metals, sediments, nodule growth rates and mineralogy, and bacterial
activity.

## ACKNOWLEDGMENTS

The list of colleagues with whom I have discussed and argued the origin of "manganese" nodules is far too long to include here. I am particularly grateful, however, for stimulating conversations with M. Bender, W. Broecker, S. Calvert, J. Dymond, M. Lyle, and T. Moore, as well as with the other members of MANOP. My recent studies of nodules have been supported through the NSF-IDOE Manganese Nodule Program (MANOP).

## REFERENCES

Adelseck, C. G., and Berger, W. H., 1975, On the dissolution of planktonic foraminifera and associated microfossils during settling and on the sea floor: *Cushman Found. Foram. Res., Spec. Publ. 13*, p. 70–81.

Andersen, M. E., and Macdougall, J. D., 1977, Accumulation rates of manganese nodules and sediments—an alpha-track method: *Geophys. Res. Lett.*, v. 4, p. 351–353.

Andrews, J. R., Callender, E., Bowser, C. J., Mero, J. L., Gauthier, M., Meylan, M. A., Craig, J. D., Binder, K., Volk, P., Chave, A., and Bachman, W., 1974, Ferromanganese deposits of the ocean floor: *Hawaii Inst. of Geophys. Rpt. HIG-74-9*, Honolulu, 194 p.

Arrhenius, G. O. S., 1954, Origin and accumulation of aluminosilicates in the ocean: *Tellus*, v. 6, p. 215–220.

——, 1967, Deep-sea sedimentation—a critical review of U.S. work: *Amer. Geophys. Un. Trans.*, v. 48, p. 604–631.

Barnes, S. S., 1967, The formation of oceanic ferromanganese nodules: Ph.D. Thesis, Scripps Inst. Oceanography, UCSD, La Jolla, Calif., 68 p.

Bender, M. L., Fanning, K. A., Froelich, P. E., Heath, G. R., and Maynard, V., 1977a, Interstitial nitrate profiles and the oxidation of sedimentary organic matter: *Science*, v. 198, p. 605–609.

——, and Gagner, C., 1976, Dissolved copper, nickel, and cadmium in the Sargasso Sea: *J. Mar. Res.*, v. 34, p. 327–339.

——, Klinkhammer, G. P., and Spencer, D. W., 1977b, Manganese in sea water and the marine manganese balance: *Deep-Sea Res.*, v. 24, p. 799–812.

——, Ku, T. L., and Broecker, W. S., 1966, Manganese nodules—their evolution: *Science*, v. 151, p. 325–328.

Bischoff, J. L., and Dickson, F. W., 1975, Seawater-basalt interaction at 200°C and 500 bars—implications for origin of seafloor heavy metal deposits and regulation of seawater chemistry: *Earth Planet. Sci. Lett.*, v. 25, p. 385–397.

Bostrom, K., Lysen, L., and Moore, C., 1978, Biological matter as a source of authigenic matter in pelagic sediments: *Chem. Geol.*, v. 23, p. 11–20.

Boyle, E., Sclater, F., and Edmond, J. M., 1977, The distribution of dissolved copper in the Pacific: *Earth Planet. Sci. Lett.*, v. 37, p. 38–54.

Bricker, O. P., 1965, Some stability relations in the system $Mn-O_2-H_2O$ at 25° and one atmosphere total pressure: *Amer. Mineral.*, v. 50, p. 1296–1354.

Bruland, K. W., Knauer, G. A., and Martin, J. H., 1978, Zinc in northeast Pacific water: *Nature*, v. 271, p. 741–743.

DEEP-SEA FERROMANGANESE NODULES

Burns, R. G., and Burns, V. M., 1977a, Mineralogy of manganese nodules, *in* Glasby, G. P., ed., *Marine Manganese Deposits*: New York, Elsevier, p. 185–248.

——, and Burns, V. M., 1977b, The mineralogy and crystal chemistry of deep-sea manganese nodules, a polymetallic resource of the twenty-first century: *Phil. Trans. Royal Soc. London*, v. A285, p. 249–258.

——, and Burns, V. M., 1981, Authigenic oxides, *in* Emiliani, C., ed., *The Sea, Vol. 7*: New York, Wiley, *in press*.

——, Burns, V. M., Stockman, H. W., Stockman, C. T., and Benson, M. D., 1978, Deployment of long-term seafloor mineral experiments to measure changes in mineralogy and composition of manganese nodules: *MTS–IEEE Oceans 78 Proc.*, p. 662–667.

Burns, V. M., and Burns, R. G., 1976, Evidence of diagenesis and growth in manganese nodules from the north equatorial Pacific by scanning electron microscopy: *Geol. Soc. Amer. Abst. Prog.*, v. 8, p. 796–797.

——, and Burns, R. G., 1978a, Diagenetic features observed inside deep-sea manganese nodules from the north equatorial Pacific: *Scanning Electron Microscopy 1978*, v. I, p. 245–252.

——, and Burns, R. G., 1978b, Post-depositional metal enrichment processes inside manganese nodules from the north equatorial Pacific: *Earth Planet. Sci. Lett.*, v. 39, p. 341–348.

Callender, E., and Bowser, C. J., 1980, Manganese and copper geochemistry of interstitial fluids from manganese nodule-rich pelagic sediments of the northeastern equatorial Pacific Ocean: *Amer. J. Sci.*, v. 280, p. 1063–1096.

Calvert, S. E., and Price, N. B., 1977, Geochemical variations in ferromanganese nodules and associated sediments from the Pacific Ocean: *Mar. Chem.*, v. 5, p. 43–74.

——, Price, N. B., Heath, G. R., and Moore, T. C., Jr., 1978, Relationship between ferromanganese nodule compositions and sedimentation in a small survey area of the equatorial Pacific: *J. Mar. Res.*, v. 36, p. 161–183.

Chester, R., and Hughes, M. J., 1967, A chemical technique for the separation of ferromanganese minerals and adsorbed trace elements from pelagic sediments: *Chem. Geol.*, v. 2, p. 249–262.

Chukhrov, F. V., Gorshkov, A. I., Sivtsov, A. B., and Berezovskaya, V. V., 1979, New data on natural todorokite: *Nature*, v. 278, p. 631–632.

Cobler, R., and Dymond, J., 1980, Sediment trap experiment on the Galapagos Spreading Center, equatorial Pacific: *Science*, v. 209, p. 801–803.

Corliss, J. B., 1971, The origin of metal-bearing hydrothermal solutions: *J. Geophys. Res.*, v. 76, p. 8128–8138.

——, 1973, The sea as alchemist: *Oceanus*, v. 16, p. 38–43.

——, Dymond, J., Lyle, M., and Crane, K., 1978, Sediment mounds—hydrothermal ferromanganese deposits near the Galapagos Rift: *Earth Planet. Sci. Lett.*, v. 40, p. 12–24.

——, Dymond, J., Gordon, L. I., Edmond, J. M., von Herzen, R. P., Ballard, R. D., Green, K., Williams, D., Bainbridge, A., Crane, K., and van Andel, T. H., 1979a, Submarine thermal springs on the Galapagos Rift: *Science*, v. 203, p. 1073–1083.

——, Edmond, J. M., and Gordon, L. I., 1979b, Some implications of heat/mass ratios in Galapagos Rift hydrothermal fluids for models of seawater-rock interaction and the formation of oceanic crust: *Amer. Geophys. Un., Maurice Ewing Ser. 2*, p. 391–402.

Cronan, D. S., 1973, Manganese nodules in sediments cored during Leg 16, Deep Sea Drilling Project, *in* van Andel, T. H., Heath, G. R., *et al., Initial Reports of the Deep Sea Drilling Project*: U.S.G.P.O., Washington, D.C., v. 16, p. 605–608.

——, 1976, Manganese nodules and other ferromanganese oxide deposits, *in* Riley, J. P., and Chester, R., eds., *Chemical Oceanography, 2nd Ed., Vol. 5*: New York, Academic, p. 217–263.

——, 1977, Deep-sea nodules—distribution and geochemistry, *in* Glasby, G. P., ed., *Marine Manganese Deposits*: New York, Elsevier, p. 11–44.

——, and Tooms, J. S., 1967a, Sub-surface concentrations of manganese nodules in Pacific sediments: *Deep-Sea Res.*, v. 14, p. 117–119.

——, and Tooms, J. S., 1967b, Geochemistry of manganese nodules from the NW Indian Ocean: *Deep-Sea Res.*, v. 14, p. 239–249.

——, and Tooms, J. S., 1969, The geochemistry of manganese nodules and associated pelagic deposits from the Pacific and Indian Oceans: *Deep-Sea Res.*, v. 16, p. 335–359.

Dugolinsky, B. K., Margolis, S. V., and Dudley, W. C., 1977, Biogenic influence on the growth of manganese nodules: *J. of Sedimentary Petrology*, v. 47, p. 428–445.

Edmond, J. M., Corliss, J. B., and Gordon, L. I., 1979a, Ridge crest hydrothermal metamorphism at the Galapagos Spreading Center and reverse weathering: *Amer. Geophys. Un., Maurice Ewing Ser. 2*, p. 383–390.

——, Measures, C., McDuff, R. E., Chan, L. H., Grant, B., Gordon, L. I., and Corliss, J. B., 1979b, Ridge crest hydrothermal activity and the balances of the major and minor elements in the ocean—the Galapagos data: *Earth Planet. Sci. Lett.*, v. 46, p. 1–18.

——, Measures, C., Magnum, B., Grant, B., Sclater, F. R., Collier, R., Hudson, A., Gordon, L. I., and Corliss, J. B., 1979c, On the formation of metal-rich deposits at ridge crests: *Earth Planet. Sci. Lett.*, v. 46, p. 19–30.

Ehrlich, H. L., 1963, Bacteriology of manganese nodules, I. Bacterial action on manganese in nodule enrichments: *Appl. Microbiol.*, v. 11, p. 15–19.

——, 1971, Bacteriology of manganese nodules, V. Effect of hydrostatic pressure on bacterial oxidation of $Mn^{+2}$ and reduction of $MnO_2$: *Appl. Microbiol.*, v. 21, p. 306–310.

Foster, A. R., 1970, Marine manganese nodules—nature and origin of internal features: M.S. Thesis, Washington State Univ., Pullman, Wash., 131 p.

Froelich, P. N., Klinkhammer, G. P., Bender, M. L., Luedtke, N. A., Heath, G. R., Cullen, D., Dauphin, P., Hammond, D., Hartman, B., and Maynard, V., 1979, Early oxidation of organic matter in pelagic sediments of the eastern equatorial Atlantic—suboxic diagenesis: *Geochim. Cosmochim. Acta*, v. 43, p. 1075–1090.

Glasby, G. P., 1977, Why manganese nodules remain at the sediment-water surface: *New Zealand J. Sci.*, v. 20, p. 187–190.

——, 1978, Deep-sea manganese nodules in the stratigraphic record—evidence from DSDP cores: *Mar. Geol.*, v. 28, p. 51–64.

Greenslate, J., 1974, Microorganisms participate in the construction of manganese nodules: *Nature*, v. 249, p. 181–183.

——, 1977, Manganese concretion wet density—a marine geochemistry constant: *Mar. Mining*, v. 1, p. 125–148.

Greenslate, J. L., Frazer, J. Z., and Arrhenius, G. 1973, Origin and deposition of selected transition elements in the seabed, *in* Morgenstein, M., ed., *The Origin and Distribution of Manganese Nodules in the Pacific and Prospects for Exploration*: Hawaii Inst. Geophys., Univ. Hawaii, p. 45–69.

Gundlach, H., Beiersdorf, H., Marchig, V., and Schnier, C., 1976, "Heated" bottom water and associated Mn/Fe-oxide crusts from the Clarion Fracture Zone SE of Hawaii: *Joint Oceanog. Assembly, Edinburgh, Late Abstracts*, UN FAO, Rome, p. 8.

Hajash, A., 1975, Hydrothermal processes along mid-ocean ridges—an experimental investigation: *Contrib. Mineral. Petrol.*, v. 53, p. 205–266.

Harada, K., and Nishida, S., 1976, Biostratigraphy of some marine manganese nodules: *Nature*, v. 260, p. 770–771.

Heath, G. R., 1978, Deep-sea manganese nodules: *Oceanus*, v. 21, p. 60–68.

———, 1979, Burial rates, growth rates and size distributions of deep-sea manganese nodules: *Science*, v. 205, p. 903–904.

———, Moore, T. C., Jr., Somayajulu, B. L. K., and Cronan, D. S., 1970, Sediment budget in a deep-sea core from the central equatorial Pacific: *J. Mar. Res.*, v. 28, p. 225–230.

Heezen, B. C., and Fornari, D. J., 1975, Geological map of the Pacific Ocean: *Geol. World Atlas, Sheet 20*, UNESCO, Paris.

Hessler, R. R., and Jumars, P. A., 1977, Abyssal communities and radioactive waste disposal: *Oceanus*, v. 20, p. 41-46.

Heye, D., and Beiersdorf, H., 1973, Radioaktive und magnetische untersuchungen an manganknollen zur ermittlung der wachstumsgeschwindigkeit bzw. zur altersbestimmung: *Z. Geophys.*, v. 39, p. 703–726.

———, and Marchig, V., 1977, Relationship between the growth rate of manganese nodules from the central Pacific and their chemical constitution: *Mar. Geol.*, v. 23, p. M19–M25.

Hinga, K. R., Sieburth, J. M., and Heath, G. R., 1979, The supply and use of organic material at the deep-sea floor: *J. Mar. Res.*, v. 37, p. 355–379.

Honjo, S., 1976, Coccoliths—production, transportation and sedimentation: *Mar. Micropaleo.*, v. 1, p. 65–79.

———, 1978, Sedimentation of materials in the Sargasso Sea at a 5,367 m deep station: *J. Mar. Res.*, v. 36, p. 469–492.

Horn, D. R., Horn, B. M., and Delach, M. N., 1972, Ferromanganese deposits of the North Pacific: *National Science Foundation, International Decade of Ocean Exploration, Technical Report No. 1*, 78 p.

———, Horn, B. M., and Delach, M. N., Factors which control the distribution of ferromanganese nodules: *National Science Foundation, International Decade of Ocean Exploration, Technical Report No. 8*, p. 1–15.

Karas, M. C., 1978, Studies of manganese nodules using deep-tow photographs and side-looking sonars: Scripps Inst. Oceanog., La Jolla, Calif., Ref. 78-20, 38 p.

Klinkhammer, G. P., 1980a, Early diagenesis in sediments from the eastern equatorial Pacific—pore water metal results: *Earth Planet. Sci. Lett.*, v. 49, p. 81–101.

———, 1980b, Observations on the distribution of manganese over the East Pacific Rise: *Chem. Geol.*, v. 29, p. 211–226.

Krishnaswami, S., and Cochran, J. K., 1978, Uranium and thorium series nuclides in oriented ferromanganese nodules—growth rates, turnover times and nuclide behavior: *Earth Plant. Sci. Lett.*, v. 40, p. 45–62.

———, Cochran, J. K., Turekian, K. K., and Sarin, M. M., 1979, Time scales of deep-sea ferromanganese nodule growth based on 10–Be and alpha-track distributions and their relation to uranium decay series measurements, *in Sur la Genese des Nodules de Manganese*: Col. Int. du C.N.R.S., No. 289, Gif-sur-Yvette, p. 251–260.

Ku, T. L., 1977, Rates of accretion: *in* Glasby, G. P., ed., Marine Manganese Deposits, New York, Elsevier, p. 249–267.

———, and Broecker, W. S., 1967, Uranium, thorium and protactinium in a manganese nodule: *Earth Planet. Sci. Letter*, v. 2, p. 317–320.

———, Hug, C. A., and Macdougall, J. D., 1979a, Pleiades "apple nodule"–radiochemical observations: *EOS*, v. 60, p. 850.

———, Omura, A., and Chen, P. S., 1979b, Be[10] and U-series isotopes in manganese nodules from the central North Pacific, *in* Bischoff, J. L., and Piper, D. Z., eds., *Marine Geology and Oceanography of the Pacific Manganese Nodule Province*: New York, Plenum, p. 791–814.

Lalou, C., and Brichet, E., 1972, Signification des mesures radiochimiques dans l'evaluation de la vitesse de croissance des nodules de manganese: *C. R. Acad. Sci. Paris, Ser. D., no. 275*, p. 815–818.

———, Brichet, E., Poupeau, G., Romary, P., and Jehanno, C., 1979, Growth rates and possible age of a North Pacific manganese nodule: *in* Bischoff, J. L., and Piper, D. Z., eds., *Marine Geology and Oceanography of the Pacific Manganese Nodule Province*: New York, Plenum, p. 815–834.

Love, C. M., ed., 1975, *Eastropac Atlas, v. 9*: Natl. Mar. Fish. Serv., Circular 330, Washington, D.C.

Lowenstein, C., Lyle, M., and Rosson, 1979, *MANOP Cruise Report Areas C and S, R/V Knorr, K79–05, May 1979*: MANOP Office, Univ. Rhode Island, 32 p.

Lupton, J. E., and Craig, H., 1978, [3]He in the Pacific Ocean–injection at active spreading centers and applications to deep circulation studies: *EOS*, v. 59, p. 1105–1106.

Lyle, M., 1978, The formation and growth of ferromanganese oxides on the Nazca Plate: Ph.D. Dissertation, School of Oceanography, Oregon State Univ., 172 p.

———, Dymond, J., and Heath, G. R., 1977, Copper-nickel-enriched ferromanganese nodules and associated crusts from the Bauer Basin, northwest Nazca Plate: *Earth Planet. Sci. Lett.*, v. 35, p. 55–64.

Lynn, D. C., and Bonatti, E., 1965, Mobility of manganese in diagenesis of deep-sea sediments: *Mar. Geol.*, v. 3, p. 457–474.

Macdougall, J. D., 1979, The distribution of total alpha radioactivity in selected manganese nodules from the North Pacific–implications for growth processes: *in* Bischoff, J. L., and Piper, D. Z., eds., *Marine Geology and Oceanography of the Central Pacific Manganese Nodule Province:* New York, Pergamon, p. 775–789.

Margolis, S. V., and Glasby, G. P., 1973, Microlaminations in marine manganese nodules as revealed by scanning electron microscopy: *Geol. Soc. Amer. Bull.*, v. 84, p. 3601–3610.

Martin, J. H., and Knauer, G. A., 1973, The elemental compositions of plankton: *Geochim. Cosmochim. Acta*, v. 37, p. 1639–1653.

———, and Knauer, G. A., 1980, Manganese cycling in northeast Pacific waters: *Earth Planet. Sci. Lett.*, v. 51, p. 266–274.

McCave, I. N., 1975, Vertical flux of particles in the ocean: *Deep-Sea Res.*, v. 22, p. 491–502.

McKelvey, V. E., Wright, N. A., and Rowland, R. W., 1979, Manganese nodule resources in the northeastern equatorial Pacific, *in* Bischoff, J. L., and Piper, D. Z., eds.,

*Marine Geology and Oceanography of the Central Pacific Manganese Nodule Province*: New York, Pergamon, p. 747–762.

Menard, H. W., 1964, *Marine Geology of the Pacific:* New York, McGraw-Hill, 271 p.

——, 1976, Time, chance and the origin of manganese nodules: *Amer. Scientist*, v. 64, no. 5, p. 519–529.

——, and Frazer, J. L., 1978, Manganese nodules on the sea floor—inverse correlation between grade and abundance: *Science*, v. 199, p. 969–971.

Menzies, M., and Seyfried, W. E., Jr., 1979, Basalt-seawater interaction—trace element and strontium isotope variations in experimentally altered basalt and periodotite: *Earth Planet. Sci. Lett.*, v. 44, p. 463–473.

Mero, J. L., 1965, *The Mineral Resources of the Sea:* Amsterdam, Elsevier, 312 p.

Monget, J. M., Murray, J. W., and Mascle, J., 1976, A world wide compilation of published, multicomponent/analyses of ferromanganese concretions: *NSF-IDOE Manganese Nodule Project Tech. Report No. 12*, 127 p.

Moore, T. C., Jr., and Heath, G. R., 1966, Manganese nodules, topography, and thickness of Quaternary sediments in the central Pacific: *Nature*, v. 212, p. 983–985.

——, and Heath, G. R., 1967, Abyssal hills in the central equatorial Pacific—detailed structure of the sea floor and sub-bottom reflectors: *Mar. Geol.*, v. 5, p. 161–179.

——, and Heath, G. R., 1978, Sea-floor sampling techniques, *in* Riley, J. P., and Chester, R., eds., *Chemical Oceanography, Vol. 7*: New York, Academic Press, p. 75–126.

Moore, W. S., Ku, T. L., Macdougall, J. D., Burns, V. M., Burns, R., Dymond, J., Lyle, M., and Piper, D. Z., 1980, Fluxes of metals to a manganese nodule—Radiochemical, chemical, structural and mineralogical studies: *Earth Planet. Sci. Lett.*, v. 52, p. 151–171.

Moritani, T., Mochizuki, T., Terashima, S., and Maruyama, S., 1979, Metal contents of manganese nodules from the GH77-1 area, *in* Moritani, T., ed., *Deep Sea Mineral Resources Investigation in the Central-Western Part of Central Pacific Basin January-March 1977 (GH77-1 Cruise)*: Geol. Survey Japan, p. 206–217.

Mottl, M. J., and Holland, H. D., 1978, Chemical exchange during hydrothermal alteration of basalt by seawater, I. Experimental results for major and minor components of seawater: *Geochim. Cosmochim. Acta*, v. 42, p. 1103–1115.

Murray, J., and Renard, A. F., 1891, *Deep sea deposits—Report of the scientific results of the H.M.S. Challenger, 1873-1876*: Her Majesty's Government Printing Office, 525 p.

Nelson, D. E., 1977, Carbon-14—direct detection at natural concentrations: *Science*, v. 198, p. 507–508.

Paul, A. Z., 1976, Deep-sea bottom photographs show that benthic organisms remove sediment cover from manganese nodules: *Nature*, v. 263, p. 50–51.

——, Thorndike, L. G., Sullivan, L. G., Heezen, B. C., and Gerard, R. D., 1978, Observations of the deep-sea floor from 202 days of time-lapse photography: *Nature*, v. 272, p. 812–814.

Payne, R. R. and Conolly, J. R., 1972, Pleistocene manganese pavement production—its relation to the origin of manganese in the Tasman Sea, *in* Horn, D., ed., *Ferromanganese Deposits of the Ocean Floor*: NSF-IDOE, Washington, D.C., p. 81–92.

Piper, D. Z. and Williamson, 1977, Composition of Pacific Ocean ferromanganese nodules: *Mar. Geol.*, v. 32, p. 285–303.

Raab, W., 1972, Physical and chemical features of Pacific deep sea manganese nodules and their implications to the genesis of nodules: Papers from a Conference on Ferromanganese Deposits on the Ocean Floor, Horn, D. R., ed., National Science Foundation, Washington, D.C., p. 31-49.

Reid, J. L., Jr., 1965, Intermediate waters of the Pacific Ocean: *The Johns Hopkins Oceanographic Studies, No. 2*, Baltimore, 85 p.

Rosson, R. A., Tebo, B. M., Nealson, K. H., and Emerson, S., 1979, Vertical distribution and potential oxidative activity of manganese bacteria in Saanich Inlet—implications for the deep sea: *EOS*, v. 60, p. 850.

Schwertmann, V., 1964, Differenzierung der eisenoxide des bodens durch photochemische extraktion mit saurer ammoniumoxalat—Lösung: Zeitchr, Fflanzenernähr, Düng, Bodenkunde, v. 105, p. 194-202.

Sclater, F. R., Boyle, E., and Edmond, J. M., 1976, On the marine geochemistry of nickel: *Earth Planet. Sci. Lett.*, v. 31, p. 119-128.

Sessions, M. H., Isaacs, J. D., and Schwartzlose, R. A., 1968, A camera system for the observation of deep-sea marine life: Soc. Photo-Optical Instr. Eng., Underwater Photo-Optical Instr. Applic. Sem., San Diego, p. 1-5.

Seyfried, W. E., Jr., and Bischoff, J. L., 1977, Hydrothermal transport of heavy metals by seawater—the role of seawater basalt ratio: *Earth Planet. Sci. Lett.*, v. 34, p. 71-78.

——, and Bischoff, J. L., 1979, Low temperature basalt alteration by seawater—an experimental study at $70°C$ and $150°C$: *Geochim. Cosmochim. Acta*, v. 43, p. 1937-1947.

——, and Mottl, M. J., 1977, Origin of submarine metal-rich hydrothermal solutions—experimental basalt-seawater interaction in a seawater dominated system at $300°C$, 500 bars, *in Proc. Second Internat. Sympos. on Water-Rock Interaction, I.A.G.C.*: Strasbourg, France, v. IV, p. 173-180.

Skornyakova, N. S., 1979, Zonal regularities in occurrence, morphology and chemistry of manganese nodules of the Pacific Ocean, *in* Bischoff, J. L., and Piper, D. Z., eds., *Marine Geology and Oceanography of the Pacific Manganese Nodule Province*: New York, Plenum, p. 699-727.

Smith, K. L., Jr., 1978, Benthic community respiration in the NW Atlantic Ocean—*in situ* measurements from 40 to 5200m: *Mar. Biol.*, v. 47, no. 4, p. 337-347.

——, White, G. A., and Lavery, M. B., 1979, Oxygen uptake and nutrient exchange of sediments measured *in situ* using a free vehicle grab respirometer: *Deep-Sea Res.*, v. 27, p. 337-346.

——, White, G. A., Laver, M. B., and Haugsness, J. A., 1978, Nutrient exchange and oxygen consumption by deep-sea benthic communities—preliminary *in situ* measurements: *Limnol. Oceanog.*, v. 23, p. 997-1005.

Sorem, R. K., 1973, Manganese nodules as indicators of long-term variations in sea floor environment, *in* Morgenstein, M., ed., *The Origin and Distribution of Manganese Nodules in the Pacific and Prospects for Exploration*: Hawaii Inst. Geophys., Honolulu, p. 151-164.

——, and Fewkes, R. H., 1977, Internal characteristics, *in* Glasby, G. P., ed., *Marine Manganese Deposits*: New York, Elsevier, p. 147-183.

——, and Fewkes, R. H., 1979, *Manganese Nodules*: New York, Plenum, 723 p.

Spencer, D. W., Brewer, P. G., Fleer, A., Honjo, S., Krishnaswami, S., and Nozaki, Y., 1978, Chemical fluxes from a sediment trap experiment in the deep Sargasso Sea: *J. Mar. Res.*, v. 36, p. 493-523.

Spiess, F. N., and Greenslate, J., 1976, Pleiades Expedition, Leg, 04, Mn76–01, R/V Melville, Preliminary Cruise Report: *Manganese Nodule Project, Tech. Rpt. No. 15*, National Science Foundation, Washington, D.C., 87 p.

Spiess, F., Lonsdale, P., Bender, M., Kadko, D., Zampal, J., and Ford, I., 1977, MANOP Cruise Report, Site Survey—Areas M and H, R/V, Melville, Indomed (Leg 1)– Sept.–Oct. 1977: MANOP Office, Univ. Rhode Island, 31 p.

Trimble, R. B., and Ehrlich, H. L., 1968, Bacteriology of manganese nodules, III. Reduction of $MnO_2$ by two strains of nodule bacteria: *Appl. Microbiol.*, v. 16, p. 695–702.

Turekian, K. K., Cochran, J. K., Krishnaswami, S., Lanford, W. A., Parker, P. D., and Bauer, K. A., 1979, The measurement of [10]Be in manganese nodules using a tandem Van de Graaff accelerator: *Geophys. Res. Lett.*, v. 6, p. 417–420.

Turner, S., and Buseck, P. R., 1979, Todorokites—Mineral or mixture? Layer or tunnel structure? High resolution TEM evidence: *Geol. Soc. Amer. Abstr. Prog.*, v. 11, p. 531.

Volkov, I. I., 1977, The mode of formation of Fe-Mn nodules in recent sediments: *Geochem. Internat.*, v. 6, p. 150–156.

Watkins, N. D., and Kennett, J. P., 1971, Antarctic bottom water—major change in velocity during the late Cenozoic between Australia and Antarctica: *Science,* v. 173,

Weiss, R. F., 1979, The MANOP bottom lander: *EOS*, v. 60, p. 851.

Robert A. Berner
Department of Geology and Geophysics
Yale University
New Haven, Connecticut 06520

# 7

# CHEMISTRY OF BIOGENIC MATTER AT THE DEEP-SEA FLOOR

Some of the most important chemical reactions taking place at the deep-sea floor are brought about, directly or indirectly, by biological activity. Organic matter, $CaCO_3$, and opaline silica secreted in the overlying water by plankton upon death, fall to the bottom where the organic matter undergoes microbiological decomposition and the $CaCO_3$ and opaline silica are dissolved. Organic matter decomposition follows a sequence of processes depending upon the free energy yield of each overall process. This gives rise to the successive utilization of $O_2$, $NO_3^-$, $MnO_2$, $Fe(OH)_3$, and $SO_4^{--}$ for oxidative destruction followed ultimately by the production of methane. Because of slow rates of deposition, organic matter decomposition in the deep sea is more complete prior to burial than in near-shore sediments. As a result, less total organic carbon is buried and the oxidizing agents higher in the list (e.g., $O_2$, $NO_3^-$) are the ones mainly used. Sulfate reduction is more characteristic of shallower water sediments but is also found in deep-sea hemipelagic sediments. The rate of organic matter decomposition via sulfate reduction has been found, like organic content, to correlate positively with sedimentation rate. This correlation can be explained theoretically through the use of diagenetic modeling.

The rate of dissolution of planktonic $CaCO_3$ in deep-sea sediments is a high order function of the degree of undersaturation of deep seawater. As a result, $CaCO_3$ disappears rapidly once the undersaturation attains a sufficiently high value. This gives rise to the carbonate lysocline (zone of rapid removal) and carbonate compensation depth (depth below which there is complete removal). Carbonate dissolution is not simply a function of the thermodynamic saturation state of the water or its degree of turbulence. Dissolution involves both calcite and aragonite, with aragonite accounting for as much as 50% of the total. Because of its higher solubility, aragonite disappears at much shallower depths than calcite.

Opaline silica dissolution occurs both at and below the deep-sea floor. It takes place by very slow detachment of $SiO_2$ from surfaces of biogenic particles, and because of this slowness, more resistant species of diatoms and radiolaria are preserved and buried to form siliceous ooze. Diagenetic modeling indicates the important role that bioturbation plays in the process of biogenic silica dissolution.

# INTRODUCTION

Life in the oceans is strongly dependent on the chemical composition of seawater. The composition of seawater, in turn, depends upon inputs to it from rivers, volcanism, glaciers, and the atmosphere and upon outputs to bottom sediments and to volcanic rocks undergoing alteration by seawater. Much of the output is accomplished by the sedimentation of particles, but upon reaching the sea floor these particles are not all simply buried; instead, some may undergo chemical reactions with bottom water with consequent alteration of the composition of both particles and water. Alteration of composition is especially true of that water buried with the particles, the so-called interstitial or porewater. Porewaters are sensitive indicators of sea-floor chemical reactions and may exhibit concentrations vastly different than the overlying bottom water. This is so because

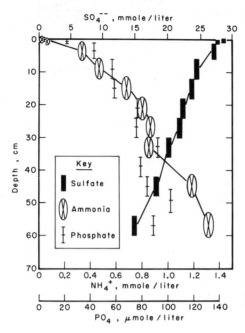

Fig. 7-1. Plots of dissolved sulfate, ammonium, and phosphate versus depth for sediments of the Santa Barbara Basin, California. Data from Sholkovitz, 1973, replotted by Berner, 1976.

the porewaters are trapped in the sediments where reaction products can build up, in contrast to the bottom water, which is free to mix with the vast volume of the oceans and thereby lose all evidence of the reactions. Some examples of porewater compositions are shown in Figs. 7-1, 7-2, and 7-3. (For a detail discussion of porewater chemistry, consult Manheim, 1976.)

As a result of concentration gradients between porewaters and bottom seawater, dissolved materials are transferred to or from the sediments by means of molecular dif-

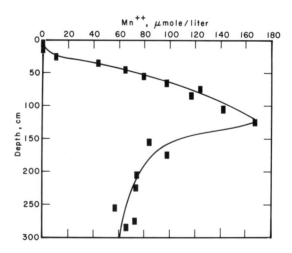

Fig. 7-2. Plot of dissolved manganese versus depth in deep-sea sediment from the Arctic Basin at 82°N, 156°W. Data from Li et al., 1969a, replotted by Berner, 1976.

CHEMISTRY OF BIOGENIC MATTER AT THE DEEP-SEA FLOOR

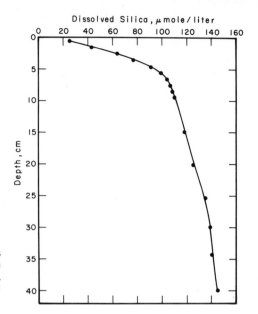

Dissolved Silica, μmole/liter

Fig. 7-3. Plot of dissolved silica versus depth for the deep-sea sediment from the Bermuda Rise of the North Atlantic Ocean. After Fanning and Pilson, 1974, replotted by Berner, 1976.

fusion, which in many sediments is enhanced by the stirring and irrigation of the sediments by bottom-dwelling organisms (benthic macrofauna). Mixing of sediment, both particles and water, is called bioturbation and it constitutes an important mechanism for the transport of products and reactants at the sediment-water interface. A summary of major substances transported to and from sediments, on a worldwide average basis, is shown in Fig. 7–4.

Many of the chemical reactions which occur on the sea floor and within the immediately underlying sediments are brought about ultimately by biological activity in the overlying water. For example, in shallow depths of the ocean, plankton, as a result of photosynthesis, produce organic compounds which are otherwise thermodynamically unstable in seawater. Upon death, some of the organic matter escapes destruction in the water column and falls to the bottom where it undergoes decomposition by microorganisms. As a result, dissolved oxygen is consumed and carbon dioxide and the nutrient

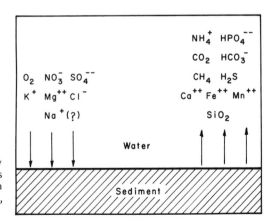

Fig. 7-4. Direction of fluxes (shown by arrows) expected for dissolved constituents resulting from chemical reactions between seawater and sediments. After Berner, 1976.

elements, nitrogen and phosphorus, are regenerated back to the oceans. Nutrient regeneration (both in the water column and at the sea floor) is important to the maintenance of photosynthesis, and thus of life, in the sea.

Burial of organic matter in the sediment can also result, via microbiological activity, in the reduction of important inorganic compounds to new substances, for example the reduction of nitrate to $N_2$ and sulfate to $H_2S$. Combining nutrient and $CO_2$ regeneration with this reduction, it can be shown that most of the dissolved species undergoing transfer across the sediment-water interface shown in Fig. 7-4 are involved in reactions accompanying the microbiological decomposition of organic matter. This includes $O_2$, $CO_2$, $HPO_4^{--}$, $NH_4^+$, $NO_3^-$, $H_2S$, $CH_4$, $Fe^{++}$, and $Mn^{++}$.

Photosynthesis and the food chain also result in the production of calcium carbonate and opaline silica, chiefly by planktonic plants and animals living in shallow water. (Benthic secretion of $CaCO_3$, e.g., by corals, will not be discussed here.) The silica and calcium carbonate form as skeletal hard parts and the chief culprits are coccolithophorids (microscopic plants which secrete $CaCO_3$ as the mineral calcite), foraminifera (unicellular animals which secrete calcite), pteropods (small swimming snails which secrete the aragonite form of calcium carbonate), diatoms (microscopic plants which secrete opaline silica), and radiolaria (unicellular animals which secrete opaline silica). Upon death and falling to the bottom, these substances also undergo liberation to seawater. Much of the deep sea is undersaturated with respect to calcium carbonate and the entire ocean is undersaturated with respect to opaline silica. Thus, upon falling to the bottom the coccolithophorids, forams, pteropods, diatoms, and radiolaria are dissolved and $Ca^{++}$, $HCO_3^-$, and $SiO_2$ are liberated to solution (see Fig. 7-4). In many parts of the oceans, dissolution is incomplete and, as a result, the remaining $CaCO_3$ and silica accumulate to form biogenic calcareous and siliceous sediments which in the deep oceans cover vast areas of the sea floor.

Other chemical reactions occur at the sea floor that are not affected by biological activity. The most important of these is the reaction of volcanic rocks and volcanogenic sediment (e.g., ash) with seawater to form new silicate minerals, chiefly clays and zeolites. Also, dissolved manganese is precipitated to form the well-known manganese nodules, and this process may or may not be strictly inorganic. Although such reactions as these are of considerable interest as controls on the composition of seawater, detailed discussion of them is beyond the scope of the present chapter. Instead we will be concerned here only with sea-floor chemical reactions which result directly or indirectly from biological activity. Specifically, we will discuss the decomposition of organic matter and the dissolution of biogenic silica and calcium carbonate. Before taking up these topics, however, it is necessary to present to the reader a brief introduction to the principles of diagenetic modeling.

## DIAGENETIC MODELING

Many of the chemical reactions occurring in sediments are sufficiently slow that they can be evidenced by changes with depth in the composition of both solids and interstitial water in the sediment. These changes are referred to collectively by the term *diagenesis*. Continually deposited sediments provide a unique opportunity for deducing rates of reactions and the dependence of these rates on various factors such as the concentrations of reactants. This can be achieved through the use of diagenetic modeling (e.g., see Berner, 1980). In diagenetic modeling, the various factors affecting the concentration of a sedimentary component at a given depth in the sediment are stated in terms of rate expressions, and the sum of these expressions (the diagenetic equation) represents all processes affecting the component. In surficial sediments the principal processes, besides chemical reaction, are mixing of solids and porewater by bioturbation, molecular diffusion of dissolved constituents in the interstitial water, and burial of water and solids by deposition. (In deep-sea sediments, molecular diffusion is far more important than bioturbation—see discussion below.) Thus, the diagenetic equations for a surficial deep-sea sediment can be expressed in the form:

## For a solid component (with concentration $C_s$)

$$\partial C_s/\partial t = R_{\text{bioturb}} + R_{\text{burial}} + \Sigma R_{\text{chem}} \qquad (1)$$

## For a dissolved component (with concentration $C$)

$$\partial C/\partial t = R_{\text{diffus}} + R_{\text{burial}} + \Sigma R_{\text{chem}} \qquad (2)$$

Here the derivatives refer to a fixed depth in the sediment, $R$ refers to the rate of change of concentration due to each process, and the summation sign, $\Sigma$, refers to all chemical reactions affecting the constituent under study.

Solution of the diagenetic equations is normally accomplished by assuming that steady state at a fixed depth is attained. This is referred to as *steady state diagenesis*. It means that the rate of all processes are so adjusted that they sum to zero and as a result $\partial C_s/\partial t = 0$ and $\partial C/\partial t = 0$. In other words, the shape of the curve of concentration versus depth does not change with time as sediment is deposited. This is shown in Fig. 7-5. Note that, although no change occurs at a fixed *depth,* there are distinct changes in a given *layer* undergoing burial. The changes within the layer (for example, the dissolution of silica) is what we normally think of when we use the word diagenesis.

If there is steady state diagenesis and only one chemical reaction of interest, we can solve the above equations for the rate of chemical reaction, providing we have a means of

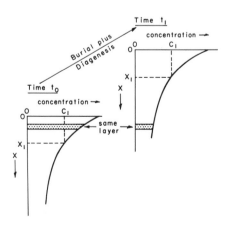

Fig. 7-5. Steady state diagenesis represented diagramatically. Note that at a fixed depth $x_1$ concentration remains constant with time, but that within a given layer it changes during burial. After Berner, 1976.

ROBERT A. BERNER

quantifying bioturbation, molecular diffusion, and burial. Burial rate can be quantified by radioactive dating of sediment layers, and molecular diffusion rates in porewaters of various types of sediments are now fairly well known (e.g., see Li and Gregory, 1974). Bioturbation is a bigger problem but several studies, using radionuclides or dated depositional events (Goldberg and Koide, 1962; Guinasso and Schink, 1975; Nozaki *et al.,* 1978, Peng *et al.,* 1977), have shown that particle bioturbation in deep-sea sediments can be treated in terms of random mixing and in many cases the value of a "biodiffusion" coefficient has been deduced. The results indicate that rates of biodiffusion are about three orders of magnitude smaller than rates of molecular diffusion and that appreciable mixing by bioturbation is confined to the top $\sim$10 cm. Thus, bioturbation of porewater can be ignored when discussing dissolved constituents of deep-sea sediments.

By expressing rates of chemical reaction in terms of concentrations of reactants and characteristic rate constants, one can solve steady state diagenetic equations, and the values obtained for the rate constants can be used to characterize the reactivity of the various sedimentary compounds. This will be demonstrated in this chapter for the case of organic matter decomposition by sulfate reduction.

For most diagenetic chemical reactions, we are dealing with simultaneous changes in the composition of both solids and interstitial water. In other words, the disappearance, for example of biogenic silica particles, with depth due to dissolution is reflected by a corresponding increase of dissolved silica in the porewater. Thus, the reactions can be studied by means of analyses for either dissolved or solid constituents. Normally dissolved constituents are determined because the techniques for analysis are better developed and results are less ambiguous. For instance, in the case of biogenic silica, there are problems in chemically distinguishing this form of silica from silica contained in clay minerals (e.g., see Demaster, 1979). Also, monotonic changes in concentration with depth due to dissolution may be masked by fluctuations due to historical changes in the amount of biogenic silica buried at the time of deposition. Interstitial water chemistry, due to smoothing by molecular diffusion, does not exhibit such sharp historically induced fluctuations in concentration. Also, water chemistry is a more sensitive indicator of diagenetic changes. For example, dissolution of only one ten-thousandth of the solid biogenic silica in a pelagic siliceous ooze would result (in the absence of molecular diffusion) in a *tenfold* increase (over that in the overlying seawater) in dissolved silica concentration in the interstitial water.


## ORGANIC MATTER DECOMPOSITION


After death, most organic matter synthesized by plankton is broken down by macro- or microorganisms living in the water column (Menzel, 1974). A small amount, however, often in the form of fecal material, escapes destruction and falls to the bottom. Further destruction then continues in the sediments and the resulting metabolic products accumulate in the porewaters. This is well illustrated by the nutrients phosphate and ammonia shown in Fig. 7-1. The microbial processes by which the organic matter is oxidized to simple ions and molecules follows a general sequence. The oxidizing agents used by bacteria are successively $O_2$, $NO_3^-$, Fe and Mn oxides, $SO_4^{--}$, and $CO_2$, and in general each oxidant is used only when its predecessors have been consumed. Overall reactions are summarized in Table 7-1. It should be emphasized that these reactions are highly generalized and involve the summation of many intermediate steps involving different microorganisms and different organic compounds. The formula $CH_2O$ refers only to the starting material, which consists of high molecular weight and insoluble substances, and which undergoes a complex series of intermediate reactions before being ultimately oxidized to $CO_2$ and/or $HCO_3^-$. The usual reason advanced for the succession shown in Table 7-1 (e.g., see Claypool and Kaplan, 1974) is that the reactions occur according to the order of energy release. In other words, if a variety of oxidizing agents are present, the one chosen is that which yields the most energy to the microorganisms. This idea is corroborated by the calculated standard Gibbs free energy changes for the reactions shown in Table 7-1. There is a definite decrease as one goes down the list.

Successive utilization of the various electron acceptors (oxidizing agents) during diagenesis gives rise to their successive disappearance with depth in the sediment. Thus, in descending through the sediment column we encounter, in order, zones of $O_2$ consumption, nitrate reduction, $Fe^{++}$ and $Mn^{++}$ production, sulfate reduction, and methane formation. The thickness of the zone dominated by each process depends upon the rate of reaction and the rate by which concentrations of the oxidants are disturbed by bioturbation or diffusion.

In shallow-water sediments, a considerable proportion of freshly killed planktonic material falls to the bottom before it can be eaten or otherwise decomposed in the water column. As a result, highly reactive organic compounds are available for bacterial activity in the sediment; consequently organic matter decomposition is rapid and electron acceptors are rapidly consumed. The utilization of $O_2$ occurs in only the top few cm of sediment (or along the walls of worm burrows), where oxygenated seawater is constantly supplied to the sediment by bioturbation. (Without bioturbation the penetration of oxygen into organic-rich, near-shore sediments by molecular diffusion is very small, on the order of about 1 mm.) The zone of nitrate reduction is also very thin (a few centimeters at most) and in many cases unmeasurable by the normal techniques of porewater sampling. By contrast, the zone of sulfate reduction is prominent and can be several meters thick. Consequently, most shallow-water organic-rich sediments are characterized by concentration gradients in dissolved sulfate and the smell of $H_2S$. A typical example is shown

TABLE 7-1. Standard State Free Energy Changes for Some Bacterial Reactions[a]

| Reaction | $\Delta G^0$ (kj $M^{-1}$ of $CH_2O$) |
|---|---|
| $CH_2O + O_2 \rightarrow CO_2 + H_2O$ | −475 |
| $5CH_2O + 4NO_3^- \rightarrow 2N_2 + 4HCO_3^- + CO_2 + 3H_2O$ | −448 |
| $CH_2O + 3CO_2 + H_2O + 2MnO_2 \rightarrow 2Mn^{++} + 4HCO_3^-$ | −349 |
| $CH_2O + 7CO_2 + 4Fe(OH)_3 \rightarrow 4Fe^{++} + 8HCO_3^- + 3H_2O$ | −114 |
| $2CH_2O + SO_4^{--} \rightarrow H_2S + 2HCO_3^-$ | −77 |
| $2CH_2O \rightarrow CH_4 + CO_2$ | −58 |

[a]After Berner, 1980. Data for $CH_2O$ and $MnO_2$ are for sucrose and fine-grained birnessite, respectively.

in Fig. 7-1. In the most organic-rich sediments, one encounters complete sulfate reduction with the resulting buildup of dissolved methane at greater depths shown in Fig. 7-6.

In deep-sea sediments, organic matter synthesized by plankton in shallow water undergoes greater decomposition prior to burial. This is due partly to the greater distance of fall through seawater, but, more importantly, it comes about because of the very low rates of sedimentation found in the deep sea. Sedimentation rates there are about a thousand times lower than they are for shallow-water sediments. Slow sedimentation rate results in the exposure at the sediment-water interface of organic compounds to oxygenated seawater for long periods of time and to the consequent removal, via oxic degradation, of the more reactive compounds. Thus, upon burial, the remaining organic matter is less reactive and decomposition rates are correspondingly slow. This results in thick zones of nitrate reduction, $Fe^{++}$ and $Mn^{++}$ production, and so forth. In fact, the zones of sulfate reduction and methane formation are either absent or so deep that they generally are not reached by conventional coring techniques, which sample only the top 1-10 m. The general situation found in pelagic sediments is shown in Fig. 7-7 (Froelich *et al.*, 1979).

Besides being less reactive, the organic matter in deep-sea sediments is also less abundant than it is in shallow-water sediments. This is shown in Table 7-2. This situation holds true, however, only for the oxygenated waters of the open ocean. In some unusual areas, such as the Black Sea, the bottom water is anoxic and organic matter deposited slowly in deep-water sediments is not as rapidly destroyed as it is in normal pelagic sediments; as a result, more organic matter accumulates.

The difference between deep-sea sediments and those deposited more rapidly in shallow water suggests that one might expect to find a correlation between the concentration or reactivity of organic matter and the rate of deposition. Such correlations have been found and, in general, they are very good. For a wide variety of fine-grained marine

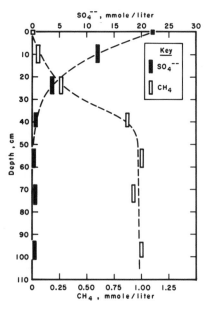

Fig. 7-6. Plots of dissolved sulfate and methane versus depth in sediments from Long Island Sound (USA). Note the large increase in methane as sulfate is depleted. After Berner, 1976; Martens and Berner, 1977.

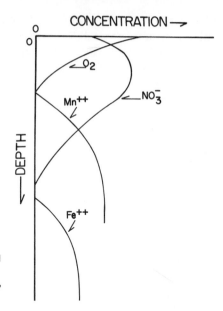

Fig. 7-7. Generalized plots of dissolved $O_2$, $NO_3^-$, $Mn^{++}$, and $Fe^{++}$ versus depth in typical deep-sea sediments. After Froelich, *et al.*, 1979.

sediments, Heath *et al.* (1977), have shown that organic matter concentration and sedimentation rate are moderately well correlated. Müller and Suess (1979) more recently have found that this correlation can be considerably improved by correcting for differing rates of organic matter production in the overlying water.

The reactivity, or rate of decomposition, of organic matter can be shown also to correlate well with the rate of sedimentation. This is especially true for organic matter decomposition via sulfate reduction (Goldhaber and Kaplan, 1975). Application of diagenetic modeling has been made to profiles of dissolved sulfate versus depth (below the zone of bioturbation) in sediments undergoing active bacterial sulfate reduction, on the assumption that organic matter decomposition occurs according to first order

TABLE 7-2. Organic Carbon Concentrations of Fine-Grained Marine Sediments[a]

| Sediment Type | % (dry wt) Organic Carbon |
| --- | --- |
| Pelagic (deep-sea) sediments | |
| Eupelagic | <0.3 |
| Hemipelagic (and Pacific equatorial divergence) | 0.3 – 2 |
| Average for all types | 0.25 |
| Shelf and slope muds | |
| Upwelling regions | 1 – 5 |
| Estuaries and closed basins | 1 – 10 |
| Other areas | 0.5 – 2 |
| Average for all types | 0.99 |

[a]After Berner, 1979.

kinetics (Berner, 1974, 1978a; Toth and Lerman, 1977). In other words, the rate of decomposition is assumed to be directly proportional to the concentration of the organic material undergoing decomposition. Mathematically this is expressed as

$$\frac{dG}{dt} = -kG \qquad (3)$$

where $G$ = concentration of organic matter undergoing decomposition (oxidation to $HCO_3^-$) by the overall process of bacterial sulfate reduction (not the total organic matter present in the sediment)

$k$ = rate constant for sulfate reduction

$t$ = time

Here the parameter $G$ represents the *concentration* of decomposing organic matter and the parameter $k$ represents its *reactivity*. Toth and Lerman (1977) have shown that there is a very good correlation between $k$ and the rate of sedimentation for a wide variety of sediments ranging from deep-sea clays deposited at only a few millimeters per thousand years to coastal muds deposited at over 1 cm/y. (Perceptible sulfate reduction in deep-sea sediments occurs only over large depth intervals—hundreds of meters—which can be seen only by deep drilling.) This correlation has been further demonstrated for additional sediments by the writer (Berner, 1978a). It is illustrated in Fig. 7–8.

In Fig. 7–8 the slope of the lines is two, which means that

$$k = A\omega^2 \qquad (4)$$

where $\omega$ = rate of deposition (length per time)

$A$ = empirical constant

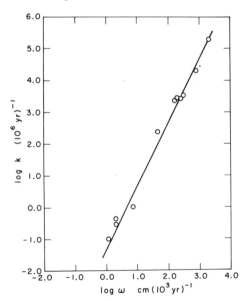

Fig. 7-8. Log-log plot of the first-order rate constant for organic matter decomposition via sulfate reduction versus rate of sedimentation for sediments showing evidence of bacterial sulfate reduction. Data are for depths below the zone of bioturbation only. After Berner, 1978a; Toth and Lerman, 1977.

CHEMISTRY OF BIOGENIC MATTER AT THE DEEP-SEA FLOOR

This relation, when combined with additional observations of an excellent correlation between sedimentation rate and the initial slope of dissolved sulfate-versus-depth (Berner, 1978a), can be used to show that the reactivity of organic matter undergoing decomposition via sulfate reduction varies greatly from one sediment to another, whereas its concentration is relatively uniform. This occurs because sulfate in most sediments is exhausted with depth before all potentially reactive organic compounds can be consumed.

It would be interesting to apply arguments similar to those presented above for sulfate reduction, to the decomposition of organic matter by other processes, namely $O_2$, $NO_3^-$, and $MnO_2$ reduction, especially since these are the dominant processes in shallowly buried deep-sea sediments (see Fig. 7-7). Unfortunately, insufficient data exist for this purpose. Once such calculations have been performed, it should be possible to characterize better the reactivity of organic materials deposited in the deep sea.

## CALCIUM CARBONATE DISSOLUTION

As stated in the introduction, calcium carbonate is precipitated in shallow waters of the ocean by planktonic coccolithophorids, foraminifera, and pteropods. Here the $CaCO_3$ does not undergo dissolution, because near the surface of the ocean, the water is supersaturated. Upon death, the calcareous skeletal remains of these organisms fall to the sea floor and, if the water is sufficiently deep, they undergo dissolution. Dissolution at depth is brought about by higher pressures, higher $CO_2$ concentrations, and lower temperatures, all of which lead to an increase in the solubility of $CaCO_3$. Removal by dissolution gives rise to the phenomenon known as the carbonate compensation depth or CCD. The CCD represents the depth at which the rate of dissolution equals the rate of fallout of $CaCO_3$ so that no $CaCO_3$ accumulates at greater depths. In a given region this situation can be visualized as analogous to snow on a mountain range. At a good distance above the CCD (snowline), appreciable concentrations of $CaCO_3$ (snow) are found in the sediments. With increasing depth the concentration of calcium carbonate (snow) drops abruptly and goes to zero when the CCD is reached. A typical plot of $CaCO_3$ content versus depth, illustrating the CCD, is shown in Fig. 7-9 (for further details on the CCD, see Berger, 1976). The term CCD is normally used to refer to calcite, which is far more abundant in deep-sea sediments than aragonite. Aragonite, because it is more soluble than calcite, disappears at much shallower depths with its own compensation depth (ACD). Thus, using the mountain analogy, aragonite can be thought of as a lower melting-point "snow," consisting of a substance other than $H_2O$, which disappears everywhere except at the very tops of the highest mountain peaks.

Originally it was presumed that the CCD simply represented the depth where seawater passes downward from a state of supersaturation to one of undersaturation with respect to calcium carbonate. In this case the CCD would represent the depth of calcite saturation. However, considerable work over the past 15 years has shown that, in general, the depth where water attains saturation with calcite is distinctly shallower than the CCD.

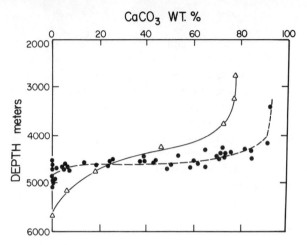

Fig. 7-9. Plot of weight per cent calcium carbonate versus water depth for deep-sea sediments of the Pacific Ocean. The dashed curve represents sediments of the central Pacific located between 15°N and 5°S, and 120°–135°W. The full line represents average values for each 500 m interval for all Pacific pelagic sediments. After Bramlette, 1961 (*in* M. Sears, ed., Amer. Assoc. Adv. Sci., Publ. 67, p. 345–366, copyright 1961 by Amer. Assoc. Adv. Sci.).

Examples are shown in Figs. 7-10 and 7-11. This finding is bolstered by the discovery that partial dissolution, manifested by the disappearance of solution-susceptible forams and coccoliths (both calcite), occurs at a depth considerably above the CCD. In fact, dissolutive removal actually accelerates at a depth, known as the lysocline (Berger, 1970), which is on the average about 1000 m shallower than the CCD. The depth where evidence

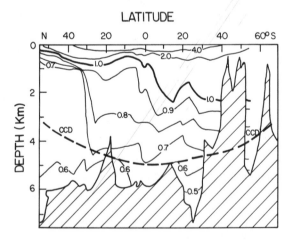

Fig. 7-10. Degree of saturation, with respect to calcite, of central Pacific Ocean water as a function of latitude. Contours are for constant values of $\Omega$. ($\Omega$ is the concentration product for calcium and carbonate ions divided by the same product for saturation equilibrium with calcite.) Also plotted is the CCD. Data from Berger, 1976; Takahashi, 1975.

CHEMISTRY OF BIOGENIC MATTER AT THE DEEP-SEA FLOOR

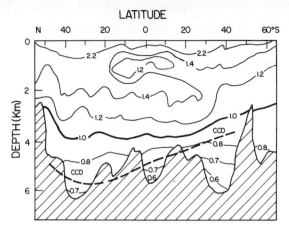

Fig. 7-11.    Degree of saturation, with respect to calcite, of the western Atlantic Ocean as a function of latitude. Contours are for constant values of $\Omega$. After Takahashi, 1975; CCD data from Berger, 1976.

of dissolution is first encountered is referred to as the $R_0$ level (Berger, 1976, 1977) and in general it coincides with the depth of saturation. This, and its relation to the lysocline and CCD, for sediments of the Atlantic Ocean are shown in Fig. 7–12.

There has been considerable controversy regarding the origin of the lysocline. One hypothesis (Li *et al.,* 1969b) is that it represents the depth where water passes downward from supersaturation to undersaturation. This is not in keeping, however, with the many observations and calculations summarized in Fig. 7–12. Another hypothesis is that forams and coccoliths undergo enhanced dissolution below a certain depth because of increased turbulence in deeper parts of the oceans. This is the so-called hydrodynamic theory of the lysocline (Edmond, 1974). Increased turbulence is believed to accelerate the rate of

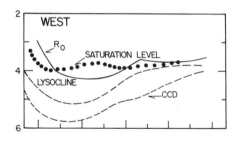

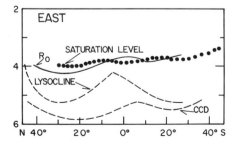

Fig. 7-12.    Plots of the saturation level, $R_0$ level, lysocline, and CCD for the Atlantic Ocean. After Berger, 1977.

dissolution by bringing about resuspension of bottom particles and by increasing the rate of migration of calcium and carbonate ions away from the surfaces of dissolving particles. (This is analogous to increased dissolution of sugar when a cup of coffee is stirred.)

Several observations furnish evidence against the hydrodynamic hypothesis. First of all, recent dissolution experiments have been conducted by immersing samples of coccoliths and forams directly in seawater on fixed moorings at a series of depths and by artificially pumping water past them at the same rate at all depths (Honjo and Erez, 1978). Rates of dissolution were found to accelerate greatly at a depth approximately equal to the lysocline, even though the stirring rate was held constant at all depths. Secondly, calculations of the rate by which calcite dissolves, at the saturation states found in the oceans, indicate that the rates are too slow to be affected by stirring (Berner and Morse, 1974). In other words, dissolution is controlled by the rate at which calcium and carbonate ions are detached from the calcite surfaces, and not by the rate of transport of the ions away from the surfaces. Thus, enhanced stirring should have little or no effect on the rate of dissolution. Finally, there is no established evidence that increased turbulence occurs everywhere in the ocean at the same depth as the lysocline.

The third theory is that of the chemical kinetic origin of the lysocline (Morse and Berner, 1972). According to this theory, the lysocline comes about because of a highly nonlinear rate law for calcite dissolution in seawater. In other words, the rate of dissolution does not smoothly parallel the degree of undersaturation but rather increases greatly for small increases in undersaturation. This idea is backed by the *in situ* dissolution experiments of Peterson (1966), Berger (1967), Milliman (1977), and Honjo and Erez (1978), and by laboratory experiments (Berner and Morse, 1974; Keir, 1979; Morse, 1978; Morse and Berner, 1972; Sjöberg, 1978) (see Fig. 7-13). It has been found in the laboratory

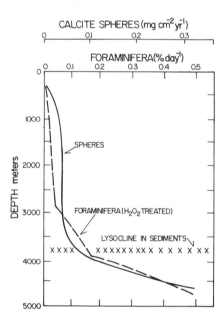

Fig. 7-13. Rates of dissolution as a function of water depth of calcite spheres (Peterson, 1966) and foraminifera (Berger, 1967) suspended in the central Pacific Ocean for four months.

that the rate of calcite dissolution in seawater, over the range of undersaturation found in the deep oceans, follows the expression

$$R_{diss} = k_d (1 - \Omega)^n \tag{5}$$

where $R_{diss}$ = rate of dissolution

$\Omega$ = ratio of the concentration product of $Ca^{++}$ and $CO_3^{--}$ divided by the same product at equilibrium saturation

$k_d$ = dissolution rate constant

The value of $n$ ranges from 3.5 to 5.5. Typical results, taken from the study by Keir (1979) are shown in Fig. 7-14. At high but still undersaturated values of $\Omega$ (low values of $1 - \Omega$), dissolution becomes so slow that it is unmeasurable both in the laboratory and in the field. This strong dependence of dissolution rate on degree of saturation explains why the lysocline exists. It occurs at a degree of undersaturation where $R_{diss}$ is so high that particles begin to undergo extensive dissolution before they can be buried in the sediment and protected from further dissolution.

So far little has been said concerning how and where dissolution occurs within a single deep-sea sediment. One would expect that upon burial the interstitial water, due to dissolution, would approach saturation with respect to calcite because of a high ratio of solids-to-water in the sediment and a lack of rapid mixing of porewater with overlying undersaturated seawater. As saturation is approached, rates of dissolution would be expected to drop considerably because of the strong dependence of $R_{diss}$ on $\Omega$ [see Eq. (5)]. In this way, dissolution at depth in the sediment would be precluded. Use of laboratory data along with diagenetic modeling (assuming bioturbation and dissolution

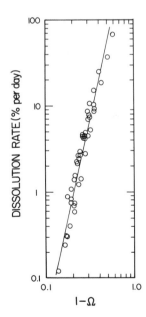

Fig. 7-14.  Dissolution rates of calcite, as measured in the laboratory, plotted as a function of saturation state in terms of $(1-\Omega)$. The steep slope on this log-log plot indicates that the rate of dissolution is strongly dependent upon $\Omega$ and therefore the kinetics are highly nonlinear. After Keir, 1980.

of solid $CaCO_3$ and molecular diffusion and addition-via-dissolution of dissolved $CaCO_3$) by Schink and Guinasso (1977, 1978), Takahashi and Broecker (1977), and Keir (1979) indicates that most dissolution should occur only within the top $\sim 2$ cm of sediment. Unfortunately, present methods of porewater sampling are not sufficiently precise to test these predictions.

Laboratory studies (Keir, 1979) have also shown that dissolution rates calculated from diagenetic modeling [using Eqs. (1) and (2)] are distinctly lower than those measured in the laboratory at the same degree of undersaturation. One explanation offered by Keir and by Schink and Guinasso is that aragonite is present due to sedimentation of pteropods and their downmixing into the sediment by bioturbation. This aragonite, because of its high solubility, is completely dissolved and adds extra dissolved $CaCO_3$ to interstitial solution, which in turn raises the value of $\Omega$ and thereby slows calcite dissolution. This is a reasonable hypothesis in that sedimentation of pteropods to the sea floor is important (see below) and complete dissolution of aragonite occurs at all depths where calcite dissolution (partial or complete) also takes place.

Aragonite not only affects the rate of dissolution of calcite but also is important in its own right. It is normally assumed that the sedimentation of aragonite to the sea floor as pteropods is very minor compared to that for calcitic coccoliths and forams. This idea has arisen because of the rarity of pteropod remains in most deep-sea sediments. However, aragonite completely dissolves away below about 500 m in the Pacific Ocean and below about 2500 m in the Atlantic Ocean (e.g. see Fig. 7-15). If one looks at sediments at sufficiently shallow depths, one occasionally encounters sediments sufficiently high in pteropod shells that the term *pteropod ooze* has been coined. Unfortunately, such shallow depths are rare in the open ocean and are usually confined to slopes near conti-

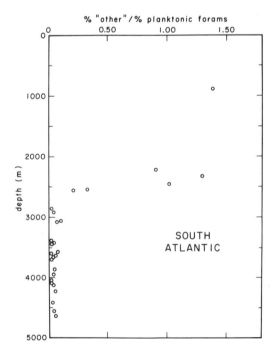

Fig. 7-15. Plot of the ratio of pteropods-to-forams as a function of water depth for pelagic sediments of the South Atlantic Ocean. Samples were taken at least 1000 km from the nearest land. Material denoted as "other" by Murray and Chumley (1924) must, under these circumstances, represent almost entirely pteropods. After Berner, 1977.

CHEMISTRY OF BIOGENIC MATTER AT THE DEEP-SEA FLOOR

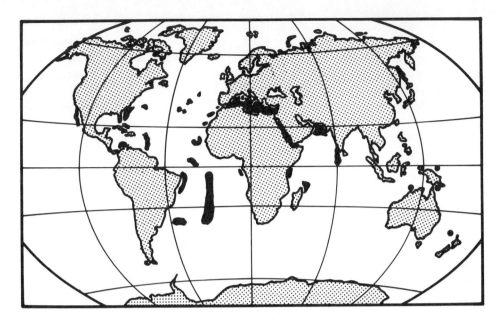

Fig. 7-16.  Worldwide distribution of pteropod ooze. Areas of pteropod ooze are denoted in black. After Berner, 1977.

nents or islands where pteropods are then masked by large influxes of land-derived detritus or by the shallow-water skeletal material secreted by benthic organisms (e.g., corals, algae, clams, snails, etc.). Only in a few locations, far from sediment sources, has pteropod-rich sediment been found. Notable locations are the deeper portions of the Mediterranean and Red Seas and shallower portions of the South Atlantic mid-ocean ridge. A location map for pteropod ooze is shown in Fig. 7–16.

Berner (1977) has calculated the approximate aragonite:calcite ratio in pteropod ooze and in shallow (700–1500 m) pelagic (far from land) sediments by examining published data for pteropod, foram, and coccolith abundances. Most of the data on pteropod and foram abundance were taken from the compilations of Murray and Chumley (1924) for 1426 sediment localities from the Atlantic Ocean, many of which represent samples taken during the famous *Challenger* expedition during the latter part of the 19th century. If shallow pelagic sediments and pteropod ooze can be considered to sample entirely undissolved carbonate particles (both calcite and aragonite) falling to the bottom, then, from the relative abundance of coccoliths, forams, and pteropods, one may calculate that about 50% of the $CaCO_3$ delivered to the sea floor is present as aragonite. This value agrees with results for Mediterranean and Red Sea sediments and with estimates of aragonite-versus-calcite production in calcareous plankton; it is much larger than previously assumed.

The reason aragonite sedimentation has been neglected (as stated above) is that pteropods are not found in most deep-sea sediments. However, this is due partly to methods of sampling. Examination of box cores from deep water, which are carefully taken to preserve the sediment-water interface, has allowed observations to be made of scattered pteropod shells which are confined to the surface of the sediment. This is believed to represent aragonite in the act of dissolution (e.g., see Morse, 1977), This, combined with the finding of appreciable aragonite (as shells of both pteropods and bivalve larvae) in deep-water plankton tows and in deep-water sediment traps, supports the idea that aragonite fallout to the bottom is important. Inclusion of pteropod dissolution in previous models of $CaCO_3$ dissolution in the deep sea (e.g., Broecker, 1971) gives much better agreement between calculated and measured rates of accumulation and dissolution than if pteropods are ignored. Thus, it is important that both aragonite and calcite be considered when discussing the dissolution of calcium carbonate in the deep sea.


## BIOGENIC SILICA DISSOLUTION

Planktonic diatoms and radiolaria are able to precipitate $SiO_2$ from seawater to form skeletons consisting of a form of disordered cristobalite known as opaline silica, even though the surrounding water, as well as the rest of the oceans, is undersaturated with respect to opaline silica. This is accomplished by excess energy ultimately supplied by sunlight. Upon death, the biogenic silica immediately begins to redissolve and most returns to solution during the fall of the skeletal material to the sea floor. However, some of the diatom and radiolarian debris arrives at the bottom, where it undergoes further dissolution. This dissolution continues during burial and gives rise to increased concentrations of dissolved silica in the interstitial water (see Fig. 7-3).

Both diatoms and radiolaria have very high surface areas (Lawson *et al.*, 1978) and reasonably high solubility in deep-ocean water ($\sim$1 m$M$/l). Thus, they would be expected to dissolve rapidly in the deep sea. Through the use of diagenetic modeling and the assumption that the rate of opaline silica dissolution is controlled by transport of dissolved silica away from the surfaces of the dissolving particles (Berner, 1978b), one may predict that all opaline silica falling to the bottom should dissolve away before it can be buried in the sediment. However, this does not happen. The existence of siliceous oozes consisting of high proportions of diatoms and radiolaria, and the presence of concentration gradients of dissolved silica in porewaters (Fig. 7-3), both point to much slower rates of dissolution. Rates calculated via diagenetic modeling, using Eqs. (1) and (2) (Hurd, 1973) are several orders of magnitude lower than predicted for simple transport-controlled dissolution. Thus, the rates of both biogenic silica and calcium carbonate dissolution are controlled by the detachment of ions, molecules, etc., from the particle surfaces, and not by transport away from the surfaces.

Dissolution of silica is so slow that the interstitial waters of deep-sea sediments containing excess biogenic silica do not attain saturation with respect to the opaline silica. The usual explanation for this (e.g., Hurd, 1973; Wollast, 1974) is that the surface of siliceous particles becomes poisoned by ions, such as $Mg^{++}$, which react with the surface and convert it to a less soluble substance (for example, sepiolite). However, if this were true, one would expect to find essentially the same asymptotic concentration of dissolved silica at a depth representing equilibrium with sepiolite or some other siliceous substance in most sediments. Instead, one finds a correlation between the asymptotic concentration and the rate of plankton production (as a measure of the rate of supply of biogenic silica to the bottom) in surface waters overlying the site of sedimentation (Schink *et al.*, 1973). This suggests that the biogenic silica added to the sediment exhibits a range of reactivity with some material dissolving so slowly that its effect on porewater concentration cannot be discerned over the depths normally sampled by coring. In this case, the asymptotic dissolved silica concentration merely represents the total consumption of reactive siliceous particles. Such an explanation has been offered by Schink *et al.* (1975) and used, along with diagenetic modeling, to explain the depth distribution of dissolved silica in deep-sea sediments.

One problem with the "reactive silica" theory is that in sediments consisting of very high proportions of biogenic silica, such as occur in the region surrounding Antarctica, dissolved silica concentrations still do not attain saturation with respect to opaline silica (Demaster, 1979). This poses a problem in that the vast majority of the silica present would have to be considered as being nonreactive. Perhaps the best explanation of the silica dissolution problem is a combination of theories. In other words, a large variation in reactivity exists, but surface poisoning and some reduction in solubility also occur. The laboratory studies of Johnson (1975) indicate that differential reactivity of different species of diatoms and radiolaria is a fact and cannot be ignored.

If a range of reactivity is truly present for biogenic silica buried in deep-sea sediments, then some interesting effects can occur as a result of bioturbation. Schink *et al.* (1975), by means of diagenetic modeling, have shown that bioturbation should result in the downmixing of reactive silica particles which otherwise would be dissolved away near the sediment-water interface. This downward transport enables excess silica to be liberated to solution at depth and, thus, to build up in concentration to values higher than would be present in the absence of bioturbation. This result, shown in Fig. 7–17, is not intuitively obvious. Normally, bioturbational mixing is thought of as an homogenizing process whereby concentration gradients of dissolved constituents are reduced. However, in deep-sea sediments, ionic diffusion is much faster than porewater bioturbation and, as a result, only solids are considered as being mixed by bioturbation. Thus, injection of dissolved silica at depth, from super-reactive particles carried downward by bioturbation, gives rise

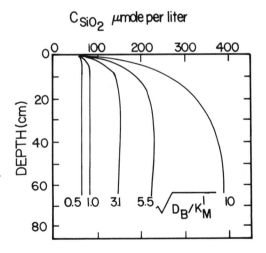

Fig. 7-17. Computer-generated plots of dissolved silica $C_{SiO_2}$ versus depth in a typical deep-sea sediment according to the diagenetic model of Schink *et al.*, 1975. $D_B$ = bioturbation coefficient for solids, $k_m'$ = rate constant for dissolution of biogenic silica. Note the increase in the asymptotic value of dissolved silica concentration as the rate of bioturbation (at fixed dissolution rate constant) is increased. After Schink *et al.*, 1975.

to increases in concentration gradients of dissolved silica. This is yet another example of how diagenetic modeling has been used to study chemical reactions occurring on the sea floor, in this case the dissolution of biogenic silica. Hopefully, more of such modeling will be done in the future.

# REFERENCES

Berger, W. H., 1967, Foraminiferal ooze—solution at depths: *Science,* v. 156, p. 383–385.

——, 1970, Planktonic foraminifera—selective solution and the lysocline: *Mar. Geol.,* v. 8, 111–138.

——, 1976, Biogenous deep sea sediments—production, preservation, and interpretation, *in* Riley, J. P., and Chester, R., eds., *Chemical Oceanography, Vol. 5:* New York, Academic, p. 265–387.

——, 1977, Carbon dioxide excursions and the deep-sea record aspects of the problem, *in* Andersen, N. R., and Malahoff, A., eds., *The Fate of Fossil Fuel* $CO_2$ *in the Oceans:* New York, Plenum, p. 505–542.

Berner, R. A., 1974, Kinetic models for the early diagenesis of nitrogen, sulfur, phosphorus, and silicon in anoxic marine sediments, *in* Goldberg, E. D., ed., *The Sea, Vol. 5:* New York, Wiley, p. 427–450.

——, 1976, The benthic boundary from the viewpoint of a geochemist, *in* McCave, I. N., ed., *The Benthic Boundary Layer:* New York, Plenum, p. 33–56.

——, 1977, Sedimentation and dissolution of pteropods in the ocean, *in* Andersen, N. R., and Malahoff, A., eds., *The Fate of Fossil Fuel* $CO_2$ *in the Oceans:* New York, Plenum, p. 243–260.

——, 1978a, Sulfate reduction and the rate of deposition of marine sediments: *Earth Planet. Sci. Lett.,* v. 37, p. 492–498.

——, 1978b, Rate control of mineral dissolution under earth surface conditions: *Amer. J. Sci.,* v. 278, p. 1235–1252.

——, 1979, A new look at biogenous material in deep-sea sediments: *Ambio Spec. Rpt. 6,* p. 5–10.

——, 1980, *Early Diagenesis—A Theoretical Approach:* Princeton Univ. Press, 241 p.

——, and Morse, J. W., 1974, Dissolution kinetics of calcium carbonate in seawater, IV. Theory of calcite dissolution: *Amer. J. Sci.,* v. 274, p. 108–134.

Bramlette, M. N., 1961, Pelagic sediments, *in* Sears, M., ed., *Oceanography:* Amer. Assoc. Adv. Sci. Publ. 67, p. 345–366.

Broecker, W. S., 1971, Calcite accumulation rates and glacial to interglacial changes in oceanic mixings, *in* Turekian, K. K., ed., *Late Cenozoic Glacial Ages:* Yale Univ. Press, New Haven, Conn., p. 239–265.

Claypool, G., and Kaplan, I. R., 1974, The origin and distribution of methane in marine sediments, *in* Kaplan, I. R., ed., *Natural Gases in Marine Sediments:* New York, Plenum, p. 99–139.

Demaster, D. J., 1979, The marine budgets of silica and $^{32}$Si: Unpublished Ph.D. dissertation, Yale Univ., 308 p.

Edmond, J. M., 1974, On the dissolution of carbonate and silicate in the deep ocean: *Deep-Sea Res.,* v. 21, p. 455–480.

Fanning, K. A., and Pilson, M. E. Q., 1974, The diffusion of dissolved silica out of deep sea sediments: *J. Geophys. Res.,* v. 79, p. 1293–1297.

Froelich, P. N., Klinkhammer, G. P., Bender, M. L., Luedtke, Heath, G. R., Cullen D., Dauphin, P., Hammond, D., Hartman, B., and Maynard, V., 1979, Early oxidation of organic matter in pelagic sediments of the eastern equatorial Atlantic—suboxic diagenesis: *Geochim. Cosmochim. Acta,* v. 43, p. 1075–1090.

Goldberg, E. D., and Koide, M., 1962, Geochronological studies of deep-sea sediments by the Io/Th method: *Geochim. Cosmochim. Acta,* v. 26, p. 417–450.

Goldhaber, M. B., and Kaplan, I. R., 1975, Controls and consequences of sulfate reduction rates in recent marine sediments: *Soil Sci.,* v. 119, p. 42–55.

Guinasso, N. L., and Schink, D. R., 1975, Quantitative estimate of biological mixing rates in abyssal sediments: *J. Geophys. Res.,* v. 80, p. 3032–3043.

Heath, G. R., Moore, T. C., and Dauphin, J. P., 1977, Organic carbon in deep-sea sediments, *in* Andersen, N. R., and Malahoff, A., eds., *The Fate of Fossil Fuel $CO_2$ in the Oceans:* New York, Plenum, p. 605–625.

Honjo, S., and Erez, J., 1978, Dissolution rates of calcium carbonate in the deep ocean; an *in situ* experiment in the North Atlantic Ocean: *Earth Planet. Sci. Lett.,* v. 40, p. 287–300.

Hurd, D. C., 1973, Interactions of biogenic opal, sediment, and seawater in the central equatorial Pacific: *Geochim. Cosmochim. Acta,* v. 37, p. 2257–2282.

Johnson, T. C., 1975, The dissolution of siliceous microfossils in deep-sea sediments: Unpublished Ph.D. dissertation, Univ. California, San Diego, Calif., 163 p.

Keir, R. S., 1980, The dissolution kinetics of biogenic calcium carbonates in seawater: *Geochim, Cosmochim, Acta,* v. 44, p. 241–252.

Lawson, D. S., Hurd, D. C., and Pankratz, H. S., 1978, Silica dissolution rates of decomposing phytoplankton assemblages at various temperatures: *Amer. J. Sci.,* v. 278, p. 1373–1393.

Li, Y. H., and Gregory, S., 1974, Diffusion of ions in seawater and in deep-sea sediments: *Geochim. Cosmochim. Acta,* v. 38, p. 703–714.

——, Bischoff, J., and Mathieu, G., 1969a, The migration of manganese in the Arctic basic sediments: *Earth Planet. Sci. Lett.,* v. 7, p. 265–270.

——, Takahashi, T., and Broecker, W. S., 1969b, Degree of saturation of $CaCO_3$ in the oceans: *J. Geophys. Res.,* v. 74, p. 5507–5525.

Manheim, F. T., 1976, Interstitial waters of marine sediments, *in* Riley, J. P., and Skirrow, G., eds., *Chemical Oceanography, Vol. 6:* New York, Academic Press, p. 115–186.

Martens, C. S., and Berner, R. A., 1977, Interstitial water chemistry of anoxic Long Island Sound sediments, 1. Dissolved gases: *Limnol. and Oceanog.,* v. 22, p. 10–25.

Menzel, D. W., 1974, Primary productivity, dissolved and particulate organic matter, and the sites of oxidation of organic matter, *in* Goldberg, E. D., ed., *The Sea, Vol. 5:* New York, Wiley, p. 659–678.

Milliman, J. D., 1977, Dissolution of calcium carbonate in the Sargasso Sea (northwest Atlantic) *in* Andersen, N. R., and Malahoff, A., eds., *The Fate of Fossil Fuel $CO_2$ in the Oceans:* New York, Plenum, p. 641–654.

Morse, J. W., 1977, The carbonate chemistry of North Atlantic Ocean deep-sea sediment porewater, *in* Andersen, N. R., and Malahoff, A., eds., *The Fate of Fossil Fuel $CO_2$ in the Oceans:* New York, Plenum, p. 323–344.

——, 1978, Dissolution kinetics of calcium carbonate in seawater, VI. The near equilibrium dissolution kinetics of calcium carbonate-rich deep-sea sediments: *Amer. J. Sci.,* v. 278, p. 344–353.

——, and Berner, R. A., 1972, Dissolution kinetics of calcium carbonate in seawater, II. A kinetic origin for the lysocline: *Amer. J. Sci.,* v. 272, p. 840–851.

Müller, P., and Suess, E., 1979, Productivity, sedimentation rate, and sedimentary organic matter in the oceans, I. Organic carbon preservation: *Deep-Sea Res.,* v. 26A, p. 1347–1362.

Murray, J., and Chumley, J., 1924, The deep sea deposits of the Atlantic Ocean: *Trans. Roy. Soc. Edinburgh,* v. 54, p. 1–252.

Nozaki, Y., Cochran, J. K., and Turekian, K. K., 1979, Radiocarbon and [210]Pb distribution in submersible-taken deep-sea cores from Project Famous: *Earth Planet. Sci. Lett.,* v. 34, p. 167–173.

Peng, T.-H., Broecker, W. S., Kipphut, G., and Shackleton, N., 1977, Benthic mixing in deep sea cores as determined by [14]C dating and its implications regarding climate stratigraphy and the fate of fossil fuel $CO_2$, *in* Andersen, N. R., and Malahoff, A., eds., *The Fate of Fossil Fuel $CO_2$ in the Oceans:* New York, Plenum, p. 355–374.

Peterson, M. N. A., 1966, Calcite—rates of dissolution in a vertical profile in the central Pacific: *Science,* v. 154, p. 1542–1544.

Schink, D. R., Fanning, K. A., and Pilson, M. E. Q., 1974, Dissolved silica in the upper porewaters of the Atlantic Ocean floor: *J. Geophys. Res.,* v. 79, p. 2243–2250.

——, Guinasso, N. L., and Fanning, K. A., 1975, Processes affecting the concentration of silica at the sediment-water interface of the Atlantic Ocean: *J. Geophys. Res.,* v. 80, p. 3013–3031.

——, and Guinasso, N. L., 1977, Modelling the influence of bioturbation and other processes on calcium carbonate dissolution at the sea floor, *in* Andersen, N. R., and Malahoff, A., eds., *The Fate of Fossil Fuel $CO_2$ in the Oceans:* New York, Plenum, p. 375–400.

——, and Guinasso, N. L., 1978, Possible role of aragonite in separating the calcite lysocline from the depth of calcite saturation: *EOS,* v. 59, p. 411.

Sholkovitz, E., 1973, Interstitial water chemistry of the Santa Barbara Basin sediments: *Geochim. Cosmochim. Acta,* v. 37, p. 2043–2073.

Sjöberg, E. L., 1978, Kinetics and mechanism of calcite dissolution in aqueous solutions at low temperatures: *Stockholm Contrib. Geol.,* v. 32, p. 1–92.

Takahashi, T., 1975, Carbonate chemistry of seawater and the calcite compensation depth in the oceans: *Cushman Found. Foram. Res. Spec. Publ. 13* (W. V. Sliter, A. W. H. Be, and W. H. Berger, eds.), p. 11–26.

——, and Broecker, W. S. 1977, Mechanism for calcite dissolution on the sea floor, *in* Andersen, N. R., and Malahoff, A., eds., *The Fate of Fossil Fuel $CO_2$ in the Oceans:* New York, Plenum, p. 455–478.

Toth, D. J., and Lerman, A., 1977, Organic matter reactivity and sedimentation rates in the ocean: *Amer. J. Sci.,* v. 277, p. 465–485.

Wollast, R., 1974, The silica problem, *in* Goldberg, E. D., ed., *The Sea, Vol. 5:* New York, Wiley, p. 359–394.

# II

## BIOLOGICAL
## ENVIRONMENT
## OF THE
## DEEP SEA

Kenneth H. Nealson
Scripps Institution of Oceanography
University of California
La Jolla, California 92093

# 8

# BACTERIAL ECOLOGY
# OF THE DEEP SEA

# ABSTRACT

Low temperature, high pressure, and low nutrient levels combine to make the deep sea a potentially hostile environment for bacteria that are not adapted to these conditions. Using special collection and incubation techniques developed in recent years, it has been possible to investigate the bacteria of the deep sea, and it is now clear that bacteria have adapted to conditions of the deep sea (that is, they are psychrophilic and barophilic). Both free-living sediment bacteria and those associated with higher organisms have been studied, although there is little evidence that demonstrates either large bacterial populations or activities in seawater or sediments. It may be that the association with particulate matter (either living or dead) or with higher organisms is a characteristic of deep-sea bacterial ecosystems. A model, based on analogy to shallow-water systems, is presented for the purpose of discussion of the possible ecological roles of bacteria in the deep sea and marine sediments.

# INTRODUCTION

In 1968 the research submarine *Alvin* accidentally filled with seawater and sank to a depth of 1540 m in the North Atlantic, where it remained for about one year. Upon recovery, it was observed that a boxed lunch consisting of bouillon, a bologna sandwich, and apples was remarkably well preserved. This famous inadvertent "bologna sandwich experiment" and its successors initiated a fascinating and controversial chapter in modern marine microbiology. The publication of the observations (Jannasch *et al.,* 1970) led to a number of experiments to substantiate the findings and to reexamine the activities of deep-sea microbes. Why was the lunch not degraded? Are there no functional bacteria in the depths? Do marine bacteria participate in the cycling of carbon and other elements in the deep sea? Before answering these questions, the properties of bacteria as they relate to the marine environment, and the properties of the deep sea as it constrains the activities of the organisms in it, are discussed.

# BACTERIAL PROPERTIES

The properties considered below are general bacterial properties. When the information is available, specific reference has been made to marine bacteria. However, with the exception of an obligate requirement for the sodium ion (MacLeod, 1968), there are probably few fundamental differences between marine and nonmarine bacteria. Rather, the properties discussed will serve to point out those bacterial features that may be important in the microbial ecology of the deep sea.

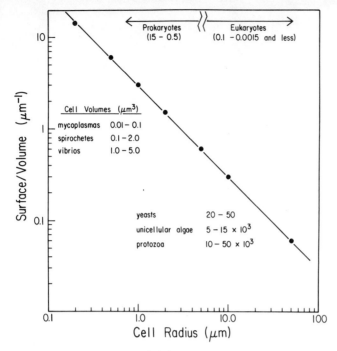

Fig. 8-1.  Surface–volume (S/V) ratios as a function of cell diameter, for spherical cells. Although the division between prokaryotes and eukaryotes is not exact, several examples are shown in the figure. The numbers under the arrows refer to the range of S to V ratios of average prokaryotes and eukaryotes.

## Size, Distribution, and Biomass

Perhaps the most overlooked property of bacteria is the most obvious one: their small size and consequent large surface to volume (S/V) ratio. The ability of an organism to interact biochemically with its environment (its biochemical potential) is at least in part dependent on the surface area of the organism, and thus on its S/V ratio. Larger organisms have specific organs and feeding habits to cope with their lowered S/V ratios. Typical bacterial cells are of the order of 1 μm in length, and usually less in diameter, while even small eukaryotes are usually 10 or more times larger. With marine bacteria, the case can be even more extreme (Azam and Hodson, 1977; Johnson and Sieburth, 1978; Watson *et al.*, 1977); the cells are often of the order of 0.2 μm in diameter, and spherical in shape. The relationship between cell diameter and S/V ratios and some representative cell volumes are shown in Fig. 8-1. A caution must be added with such considerations: Among bacteria, and even within a given species, cell size can vary considerably. For a healthy, rapidly growing *E. coli* cell, it is common to have sizes 2-3 times larger than the same species under conditions of nutrient limitation or starvation. Such changes in size may be a response to the limitation of nutrients, and the result is undoubtedly in part an increase in the efficiency of nutrient gathering. Thus, at very small sizes, the relationship between biochemical potential and cell size will not be as shown in Fig. 8-1; rather, it will be some variation of this theme that takes into consideration the physio-

## TABLE 8-1. Surface Areas of Marine Biota

*Surface Area* (SA) = 3/r [*Cell Volume* (CV)]

For bacteria:

$r = 0.2 \times 10^{-4}$ cm

CV = $10^{-4}$ cm³ (total volume per liter of seawater)[a]

SA = $3/0.2 \times 10^{-4} \times 10^{-4}$ = 15 cm²/l

For all other organisms combined:

$r = 1 \times 10^{-3}$ cm

CV = $5 \times 10^{-4}$ ml[a]

$$SA = \frac{3}{1 \times 10^{-3}} \times (5 \times 10^{-4}) = 1.5 \text{ cm}^2/l$$

[a]CV based on estimates of biomass organic carbon of bacteria and other organisms, as stated by Azam (1980).

logical changes that accompany nutrient limitation. However, for a given amount of biomass, the larger bacteria have 10-100 times greater S/V ratios than the eukaryotic organisms; this "biochemical potential," coupled with the ability to grow exponentially with doubling times of less than one hour, allow bacteria the kind of potential for which they are famous.

Table 8-1 demonstrates that the total bacterial surface area in seawater is about a factor of 10 greater than that of all other organisms combined. Thus, if reactive surfaces are considered, the bacterial component of the biomass cannot be ignored, even in seawater, where it comprises only a small portion of the total biomass.

Although the exact proportion is still not well known, the bacteria are estimated to make up 10-20% of the total oceanic biomass. Bacterial biomass was estimated by both direct count and ATP methods to be between 1.5 and 8 mg carbon per cubic meter (mg C/m³) (Siebuth, 1976). Total biomass is between 10 and 100 mg C/m³, so that the

## TABLE 8-2. Detrital Food Chain Dynamics[a]

|  | Bacteria | Protozoa | Invertebrates |
|---|---|---|---|
| Metabolic rate (kcal/m²/day) | 5.0 | 0.5 | 0.05 |
| Specific metabolic rate (kcal/g/day) | 2.0 | 0.1 | 0.5 |
| Biomass of each individual (g) | $5 \times 10^{-13}$ | $10^{-7}$ | $10^{-3}$ |
| Total biomass (g/m³) | 2.5 | 5.0 | 0.5 |
| Total number of individuals (organisms/m³) | $5 \times 10^{12}$ | $5 \times 10^{7}$ | $5 \times 10^{2}$ |

[a]These data are from Fenchel and Jorgensen, 1977, and are taken from an estuarine ecosystem.

bacteria represent a small part of the total, but can have great potential biochemically.

In the sediments, bacterial populations, although difficult to quantitate, are often orders of magnitude higher than in the open ocean (ZoBell, 1946). Very little detailed work has been done in deep-sea sediments, but estimates of bacterial biomass and activities as compared to other organisms in shallow marine sedimentary environments are shown in Table 8-2 (Fenchel and Jorgensen, 1977). These data support the notion that the bacteria have specific metabolic or biochemical potential much higher than that of other organisms, and that they may represent, at least potentially, the most biochemically important component of the biomass.

## Transport Mechanisms and Nutrition

The nearly universal rigid cell wall necessary for the structural integrity of bacteria precludes phagocytotic or pinocytotic methods of feeding common in eukaryotes, and restricts the bacteria to the use of soluble nutrients. Bacteria possess a variety of extracellular enzymes for the production of such nutrients by the solubilization of polymers (Table 8-3). Thus, the bacteria are distinct from animals and protists in their ability

TABLE 8-3.  Transport Systems and Extracellular Enzymes of Bacteria

| *Transport Systems* | | | |
|---|---|---|---|
| *System* | *Materials Transported* | *Specific Protein Required* | *Energy Required*[a] |
| Diffusion: | | | |
| Passive | Inorganic ions | – | – |
| Facilitated | Rare mechanism in prokaryotes (glycerol transport) | + | – |
| Active transport: | | | |
| Binding protein systems | Carbohydrates, amino acids, peptides, bases | + | PMF |
| Group translocation | Carbohydrates | + | PEP |

| *Exoenzymes: Their Substrates and Products*[b] | | |
|---|---|---|
| *Enzyme Group* | *Substrate Acted Upon* | *Products* |
| Amylase | Starch | Glucose, maltose, oligoglycosides |
| Cellulase | Cellulose | Glucose, cellobiose |
| Pectinase | Pectin | Galacturonic acid |
| Lysozyme | Peptidoglycan | |
| Chitinase | Chitin | Chitobiose, N-acetylglucosamine |
| Peptidases (proteases) | Proteins, peptides | Amino acids, peptides |
| Deoxyribonucleases | DNA | Deoxyribonucleosides |
| Ribonucleases | RNA | Ribonucleosides |
| Lipase | Lipids | Glycerols, fatty acids |

[a]PMF = proton motive force; PEP = phosphoenolpyruvate.

[b]From Stanier *et al.,* 1976.

to interact with and scavenge low levels of soluble nutrients. The eukaryotes, which have developed many behavioral mechanisms to increase their efficiency of feeding (browsing, filtering, chemotaxis, etc.), have little access to the soluble organic carbon other than through the bacteria that can utilize it and convert it into particulate carbon (bacterial biomass). (See Wright and Stephens, Chap. 13, this volume.)

It may be at the level of transport that a major difference exists between marine and terrestrial bacteria. Marine bacteria have a definite requirement for sodium ion (MacLeod, 1968; Reichelt and Baumann, 1974), and membrane integrity and nutrient transport probably require high sodium levels. It seems likely that the affinities, specificities, and regulatory properties of the transport systems will be of great importance in determining the activities and ecological successes of the microbes. Specific knowledge of transport systems of deep-sea bacteria is lacking, but one open-ocean isolate capable of growth on low nutrients has been found to possess a high affinity transport system that may account, at least in part, for its low nutrient character (Hodson et al., 1979).

## Bacterial Metabolism

Two uniquely prokaryotic metabolic properties, anaerobic respiration and chemolithotrophy, may also be important in marine ecosystems. The property of growth in the absence of molecular oxygen is probably of widespread importance in marine sediments, while chemolithotrophy may be restricted to more localized environments. Respiration using alternate electron acceptors (Table 8-4) permits carbon utilization and mineral solubilization to occur in sediments, even in the absence of molecular oxygen. Depending on the electron acceptor involved, the effects can be substantial; nitrate is removed, manganese is reduced (and mobilized by solubilization), iron is mobilized, sulfide is produced (by sulfate reduction), or methane is produced. Sulfate reduction is believed to be a major process in anaerobic marine sediments because of the great abundance of sulfate in seawater, but in many cases sediments remain too aerobic to allow sulfate reduction to occur.

TABLE 8-4.  Bacterial Redox Reactions Used to Oxidize Organic Carbon[a]

| Process | Electron Acceptor | Reaction Products | Eh Range (mv) |
|---------|-------------------|-------------------|---------------|
| Aerobic respiration | $O_2$ | $CO_2$, $H_2O$, biomass | +500 to +800 |
| Nitrate reduction | $NO_3^-$ | $NO_2^-$ | +300 to −500 |
| Metal reduction | $MnO_2$ | $Mn^{++}$ | +100 to +400 |
|  | $Fe_2O_3$ | $Fe^{++}$ |  |
| Sulfate reduction | $SO_4^=$ | $SH^-$, acetate | −100 to −400 |
| Methanogenesis | $CO_2$ | $CH_4$ | −800 |
| Fermentation | Organic carbon | $H_2$, $CO_2$, organic acids | [b] |

[a] For details of these reactions, see text.
[b] Fermentation begins at a variety of Eh values, and for different organisms and different substrates these may be quite different.

The contribution of anaerobic bacteria to marine anaerobic ecosystems is only now beginning to be understood. Integrated anaerobic ecosystems in the stomachs of ruminants have been described by Hungate (1963), and postulated for the Black Sea (Jannasch et al., 1974); comparable ones probably exist in marine sediments (Fenchel and Jorgensen, 1977). Such anaerobic respiration should be taken into consideration in making metabolic budgets of the environment. For instance, total oxygen uptake is often used as an indicator of total activity; if anaerobic respiration is extensive, the oxygen consumption measurements will underestimate total activity (Fenchel and Jorgensen, 1977).

Some of the relevant redox reactions that bacteria can catalyze are shown in Table 8-4. While some of these reactions occur spontaneously at Eh and pH conditions of the ocean, others do not. For example, sulfate reduction under the conditions of the deep sea is exclusively a microbial process. The reactions are arranged in order of decreasing electron potentials, so that the sequence of reactions could represent a sequence of reactions found in a sediment column as it becomes more anaerobic. Probably all the reactions are due, at least to some extent, to microbial activities. After oxygen is depleted, nitrate is consumed. Because nitrate is not abundant in seawater or oceanic porewaters, its importance has always been considered to be minimal. However, it should be stressed that flux measurements are not generally available, and it is possible that nitrate represents a dynamic component of the system that, while being maintained at low concentration, is nevertheless very important in the cycling of organic matter. Manganese and iron, on the other hand, are abundant metals in marine sediments, and under appropriate conditions might be significant electron acceptors. This possibility is not well documented, but the redox potentials of these metals, their distribution in marine sediments (Froelich et al., 1979), and the existence of organisms that can reduce them suggest that this may be the case. Sulfate reduction is the dominant reaction in anaerobic basins with good organic input and is well documented for at least two reasons in marine anaerobic environments. First, the concentration of sulfate in seawater is very high, on the order of several grams per liter, and thus sulfate supplies sufficient oxidant to dispose of large amounts of organic material. Furthermore, the product of sulfate reduction, sulfide, is toxic. The accumulation of sulfide leads to the near-exclusion of other organisms, producing an environment dominated by the few organisms that are resistant to sulfide. In metal-rich sediments, insoluble metal sulfides (primarily iron sulfides) are formed, with characteristic marine black muds resulting. Thus, the sulfate reducers (in the marine environment, the genus *Desulfovibrio*) must be regarded as very significant organisms both in the turnover of organic matter in anaerobic sediments, and in the geochemical and biological effects they exert on the sedimentary environments. It should be kept in mind that in general the oceans are aerobic, and sulfate reduction is confined to sediments with sufficient organic input. There are, however, highly productive environments where sulfate reduction occurs in the water column, and the sulfur cycle may be central to the turnover of organic matter as is postulated for the Black Sea (Jannasch et al., 1974).

Another unique prokaryotic metabolic property of potential importance to transformations in the marine environment is chemolithotrophy, the ability to utilize inorganic compounds as sources of energy. In the dark environment of the deep sea, such pathways may represent a metabolic alternative to heterotrophy. Some of the reactions that are well characterized as sources of energy for bacteria are shown in Table 8-5. In the deep ocean, where the heterotrophic nutrients are extremely dilute and probably growth-limiting, it is tempting to speculate that autotrophic prokaryotes take advantage of inorganic substrates. While these speculations are useful and entertaining, a note of caution must be added. The reduced inorganic ions are produced or available as a result of metabolic activity that is ultimately dependent on the input of organic carbon. Thus, such autotrophic activities may be confined to environments with substantial inputs of organic carbon. They may also be important in localized areas such as spreading centers, where reduced inorganics are introduced in hydrothermal fluids (see below). For instance, while the nitrifying bacteria are good candidates for such a role, growing on ammonia or nitrate and requiring oxygen for the production of nitrate, the concentrations of ammonia and nitrite in the deep ocean are usually quite low. Again, what is lacking in these studies are good indications of magnitudes of the fluxes involved. It seems likely that the nitrifiers play a major role in the oceanic nitrogen cycle, replacing the nitrate removed by the phytoplankton in the euphotic zone. Studies to determine the quantitative contribution of the nitrifiers to nitrate regeneration and to oceanic production in general have been hampered by the difficulty of growing these organisms, and the lack of *in situ* flux measurements.

Sulfur oxidizers may well be important in those instances where a ready supply of sulfide or thiosulfate is available. A striking example was recently reported in the Galapagos thermal vent area, where bacteria growing on hydrogen sulfide emanating from the vents apparently form the basis of a well-developed community (mollusks and other invertebrates) deep in the Pacific (Ballard, 1977; Corliss et al., 1979). That these bacteria are at the base of the food web is consistent with the enrichment of unusually

TABLE 8-5. Some Energy Yielding Reactions of Lithotrophic Bacteria[a]

Hydrogen oxidizing bacteria:

$$H_2 + 0.5 \, O_2 \longrightarrow H_2O$$

$$5 \, H_2 + 2 \, NO_3^- + 2H^+ \longrightarrow N_2 + 6 \, H_2O$$

Nitrifying bacteria:

$$NH_4 + 1.5 \, O_2 \longrightarrow NO_2^- + 2H^= + H_2O$$

$$NO_2^- + 0.5 \, O_2 \longrightarrow NO_3^-$$

Sulfur oxidizing bacteria:

$$S^= + O_2 \longrightarrow SO_4^=$$

$$S_2O_3 + 2O_2 + H_2O \longrightarrow 2 \, SO_4^= + 2 \, H^+$$

$$S^\circ + 1.5 \, O_2 + H_2O \longrightarrow 2 \, SO_4^= + 2 \, H^+$$

$$5 \, S^\circ + 6 \, NO_3^- + 2 \, H_2O \longrightarrow 5 \, SO_4^= + 3 \, N_2 + 4 \, H^+$$

Iron oxidizing bacteria:

$$2 \, Fe^{++} + 0.5 \, O_2 + 2 \, H^+ \longrightarrow 2 \, Fe^{+++} + H_2O$$

[a]The energy yielding reactions for these groups are well documented (see Schlegel, 1975). Another reported energy yielding reaction is the oxidation of manganese. These reports have been made by several workers and disputed by others.

BACTERIAL ECOLOGY OF THE DEEP SEA

high numbers of sulfur oxidizers.[1] Other areas where sulfur oxidizers might prevail are estuaries and anaerobic basins or seas, such as the Black Sea, where reduced sulfur is readily available due to the metabolic activities of sulfur reducers. Again, it should be stressed that these are local environments and that the contribution of these chemo-autotrophs on a global scale is not known.

Whether hydrogen and single-carbon compound oxidizing bacteria exist in marine systems is not yet known. Significant amounts of methane may be regularly put into the ocean, and it is of interest to know if methane oxidizing bacteria can take advantage of this substituent.

Finally, it has been proposed that manganese nodules are the result of the accumulation of manganese oxides formed by lithotrophic metabolism of manganese-oxidizing bacteria (Ehrlich, 1975). Quantitative evidence needed to prove this hypothesis is lacking, and the determination of *in situ* rates of manganese oxidation by bacteria are necessary to understand the role of lithotrophic bacteria in the formation of deep-sea metal deposits.

## Commensal Associations

Another prokaryotic feature to be considered is the tendency to interact with and attach to surfaces and to associate with eukaryotes. Such bacterial interaction with surfaces has been commented upon extensively (Corpe, 1970; Marshall, 1976; ZoBell, 1943a). Marine bacteria are often found associated with nutrient-rich particulate matter such as fecal pellets. In fact, it has long been known that marine microbial activities are enhanced by the addition of solid surfaces to the culture medium (ZoBell, 1943b). Particulate matter at specific nutrient levels enhances the ability of bacteria to utilize low concentrations of organic matter (Jannasch and Pritchard, 1972). Nitrate reduction, a process that is inhibited by molecular oxygen, was shown to occur in an aerated culture after the addition of chitin particles (Jannasch, 1978). By attaching to the particles, bacteria apparently form microenvironments with reduced $pO_2$ in which denitrification occurs. The understanding of such processes is obviously of great importance for accurately assessing the contribution of marine bacteria to chemical transformations of the sea.

Attachment to or association with living organisms represent other important niches for marine bacteria. Associated bacteria are required for the very existence, or at least for the well-being, of animals from terrestrial and lacustrine environments, and the same is probably true of deep-sea organisms. However, with the exception of *Photobacterium* species that populate light organs of luminous fish (Nealson and Hastings, 1979), little is known of marine bacterial associations in the deep sea. Specific enteric bacteria may aid in digestion and homeostasis, specifically chitin digestion, as proposed by Spencer (1961).

The relative quantitative contributions of marine bacteria in the free-living and associated niches is not known, and may differ from one part of the ocean to another, but the associated fraction must be considered in any formulation of the bacterial contribution to the chemistry of the sea. Bacterial/eukaryotic associations may be transient or permanent (Sieburth, 1976). The extent and functional importance of these associa-

[1] Jannasch, personal communication, 1979.

tions are not understood and represent one of the most challenging and important areas of future research. Probably the only thing that is certain is that any consideration of oceanic microbial ecology that does not include both bacterial and eukaryotic components is both naive and incomplete (Sieburth, 1976).

## LIMITS TO LIFE IN THE OCEAN

What limits or restricts the distribution or activities of deep-sea microbes? For bacteria, three properties of the deep sea loom as important limiting factors: temperature, pressure, and nutrients. A typical profile from the North Pacific is shown in Fig. 8-2. Within the

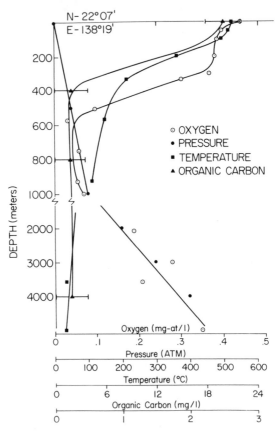

Fig. 8-2. A profile of variables with depth in the ocean. The temperature and oxygen data were gathered by the author during a cruise aboard the *R/V Kallisto* (January 1979); the pressure data were calculated. The nutrients data are averages from several reports and serve only to illustrate the fact that nutrients decrease rapidly with depth to a few hundred meters, then remain constant to the bottom. The data are of total organic carbon, which should contain POC:DOC of approximately 10:1.

BACTERIAL ECOLOGY OF THE DEEP SEA

first 1000 m, nutrient concentration and temperature change dramatically and then remain constant with depth, while pressure continues to increase as a linear function of depth. The amounts and patterns of nutrients and the location and shape of the thermocline differ with geographic location and season, although the overall pattern remains the same. Neither total dissolved organic carbon (DOC) nor particulate organic carbon (POC) give any indication of the qualitative changes that may be occurring with depth, and substantial changes occur due to recycling as the organic matter sinks (Sieburth, 1976). The organic matter that reaches the sea floor is probably quite different chemically from that which began the descent. The mechanisms that marine bacteria have evolved to cope with these factors are discussed below.

## Temperature

Although the mechanisms by which temperature inhibits growth are not understood, adaptation to specific temperature regimes apparently involves many alterations of the cell (Morita, 1975, 1976). Temperature is thus an important ecological factor that determines in part the populations of organisms that predominate in a given location at certain seasons (Kaneko and Colwell, 1973; Ruby and Nealson, 1977; Sieburth, 1967). For example, the population of luminous bacteria in coastal waters near San Diego changed seasonally over a two-year period and the species fluctuated almost exactly with water temperature (Ruby and Nealson, 1977). Growth rates as a function of temperature for a variety of marine bacteria are shown in Fig. 8-3. The results of such experiments

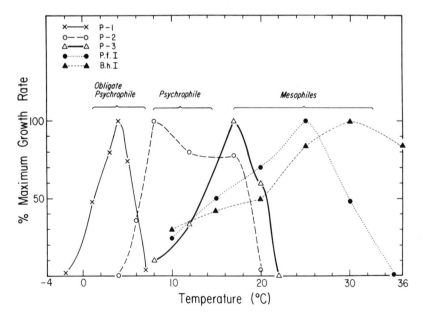

Fig. 8-3.    Growth as a function of temperature for a variety of marine bacteria from the author's laboratory. The strains labeled P are various psychrophiles isolated during a cruise to the Antarctic; P.f. is a laboratory strain of *Photobacterium fischeri*, and B.h. is a laboratory strain of *Beneckea (Vibrio) harveyi*.

support the idea that neither high nor low temperature is an unsurmountable problem for bacteria. However, adaptation to extreme temperatures is often correlated with a restricted temperature range of growth. That is, bacteria are not only tolerant of low temperature (psychrophilic), but are obligate psychrophiles, requiring low temperature for growth (Morita, 1976).

Since many other variables also fluctuate, especially in shallow, near-shore areas, seasonal changes in microbial populations cannot be attributed to temperature alone. Temperatures in the deep sea are much more constant, varying usually from 1–5°C, and nonpsychrophilic bacteria that sink to these low-temperature environments are likely to be inhibited. Indigenous deep-sea bacteria are almost certain to be psychrophilic, and are probably obligate psychrophiles. In fact, deep-sea psychrophiles have been known for many years (Morita, 1975). Some obligate psychrophiles have growth rates at low temperature comparable to surface bacteria (mesophiles) grown at moderate temperatures. Psychrophiles from Arctic and Antarctic waters display active metabolic rates under naturally prevailing low-temperature conditions (Holm-Hansen *et al.,* 1977).

## Pressure

The mechanisms by which high pressure limits growth are even less well understood than those of temperature. Although obligate barophiles (pressure-requiring bacteria) were first reported in 1957 (ZoBell and Morita), pure cultures of such forms have been achieved only recently (Yayanos *et al.,* 1979). This recent achievement is in large part due to technological and methodological developments. Whereas temperature effects can be studied easily and relatively inexpensively, pressure studies involve costly and ponderous equipment both for sample retrieval and for laboratory manipulation without decompression. Examples of pressure apparatus for the study of barophilic bacteria are shown in Fig. 8-4. While pressure chambers for the culture and laboratory study of bacteria have long been available (ZoBell, 1968), this equipment is only now being refined for field use. As a result, deep-sea bacteria have usually been collected and cultured under 1-atm pressure, without the advantage of pressurized chambers. Even so, bacteria that exhibit some degree of barotolerance are well known (Brauer, 1972; Marquis, 1976; Morita, 1976; ZoBell and Kim, 1972). In fact, it is probably fair to say that "barophilic bacteria are part of the ordinary marine psychrophilic flora" (Berger, 1974). The understanding of the extent and activities of obligate barophiles is extremely important for the understanding of the role of deep-sea bacteria in the chemistry of the oceans, and until it is possible to retrieve samples without decompression, or to do meaningful *in situ* experiments, it will not be possible to understand these problems. For example, Schwarz *et al.* (1976) reported a mixed population that grew as well at 700 atm as it did at 1 atm, but when pure cultures were obtained they did not grow at 700 atm. These problems may relate to sensitivity to decompression. It is possible that many barophiles are killed upon retrieval, as hypothesized by ZoBell (1968):

> Although many bacteria from the deep sea survived, this observation fails to prove that some bacteria, probably the most sensitive ones, were not destroyed by decom-

Fig. 8-4.    The apparatus of Jannasch, Wirsen, and Taylor (1976) for the retrieval of undecompressed water samples. Photo courtesy of Dr. H. Jannasch.

pression. Answering this question may require the examination of deep-sea bacteria at *in situ* pressure without subjecting them to decompression.[2]

The interpretation of the effects of pressure on deep-sea microbes is thus severely hampered by the fact that virtually all of the laboratory studies of the effects of pressure have been done with nonbarophilic organisms. Extensive studies (see reviews by Marquis, 1976; Morita, 1976; ZoBell, 1968; ZoBell and Kim, 1972) of the effects of pressure on the growth, viability, and activities of a variety of marine bacteria have yielded varying results. In general, however, at pressures of 100 atm or less growth has not been significantly inhibited, whereas above about 100 atm, pressure effects begin to be seen. At approximately 500 atm and above, very few bacteria survive or grow. In addition, the

[2]ZoBell, 1968, p. 87.

effects of pressure are strongly influenced by other environmental variables such as temperature, salinity, pH, and so on. For instance, while some bacteria are tolerant to 500 atm or greater at room temperature, very few are tolerant to the same pressures at 4°C.

However, barophilic bacteria with growth optima at high pressure do exist (Yayanos *et al.*, 1979; Fig. 8-5). These barophilic bacteria will need to be compared to nonbarophilic and barotolerant isolates for clues to the mechanisms of adaptation to high pressures. Furthermore, although pure cultures of the barophilic organisms have not yet been carefully studied, their existence confirms what many workers have inferred from knowledge of bacterial-animal associations. For example, midwater fish (Opisthroproctids, Macrourids, and Ceratioids) that harbor luminous symbiotic bacteria are common to depths of 1300 m and deeper (Herring and Morin, 1979; Ruby and Morin, 1978). These symbiotic bacteria are apparently metabolizing and growing well at pressures and temperatures inhibitory to normal surface bacteria. Commensal bacteria associated with the gut tracts of deep-sea animals must be thriving as well. Thus the concept of barophilic organisms as first suggested by ZoBell and Johnson (1949) may well be correct, but technological problems have limited its study.

## Nutrients

The availability of organic nutrients, both as dissolved organic carbon (DOC) and particulate organic carbon (POC), also determines distributions of heterotrophic bacteria. The ratios of dissolved to particulate to living organic carbon are estimated to be 100:10:2 (Parsons, 1963). The oceanic DOC probably represents the largest reserve of total carbon in the entire hydrosphere (Carlucci and Williams, 1978). From 10-30% of the marine primary production may end up as DOC (Riley, 1970), an input that is the major source of nutrients for the heterotrophic bacterioplankton. The low molecular weight organic

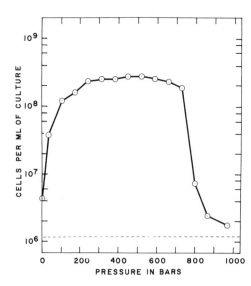

Fig. 8-5. Growth as a function of pressure for a barophilic isolate. Courtesy of Dr. A. A. Yayanos (Yayanos *et al.* 1979, *Science*, v. 205, p. 808–810, copyright 1979 by Amer. Assoc. Adv. Sci.).

fraction of the DOC probably constitutes the dynamic component of the carbon flow in shallow waters; that is, it is this component that is produced, excreted, and consumed by other organisms (Fenchel and Jorgensen, 1977). The DOC changes qualitatively with depth (Sieburth, 1976), but the nature of these changes, and whether or not smaller molecular weight carbon compounds are available to bacteria in the deeper waters, are open questions. As the POC sinks, it is solubilized and cycled (Sieburth, 1976), but the details of release with depth are not well understood. Although methods for accurate and rapid assessment of low molecular weight compounds are not available, Williams (1971) has estimated that only 10% of the DOC is in the form of "available" nutrients such as amino acids, sugars, etc. Marine bacteria are apparently adapted to low levels of nutrients, and high levels can inhibit many marine forms (Carlucci and Pramer, 1957; ZoBell, 1941), yet heterotrophic growth is apparently limited by low (micromolar or less) concentrations of nutrients. Bacteria capable of scavenging the low levels of nutrients for growth on unenriched seawater have recently been isolated (Carlucci and Shimp, 1974). These bacteria grow very slowly, doubling in 34 h; they may be similar to those grown on unenriched seawater in chemostat cultures (Jannasch, 1967, 1968, 1970).

What adaptations are required to grow in the low-nutrient regime of the deep sea? Transport systems of some marine bacteria may have a higher substrate affinity and lower capacity than those of other microbes (Hodson *et al.,* 1979). For one low-nutrient marine isolate studied, the transport system was found to be constitutive with a high degree of specificity, providing an everpresent system for the uptake of many nutrients at low levels. The regulation and characteristics of these transport systems may be key features in the ecology of the deep-sea heterotrophs.

Taken together, these studies suggest that one major role of marine bacteria is the conversion of DOC into utilizable POC as bacterial biomass. The bacterioplankton thus may represent an important link in the marine food web, providing access to the dissolved nutrients for the eukaryotes otherwise unable to utilize them.

To summarize, marine bacteria have adapted well to low temperature, high pressure, and low nutrients; in each case, the adaptations are rather extensive. However, while individual parameters can be dealt with singly, together they may impose substantial restrictions on nonindigenous organisms. Thus, the deep-sea bacteria can be viewed as a highly adapted population to an unusual environment.


## MICROBIAL ACTIVITIES IN THE DEEP SEA

The "lunch pail" observations of Jannasch *et al.* (1970) can be viewed now from the perspective developed here:

1. Surface bacteria subjected to high pressures and low temperatures to which they are not adapted should not be expected to degrade terrestrial organic matter with high efficiency.

2. Since most bacteria may be associated with sediments, the physical isolation of the lunch pail may have precluded contact of the indigenous bacteria of the deep sea with the lunch.

3. The bread, mayonnaise, bologna, and bouillon may have provided nutrients either too concentrated or too unusual to have been metabolized by deep-sea bacteria.

The first hypothesis has already been shown to be true (Jannasch *et al.*, 1976). When surface bacteria were either lowered to the deep sea or subjected to deep-sea conditions in the laboratory with a variety of nutrients, little or no growth could be detected. Furthermore, extensive experiments with a variety of radioactively labeled substrates show metabolic activity to be very low (Jannasch *et al.*, 1976; Jannasch and Wirsen, 1973; Wirsen and Jannasch, 1975, 1978).

The second hypothesis may be inadequate. Chambers containing a variety of nutrients suspended in solid substrates were placed into and above deep-sea sediments for long periods and then retrieved for analysis (Wirsen and Jannasch 1975, 1978). They observed that the samples were colonized but not degraded. The effects were similar to those measured with surface bacteria, namely that little activity was seen as compared to controls incubated at 1-atm pressure. Apparently microbes were abundant and capable of contact with the nutrients, but their metabolic rates were very slow. Thus physical isolation cannot alone have accounted for the lack of degradation of the *Alvin* lunch. When undecompressed samples of bottom water were brought from *Alvin's* surroundings to the surface, barophilic bacteria could not be isolated from the retrieved samples. Surface-water samples showed as much activity under simulated deep-sea conditions as did the deeper water samples. Regardless of the source of the samples, if the pressures were decreased to 1 atm, metabolism increased, suggesting that high pressures limit even the bacteria collected at depth. The observations suggest that bacteria abound in the deep sea in a nongrowing state. Many must be transported there as part of the sedimentary POC during sedimentation. In support of this view, it was noted that at 1-atm pressure within a few days of recovery, degradation of the *Alvin* lunch began, suggesting that the bacteria that had gone to the deep sea with the lunch had not been killed, but only severely inhibited.

An apparent paradox thus exists; on the one hand, barophilic, deep-sea tolerant bacteria have been reported, and on the other, samples of water from the deep sea show very little evidence of pressure-requiring metabolic rates. Further experiments of Jannasch and Wirsen (1973, 1977) and Yayanos *et al.* (1979) may offer an explanation. Distribution of bacteria in the deep sea may differ fundamentally from that in terrestrial or freshwater environments. The amount and quality of DOC reaching the sea floor may be too low to support the growth of heterotrophic bacteria that is typical of terrestrial environments. Barotolerant and barophilic bacteria may occur mainly in association with animals and protists. These bacteria, which may have enzymes such as chitinase or cellulase that their hosts lack, may take advantage of the behavioral adaptations of the hosts to find their way to the limited nutrients sinking to the deep sea. In the experiments of Jannasch and Wirsen, a variety of nutrients covered by membranes that excluded eukaryotes but not bacteria were placed directly into and above the sediments. The results indicate that bacterial colonization of solid surfaces occurs, but with little accompanying degradation. Alternatively, when the membranes were breached and the samples were fed on by eukaryotes, the effects were dramatic. Deep-sea animals, undoubtedly with associated bacteria, are well adapted for finding the POC near the ocean bottom, and it may be here that an important and active bacterial component resides.

To test this hypothesis, a baited trap to attract and retrieve deep-sea amphipods without decompression was designed (Yayanos, 1978; Fig. 8-6). Amphipods brought to the surface were kept alive for as long as nine days at *in situ* temperature and pressure. After they died, degradation under *in situ* conditions was observed. The amphipod carcasses were completely degraded after five months, and an abundant bacterial population was observed in the trap. A barophilic *Spirillum* species with high-pressure requirements for growth was isolated (Fig. 8-5). Its maximum growth rate is between 100 and 700 atm; outside these ranges growth is inhibited. Growth is extremely slow at 1 atm. This is the first report of a pure culture of an obligate barophile.

The notion that the distribution of bacteria in the deep sea is at least in part a function of higher-organism distribution and activity may also relate to the third hypothesis considered above, namely that low-nutrient bacteria exist and are important in the chemistry of the deep sea. If low-nutrient bacteria exist in the depths, they were not enriched in the experiments of Jannasch and Wirsen. More extensive efforts may be needed with undecompressed samples to establish whether or not such types are really present. The barophilic *Spirillum* isolated by Yayanos *et al.* (1979) was successfully cultured on a peptone medium and is not a low-nutrient bacterium. However, this may merely reflect its association with the amphipod. Low-nutrient bacteria have neither been verified nor precluded for the deep sea. The nutrient types and amounts in deep-sea experiments are thus critical to any experimental plan. For example, lithotrophic (chemoautotrophic) marine bacteria that derive energy from the oxidation of inorganic ions and that live under high pressures and low temperatures of the deep sea would not have been detected by any experiments so far reported.

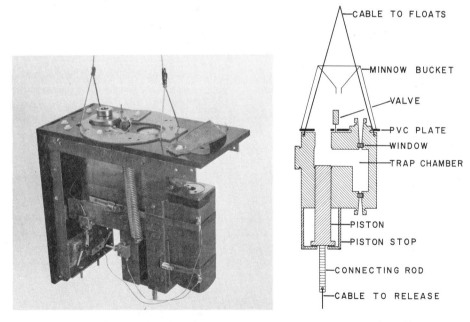

Fig. 8-6. Apparatus of Yayanos for the retrieval of undecompressed amphipods. Photo courtesy of Dr. A. A. Yayanos (Yayanos, 1978, *Science*, v. 200, p. 1056–1059, copyright 1978 by Amer. Assoc. Adv. Sci.).

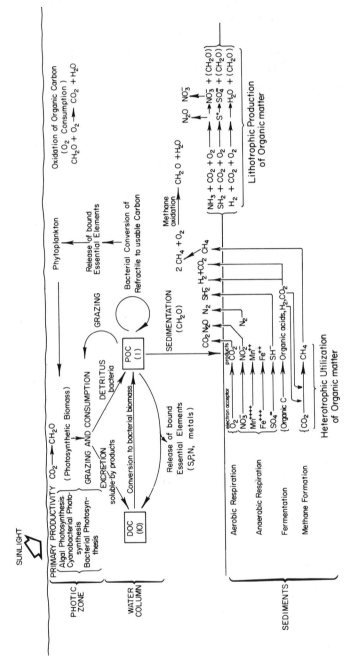

Fig. 8-7. Postulated roles for bacteria in the sea. The various bacterial reactions and reaction products are outlined. The contribution of the prokaryotes to primary productivity is thought to be extremely small in comparison with the eukaryotic photosynthesis. Probably it is insignificant in the open oceans, where nutrients are generated that directly affect the water-column dynamics. For the other reactions, while the reactions are documented, the magnitudes of the rates are generally unknown.

Although the debate concerning microbial activity in the deep sea is far from resolved, the issues are more clear. Barophilic heterotrophic bacteria, as determined by all methods used to date, are rare in samples of deep water and marine sediments. Whether there are low-nutrient bacteria or chemolithotrophic forms indigenous to the depths has not been determined. Barophilic heterotrophic bacteria have been isolated as associated forms with amphipods. Such associations may be common in the marine environment, with the microbes taking advantage of the behavioral repertoires of their hosts to locate nutrients.

The role of bacteria in forming DOC and in the conversion and cycling of nutrients in the water column is probably complicated, differing from place to place. This involves both the conversion of DOC into POC (bacterial biomass), which is further cycled, and the solubilization of POC and minerals for use by other organisms. Many animals and protists feed quite well on bacteria (ZoBell, 1946), yet the bacterial contributions to marine food webs have not been quantitated (Sieburth, 1976). Since bacteria are uniquely adapted for scavenging small amounts of nutrients in soluble form, they may be the most important factor determining the rate of exchange between the DOC and POC reservoirs and vice versa. Refractile POC such as chitin or cellulose may be converted into usable bacterial biomass. Bacteria probably accelerate the release of bioorganically bound essential elements (minerals) and particulate carbon in general. Bacteriotrophic organisms probably consume the bacteria on sinking fecal material and other POC at several levels in the water column (Sieburth, 1976). This tempting speculation is also poorly documented.

The possible functions, admittedly speculative, of deep-sea bacteria are outlined in Fig. 8-7. Neither the rate constants nor the qualitative details of the various processes are known. Quantitation requires far more extensive *in situ* experiments, of the type now being done by many workers.

# REFERENCES

Azam, F., 1980, Trophodynamic role of bacterioplankton in pelagic marine environments—a review: *Oceanologica Acta,* in press.

———, and Hodson, R. E., 1977, Size distribution and activity of marine heterotrophs: *Limnol. Oceanog.,* v. 22, p. 492–501.

Ballard, R. D., 1977, Notes on a major oceanographic find: *Oceanus,* v. 20, p. 35–44.

Berger, L., 1974, Some thoughts on the nature of barophilic bacteria, *in* Colwell, R. R., and Morita, R. Y., eds., *Effect of the Ocean Environment on Microbial Activities:* Univ. Park Press, p. 222–226.

Brauer, R. W., ed., 1972, *Barobiology and the Experimental Biology of the Deep Sea:* North Carolina Sea Grant Publ., Univ. N.C., Chapel Hill, N.C., 428p.

Carlucci, A. F., and Pramer, D., 1957, Factors influencing the plate method of determining abundance of bacteria in sea water: *Proc. Soc. Exp. Biol. Med.,* v. 96, p. 392–394.

———, and Shimp, S. L., 1974, Isolation and growth of a marine bacterium in low concentrations of substrate, *in* Colwell, R. R., and Morita, R. Y., eds., *Effect of the Ocean Environment on Microbial Activities:* Md., Univ. Park Press, p. 363–367.

——, and Williams, P. M., 1978, Simulated *in situ* growth rates of pelagic marine bacteria: *Naturwissenschaften,* v. 65, p. 541–542.

Corliss, J. B., Dymond, J., Gordon, L. I., Edmond, J. M., von Herzen, R. P., Ballard, R. D., Green, K., Williams, D., Bainbridge, A., Crane, K., and van Andel, T. H., 1979, Submarine thermal springs on the Galapagos Rift: *Science,* v. 203, p. 1073–1083.

Corpe, W. A., 1970, Attachment of marine bacteria to solid surfaces, *in* Manly, E. P., ed., *Adhesion in Biological Systems:* New York, Academic Press, p. 74–85.

Ehrlich, H. L., 1975, The formation of ores in the sedimentary environment of the deep sea with microbial participation—the case for ferromanganese concretions: *Soil Sci.,* v. 119, p. 36–41.

Fenchel, T. M., and Jorgensen, B. B., 1977, Detritus food chains of aquatic ecosystems—the role of bacteria: *Adv. Micro. Ecol.,* v. 1, p. 1–58.

Froelich, P. N., Klinkhammer, G. P., Bender, M. L., Luedtke, N. A., Heath, G. R., Cullen, D., Dauphin, P., Hammond, D., Hartman, B., and Maynard, V., 1979, Early oxidation of organic matter in pelagic sediments of the eastern equatorial Atlantic—subtoxic diagenesis: *Geochem. Cosmochem. Acta,* v. 43, p. 1075–1090.

Herring, P. J., and Morin, J. G., 1978, Bioluminescence in fishes, *in* Herring, P. J., ed., *Bioluminescence in Action:* New York, Academic Press, p. 273–329.

Hodson, R. E., Carlucci, A. F., and Azam, F., 1979, Glucose transport in a low nutrient bacterium: *Abstr. Amer. Soc. Microbiol.,* N59, p. 189.

Holm-Hansen, O., Azam, F., Carlucci, A. F., Hodson, R. E., and Karl, D. M., 1977, Microbial distribution and activity in and around McMurdo Sound: *Antartic J. U.S.,* v. 12, p. 29–32.

Hungate, R. E., 1963, The rumen bacteria, *in* Nutman, P. S., and Mosse, B., eds., *Symbiotic Associations:* New York, Cambridge Univ. Press, p. 266–297.

Jannasch, H. W., 1967, Growth of marine bacteria at limiting concentrations of organic carbon in seawater: *Limnol. Oceanogr.,* v. 12, p. 264–271.

——, 1968, Growth characteristics of heterotrophic bacteria in seawater: *J. Bacteriol.* v. 95, p. 722–723.

——, 1970, Threshhold concentrations of carbon sources limiting bacterial growth in seawater, *in* Hood, E. W., ed., *Organic Matter in Natural Waters:* Inst. Mar. Sci. College, Alaska, p. 321–328.

——, 1978, Microorganisms and their aquatic environment, *in* Krumbein, W., ed., *Env. Biogeochem. Geomicrobiol.,* v. 1: Mich., Ann Arbor Press, p. 17–24.

——, Eimhjellen, K., Wirsen, C., and Farmanfarmaian, A., 1970, Microbial degradation of organic matter in the deep sea: *Science,* v. 171, p. 672–675.

——, and Pritchard, P. H., 1972, The role of inert particulate matter in the activity of aquatic microorganisms: *Mem. 1st. Ital. Idrobiol., Suppl.,* v. 29, p. 289–308.

——, Truper, H. G., and Tuttle, J. H., 1974, Microbial sulfur cycle in the Black Sea, *in: The Black Sea—Geology, Chemistry, and Biology:* Amer. Assoc. of Petroleum Geologists, p. 419–425.

——, and Wirsen, C. O., 1973, Deep-sea microorganisms—*in situ* response to nutrient enrichment: *Science,* v. 180, p. 641–643.

——, and Wirsen, C. O., 1977, Microbial life in the deep sea: *Scient. Amer.,* v. 236, p. 42–52.

——, Wirsen, C. O., and Taylor, C. D., 1976. Undecompressed microbial populations from the deep sea: *Appl. Env. Microbiol.,* v. 32, p. 360–367.

Johnson, P. W., and Sieburth, J. McN., 1978, Morphology of non-cultured bacterio-plankton from estuarine, shelf and open ocean waters: *Abstr. Amer. Soc. Micro.,* N95, p. 178.

Kaneko, T., and Colwell, R. R., 1973, Ecology of *Vibrio parahemolyticus* in Chesapeake Bay: *J. Bacteriol.,* v. 113, p. 24–32.

MacLeod, R. A., 1968, On the role of inorganic ions in the physiology of marine bacteria: *Advan. Microbiol. Sea,* v. 1, p. 95–126.

Marquis, R. E., 1976, High pressure microbial physiology: *Adv. Microbiol. Physiol.,* v. 14, p. 159–241.

Marshall, K. C., 1976, *Interfaces in Microbial Ecology:* Cambridge, Mass., Harvard University Press, 156p.

Morita, R. Y., 1975, Psychrophilic bacteria: *Bact. Rev.,* v. 39, p. 144–167.

———, 1976, Survival of bacteria in cold and moderate hydrostatic pressure environments with special reference to psychrophilic and barophilic bacteria: *26th Symp. Soc. Gen. Microbiol.,* v. 26, p. 279–298.

Nealson, K. H., and Hastings, J. W., 1979, Bacterial bioluminescence—its control and ecological significance: *Microbiol. Rev.,* v. 43, p. 496–518.

Parsons, T. R., 1963, Suspended matter in seawater, *in* Sears, M., ed., *Progress in Oceanography, Vol. I:* New York, Pergamon, p. 203–239.

Reichelt, J. L., and Baumann, P., 1974, Effect of sodium chloride on the growth of heterotrophic marine bacteria: *Arch. Microbiol.,* v. 97, p. 329–345.

Riley, G. A., 1970, Particulate organic matter in seawater: *Adv. Mar. Biol.,* v. 8, p. 1–118.

Ruby, E. G., and Morin, J. G., 1978, Specificity of symbiosis between deep-sea fish and psychotrophic luminous bacteria: *Deep-Sea Res.,* v. 25, p. 161–171.

———, and Nealson, K. H., 1977, Seasonal changes in the species composition of luminous bacteria in nearshore water: *Limnol. Oceanog.,* v. 23, p. 530–533.

Schlegel, H. G., 1975, Mechanisms of chemoautotrophy, *in: Marine Ecology, Vol II*: London, Wiley, p. 9–60.

Schwarz, J. R., Yayanos, A. A., and Colwell, R. R., 1976, Metabolic activities of the intestinal microflora of a deep-sea invertebrate: *Appl. Env. Microbiol.,* v. 31, p. 46–48.

Sieburth, J. McNeill, 1967, Seasonal selection of estuarine bacteria by water temperature: *J. Exp. Mar. Biol. Ecol.,* v. 1, p. 98–121.

———, 1976, Bacterial substrates and productivity in marine ecosystems: *Ann. Rev. Ecol. Syst.,* v. 7, p. 259–285.

Spencer, R., 1961, Chitinoclastic activity in the luminous bacteria: *Nature,* v. 190, p. 938.

Stanier, R. Y., Adelberg, E., and Ingraham, J., 1976, *The Microbial World:* Englewood Cliffs, N. J., Prentice-Hall, 871p.

Watson, S. W., Novitsky, T. J., Quinby, H. L., and Valois, F. W., 1977, Determination of bacterial number and biomass in the marine environment: *App. Env. Microbiol.,* v. 33, p. 940–946.

Williams, P. M., 1971, The distribution and cycling of organic matter in the ocean, *in* Faust, S. J., and Hunter, F. V., eds., *Organic Compounds in Aquatic Environments:* New York, Dekker, p. 145–160.

Wirsen, C. O., and Jannasch, H. W., 1975, Activity of marine psychrophilic bacteria at elevated hydrostatic pressures and low temperatures: *Mar. Biol.*, v. 31, p. 201–208.

——, and Jannasch, H. W., 1978, Experiments in deep-sea microbiology using the DSRV *Alvin*: *Abstr. Amer. Soc. Microbiol.*, N13, p. 164.

Yayanos, A. A., 1978, Recovery and maintenance of live amphipods at a pressure of 580 bars from an ocean depth of 5700 meters: *Science*, v. 200, p. 1056–1059.

——, Dietz, A. S., and Van Boxtel, R., 1979, Isolation of a deep-sea barophilic bacterium and some of its growth characteristics: *Science*, v. 205, p. 808–810.

ZoBell, C. E., 1941, Studies on marine bacteria; 1. The cultural requirements of heterotrophic aerobes: *J. Mar. Res.*, v. 4, p. 42–75.

——, 1943a, Bacterial utilization of low concentrations of organic matter: *J. Bacteriol.*, v. 45, p. 555–564.

——, 1943b, The effect of solid surfaces upon bacterial activity: *J. Bacteriol.*, v. 46, p. 39–56.

——, 1946, *Marine Microbiology:* Waltham, Mass., Chronica Botanica, 240p.

——, 1968, Bacterial life in the deep sea: *Bull. Misaki Mar. Biol. Inst.*, Kyoto Univ., v. 12, p. 77–96.

——, and Johnson, F. H., 1949, The influence of hydrostatic pressure on the growth and viability of terrestrial and marine bacteria: *J. Bacteriol.*, v. 57, p. 179–189.

——, and Kim, J., 1972, Effects of deep-sea pressures on microbial enzyme systems, *in* Sleigh, M. A., and MacDonald, A. G., eds., *The Effects of Pressure on Organisms:* New York, Academic Press, p. 125–146.

——, and Morita, R. Y., 1957, Barophilic bacteria in some deep-sea sediments: *J. Bact.*, v. 73, p. 563–568.

Reinhardt A. Rosson[+] and Kenneth H. Nealson
Scripps Institution of Oceanography
University of California
La Jolla, CA 92093

# 9

# MANGANESE BACTERIA AND THE MARINE MANGANESE CYCLE

[+]Present address: University of Texas Marine Science Institute, Port Aransas, Texas 78373

# ABSTRACT

Modulation (mobilization and/or precipitation) of metals is a role often ascribed to bacteria. Many bacterial species are known that have the capacity to oxidize and/or reduce manganese. In many cases, laboratory studies have shown great potential for microbial catalysis of manganese chemistry. Thus, bacteria have often been implicated as agents of importance in the marine manganese cycle, and even in the formation of manganese nodules. However, proof of bacterial involvement requires that several criteria be satisfied, including (1) the presence of the causative organism, (2) the presence of structures (metal precipitates) formed by microbes, and (3) the measurement of *in situ* activities. Laboratory and field studies have seldom met all of these criteria, although in several environments it now seems clear that bacteria play important roles in manganese oxidation. Using these approaches, it should be possible to examine marine environments, including the deep sea, with the hope of distinguishing biological from chemical processes.

## MANGANESE IN THE MARINE ENVIRONMENT

Manganese has recently attracted the attention of many scientists, partly because of its abundance in ferromanganese nodules on the sea floor. The formation, distribution, and properties of these concretions is discussed in detail by Heath (this volume, Chap. 6). Manganese occurs both as the divalent, Mn(II), soluble salt, and in a variety of insoluble manganates ($MnO_x$) such as pyrolusite, birnessite, todorokite, and many others (Table 9–1). Despite its abundance in marine sediments (up to 0.08% by weight), the concentra-

TABLE 9-1. Some Common Manganese-Containing Minerals[a]

| | |
|---|---|
| Oxides and hydroxides: | |
| Birnessite (delta $MnO_2$) | $(Na,K,Ca)(Mg,Mn^{++})Mn_6O_{14} \cdot 5H_2O$ |
| Buserite | Na-Mn oxide hydrate |
| Hausmannite | $Mn_3O_4$ |
| Hollandite | $(Ba,K)_{12}Mn_8O_{16} \cdot xH_2O$ |
| Manganite | $MnOOH$ |
| Manganosite | $MnO$ |
| Psilomelane | $(Ba,K,Mn^{++},Co)_2Mn_5O_{10} \cdot xH_2O$ |
| Pyrolusite (Rhamsdellite) | $MnO_2$ |
| Pyrochroite | $Mn(OH)_2$ |
| Todorokite | $(Na,K,Ca)(Mg,Mn^{++})Mn_5O_{12} \cdot xH_2O$ |
| Iron and iron-silicates: | |
| Jacobsite | $MnFeO_4$ |
| Pyromanganite | $(Mn,Fe)SiO_3$ |
| Rhodonite | $(Mn,Fe,Ca)SiO_3$ |
| Carbonate: | |
| Rhodochrosite | $MnCO_3$ |
| Sulfide: | |
| Albandite | $MnS$ |

[a]Data from Burns and Burns, 1979; Marshall, 1979; R. Potter, personal communication.

tion of Mn(II) in seawater is very low, on the order of $10^{-9}M$, and hence the source of Mn(II) for the formation of manganese nodules and crusts has always been a matter of some speculation. Recently, however, with the discovery of the deep-sea hydrothermal vents, this dilemma may have been resolved. It is no longer necessary to hypothesize that the manganese in sediments is reduced and mobilized toward the surface where the Mn(II) is reoxidized, forming crusts, since it appears that sufficient manganese is introduced through the thermal vents to account for virtually all of the manganese precipitated in the deep sea (Edmond *et al.*, 1979).

Since manganese solubility changes as a function of oxidation state, the pH-Eh relationship of the various manganese phases are extremely important to the understanding of marine chemistry; a typical stability diagram for manganese under various pH-Eh conditions is presented in Fig. 9-1. The redox chemistry is further complicated by the tendency of Mn(II) to be actively chelated by a variety of organic and inorganic materials (Murray, 1974; Nakhshina, 1975). In fact, the manganates themselves chelate a variety of other minerals as well as adsorb Mn(II) and catalyze its further oxidation (Morgan and Stumm, 1965; Murray, 1974). Such inorganic autocatalyses have led many authors to conclude that biological oxidation is not necessary to explain manganese chemistry in the deep sea. Manganese adsorption by manganates is rapid and significant at pH values greater than 8.5 (Morgan and Stumm, 1965), and the possibility that manganese auto-oxidation is the major oxidative process in the sea must certainly be considered. However, it is also possible that the biota catalyzes some of the manganese chemistry of the sea. Some authors (Ehrlich, 1974; Silverman and Ehrlich, 1964) have proposed that bacteria are of fundamental importance in driving the chemical reactions, both oxidative and reductive, that are responsible for the marine manganese cycle. This chapter examines the properties of microbially catalyzed manganese chemistry of the sea.

## MANGANESE BACTERIA

Extensive laboratory studies of manganese bacteria, primarily from fresh water and soils, have revealed that manganese oxidation and/or reduction is catalyzed by many different bacterial genera (Table 9-2). It seems probable that like the freshwater and soil forms, many physiologically distinct taxa, utilizing a variety of different mechanisms, will be isolated from the marine environment.

Many of the bacteria, fungi, and algae that oxidize manganese act indirectly by changing the pH and Eh of their environment, resulting in conditions that favor one manganese phase over another. Algae remove $CO_2$ and produce oxygen, thus driving the oxidation of manganese, while many fungi excrete ammonia or other basic compounds, producing conditions that favor manganese precipitation. On solid culture media, indirect effects may be visualized as zones of oxidation or reduction around colonies, often at considerable distances from the colonies. Some effects are cell associated, like the production of compounds that adsorb Mn(II), raising its concentration and facilitating more rapid rates of oxidation, as postulated by Silverman and Ehrlich (1964). For example, the soil bacterium *Pedomicrobium* has been shown to excrete a surface associated acidic polysaccharide upon which the precipitated manganese is found (Ghiorse

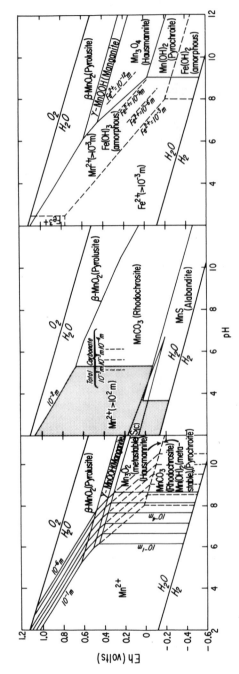

Fig. 9-1. Stability diagram for manganese under various pH-Eh conditions.

| Bacterial Group[b] | Genera | Soil | Freshwater | Marine |
|---|---|---|---|---|
| Manganese Oxidizers: | | | | |
| Pseudomonads | Pseudomonas | V | Z | N |
| Facultative gram negatives | Aeromonas | | | N |
| | Arthrobacter[c] | B,V | | E |
| Enterobacteria | Flavobacterium | | | N |
| | Oceanospirillum | | | E' |
| Sphaerotilus group | Leptothrix[d] | V | | |
| | Clonothrix | | P | |
| Prosthecate bacteria | Metallogenium | | P | |
| | Kuznetsovia | | P | |
| | Caulococcus | | P | |
| | Pedomicrobium | G | | |
| | Hyphomicrobium | | T | |
| Actinobacteria | Nocardia | S | | |
| | Streptomyces | B' | | |
| Gram-positive aerobes | Bacillus | V | | N' |
| | Micrococcus | | | E |
| Manganese Reducers: | | | | |
| Many genera of both gram-positive and gram-negative bacteria are reducers | | | T',P | |

[a]This table is not meant to be inclusive, rather it demonstrates the taxonomic and distributional versatility of manganese bacteria.

[b]Groupings of Margulis, 1974.

[c]Also called *Siderocapsa* (Perfil'ev and Gabe, 1965).

[d]*Leptothrix* and *Sphaerotilus* are used synonymously (van Veen *et al.* 1978).

[e]References in the table: B. Bromfield, 1974; B', Bromfield, 1979; E, Ehrlich, 1966; E', Ehrlich, 1978a; G, Ghiorse and Hirsch, 1979; N, Nealson, 1978; N', Nealson and Ford, 1980; P, Perfil'ev and Gabe, 1965; S, Schweisfurth, 1971; T, Tyler and Marshall, 1967; T', Troshanov, 1965; V, van Veen, 1973; Z, Zavarzin, 1962.

and Hirsch, 1979). Still other manganese oxidation is apparently directly catalyzed by enzymes or proteins (Bromfield 1979; Ehrlich, 1968; Jung and Schweisfurth, 1979). In many of these cases, the optimum pH of manganese oxidation is 7.0 or below, conditions where the reaction would not thermodynamically be favored (Fig. 9-1).

Since energy is available from the oxidation of manganese (Crerar *et al.,* 1979; Ehrlich, 1976; Ehrlich, 1978b), it has long been speculated that organisms capable of manganese oxidation should also be able to obtain energy from the oxidation, and hence there should be manganese chemolithotrophs. Ali and Stokes (1972) reported chemolithotrophic and mixotrophic growth of *Leptothrix* (*Sphaerotilus*) *discorphorus,* but these reports have been challenged (Hajj and Makemson, 1976; van Veen *et al.,* 1978). A preliminary report of mixotrophic growth of a marine isolate, Strain B345, indicated that manganese stimulated leucine uptake and also ATP synthesis (Ehrlich, 1978a). Whether or not manganese chemolithotrophs truly exist is for the moment unknown, but it is an exciting question for future research.

Ehrlich (1966, 1968) has proposed that some marine bacteria oxidize manganese only in the presence of solid $MnO_2$. A later report suggested that other surfaces, including iron hydroxides, also suffice (Ehrlich, 1978a). A variety of solid surfaces, including calcite, glass beads, and sand grains, significantly increased the rate of bacterial manganese oxidation for a different marine isolate, SG-1 (Nealson and Ford, 1980). SG-1 forms spores that have been shown to adsorb and oxidize manganese, even at very low concentrations, with no requirement for surface interaction (Rosson and Nealson, 1980). Recent evidence suggests that these surfaces concentrated growth-limiting organic nutrients, and hence stimulated SG-1 sporulation and manganese oxidation.[1] For all conditions tested, SG-1 spores, and not the vegetative cells, bound and oxidized manganese.[2] Thus, while surface interactions undoubtedly play an important role for marine manganese oxidizers, they are probably not always required.

One striking difference between marine and freshwater environments is the abundance of sheathed, filamentous, and budding bacteria in the latter environment as compared to the sea (Table 9-2). Most of the marine manganese oxidizers so far reported are morphologically simple (gram positive or gram negative rods or cocci), while those in soils and fresh water may be either simple or complex forms. The lack of identification of complex forms in seawater may simply be due to their not yet having been isolated, or there may be a fundamental difference between the manganese bacteria of the two environments. With regard to this question, Ghiorse (1980) has recently reported *Pedomicrobium*-like organisms on manganese concretions from the Baltic Sea. As studies like these proceed, it may well be that complex morphologies previously attributed just to freshwater and terrestial habitats will also be found to be abundant in the sea.

Irrespective of the energetics, mechanisms, or organisms involved, there is no doubt that under laboratory conditions, manganese-oxidizing bacteria exert substantial effects and thus have the potential to be significant in the environment. However, in many of the laboratory studies, concentrations of manganese are quite high (up to millimolar levels), and such concentrations are rarely seen in nature. Manganese oxidation that is easily demonstrated under laboratory conditions may not occur at all in nature, and experiments using low concentrations are needed to assess the ability of bacteria to function under conditions similar to those found in marine environments.

Manganate reduction by bacteria, like manganese oxidation, can either be direct or indirect. An indirect reduction as the result of oxygen removal and acid production is probably the most common mechanism, although direct reduction has been shown to proceed enzymatically, involving both an inducible enzyme ($MnO_2$ reductase) and a specific cytochrome system for electron transport (Ghiorse and Ehrlich, 1974; Trimble and Ehrlich, 1968, 1970). It was hypothesized that $MnO_2$ was used by these bacteria as an alternate electron acceptor in the absence of molecular oxygen. However, this reduction was shown to proceed equally well in the presence and absence of molecular oxygen, thus lending doubt to this speculation (Trimble and Ehrlich, 1968). An *Arthrobacter* species that can either oxidize or reduce manganese depending on the growth conditions

[1] Kepkay and Nealson, unpublished.

[2] Rosson and Nealson, unpublished.

has also been reported (Bromfield and David, 1976). Since enzymatic reactions are usually reversible *in vitro*, it is possible that an organism could use the same enzymatic system to catalyze both processes. If so, the importance of studying oxidation and reduction under conditions approximating a natural environment is further emphasized. It is possible, for example, that many so-called oxidizers are actually identified as such simply because of the great excess of Mn(II) in the growth medium. With the exception of the report mentioned above, no other studies of oxidation and reduction by the same organism have appeared.

The results of laboratory studies of bacteria isolated directly from deep-sea manganese nodules and sediments are not conclusive. In general, the bacteria isolated (cultured initially at 1 atm) from the deep sea do not grow well under the high pressure and low temperatures characteristic of the deep sea (Ehrlich *et al.*, 1972), although at least in one case, cells that were grown under low pressure are able to catalyze manganese oxidation when shifted to 400 atm. The same generalization applies to the manganese reducers isolated from nodules.

To summarize the laboratory studies, the manganese bacteria clearly have the potential to contribute substantially to the redox chemistry of manganese in the ocean. Neither the mechanism of action of the oxidizers nor of the reducers is known, and further work is needed before the effects of pressure and temperature on growth and activity are understood. This is true with regard to the formulation of models of the marine manganese cycle in the deep-sea sediments where the manganese nodules and deposits are found.

## ISOLATION AND IDENTIFICATION
## OF MANGANESE BACTERIA IN NATURE

In soils, pipelines, freshwater sediments and freshwater manganese concretions, manganese-oxidizing bacteria are abundant; convincing evidence of similar distributions of bacteria in deep-sea sediments and nodules is lacking. Manganese oxidizers have been cultured from ferromanganese nodules and deep-sea sediments (Ehrlich *et al.*, 1972; Krumbein, 1971; Nealson, 1978; Schutt and Otto, 1977, 1978; Sorokin, 1971). Of the heterotrophs that grew from nodules, up to 50% were found to be capable of manganese reduction, while up to 18% were capable of manganese oxidation (Ehrlich *et al.*, 1972). However, the populations of manganese oxidizers were small; only $10^2$ to $10^5$ heterotrophs per gram of nodule material were found, so that the manganese oxidizers rarely exceeded $10^4$ per gram. Similar studies from our laboratory, sampling nodules from the deep mid-Pacific, have yielded even lower numbers, from 10–100 heterotrophs per gram, with only 1 to 10% being capable of manganese oxidation.

In several reports, high numbers of manganese oxidizers were found associated with manganese nodules, but the sampling was not immediate (Ehrlich *et al.*, 1972; Schutt and Otto, 1978). In the former report, it was noted that bacterial growth occurred on nodules,

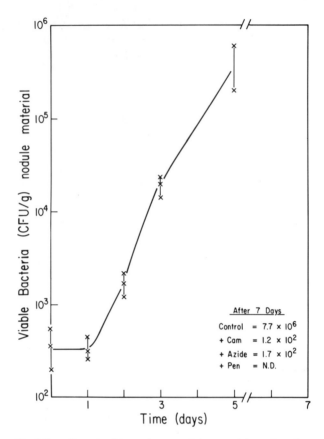

Fig. 9-2. Growth of bacteria on nodule material as a function of time, incubated at 4°C and 1 atm with and without inhibitors.

even when they were stored at 4°C. This is illustrated in Fig. 9-2[3] in which the bacteria on a stored nodule are seen to increase after a lag time of one day, by a factor of ten each day, so that after a few days the counts are substantial. If either chloramphenicol (Cam), which inhibits protein synthesis, or penicillin (Pen), which kills off growing cells, is added no increase in the bacterial population is seen, suggesting that the high numbers seen on stored nodules are the result of microbial growth. Therefore, if nodules are not sampled immediately upon retrieval, the resulting analyses of microbial populations will probably reflect neither the relative abundance of types nor numbers of bacteria originally present in samples.

Manganese-oxidizing bacteria are also found in deep-sea sediments. The bacteria and the proportion of Mn(II) oxidizers in these sediments also increase with time after sampling (Ehrlich et al., 1972; Schutt and Otto, 1978).[4] These bacteria could easily be contaminants from the water column or surface bacteria that have been carried to the bottom on particulate matter. Due to the low temperatures and high pressures of the

[3] Nealson, unpublished.

[4] Also, Nealson, unpublished.

MANGANESE BACTERIA AND THE MARINE MANGANESE CYCLE

deep sea, such bacteria might be in a resting state; growth and metabolic activity would begin only after decompression upon retrieval (Novitsky and Morita, 1977). Since few of the bacteria from manganese nodules or sediments have been shown to be capable of growth at high pressures and low temperatures, this possibility should be considered.

It is also possible that the ferromanganese nodules and deep-sea sediments are populated by bacteria that do not grow in the culture media used, or by bacteria that are obligate barophiles (Yayanos, *et al.,* 1979). If manganese oxidizers are important in the genesis and growth of nodules, then based on the rates of nodule growth (Heath, this volume, Chap 6), they should have been in the deep sea for millions of years and are likely candidates to be obligate barophiles. No nodules have been retrieved without decompression or studied *in situ.* Thus, it is difficult to ascribe a significant role to bacteria in the genesis of manganese nodules.

One possible way to resolve such problems is the microscopic examination of manganese precipitates for the presence of manganese-oxidizing organisms. This approach, however, is fraught with difficulty because manganese bacteria assume a variety of morphologies depending on their age, the growth medium, and the extent of manganese oxidation (Gabe *et al.,* 1965; Ghiorse and Hirsch, 1978, 1979; Nealson and Tebo, 1980; Tyler and Marshall, 1967). In general, as manganese is oxidized and the bacteria become coated with the resulting precipitate, they become less easily recognized as bacteria (Fig. 9-3). Therefore, the value of the examination of nodules for morphologies typical of bacteria to prove their involvement in the precipitation of manganese is questionable. However, the manganates can be reduced by chemical treatment, leaving only intact cells which can be visualized with either scanning or transmission electron microscopy techniques and may offer a valuable method for the assessment of the biological contribution to manganese precipitates.

Scanning electron microscopic analysis of nodules in the Blake Plateau revealed large populations ($6 \times 10^5/cm^2$ of nodule surface) of microorganisms (LaRock and Ehrlich, 1975). When nodules from the same area were examined for viable bacteria, only about $10^3$ cells per gram of nodule surface were obtained. The reason for this discrepancy is not known, but even if the bacteria-like bodies are indeed bacteria, they are very likely not the ones involved with manganese precipitation, as judged by lack of associated manganese. A more fruitful approach may be that of sectioning the nodules and looking under the surface for the microbes buried in precipitates of their own making. At least with laboratory cultures, this approach works; when thin sections are made of pure cultures of manganese-oxidizing bacteria and examined with transmission electron microscopy, it is obvious that microbes are associated with the precipitation of these metals (Fig. 9-4). In one case, the precipitated manganese is inside the cell, while in the others, it is outside. In most cases so far examined, it is possible to identify biologically precipitated manganese. Such studies have been recently done for natural precipitates with encouraging results[5] (Ghiorse, 1980; Ghiorse and Hirsch, 1978, 1979).

In studies of manganese and iron biochemistry in lakes, several Russian workers have reported the abundance of highly pleomorphic organisms of the genera *Metallogenium* and *Kuznetzovia.* These organisms can be identified using special techniques called over-growth methods (Kutzova, 1975) and capillary methods (Perfil'ev and Gabe, 1965).

[5] Emerson, Jacobs, Kalhorn, Tebo, Nealson, and Rosson, unpublished.

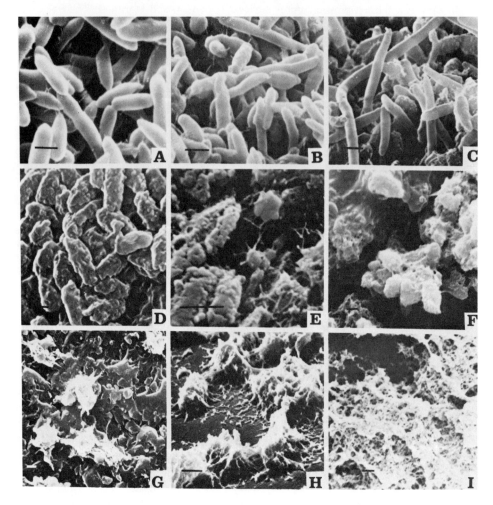

Fig. 9-3.    Scanning electron micrographs of manganese oxidizing bacillus SG-1 under different conditions. (Bars = 1.0 μm).

A. Agar medium lacking manganese, cells after one week of growth.

B. Cells after one day of growth in manganese-containing medium.

C. Cells after one week of growth in manganese-containing medium.

D. Cells taken from the top surface of a colony after one month of growth in manganese medium.

E. Cells taken from the bottom surface of the colony after one month of growth in manganese-containing medium.

F. Naturally occurring clumps of manganese coated bacteria in seawater after several months of growth.

G. Clumps of $MnO_2$ coated bacteria on sand grains in seawater after one week of growth.

H. Clumps of $MnO_2$ coated bacteria on glass slides after one week of growth.

I. Clumps of $MnO_2$ bacteria on glass slides after three months of growth.

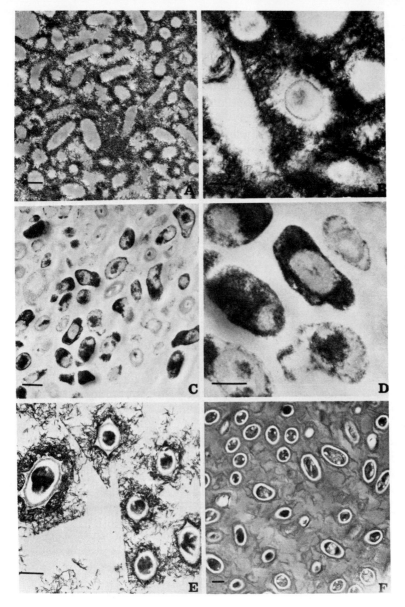

Fig. 9-4.   Transmission electron micrographs of manganese precipitation by bacteria.

A. A gram-negative rod forms an extracellular precipitate of $MnO_2$ • Bar = 0.5 μm.
B. Same as A, but at higher magnification. Bar = 0.25 μm.
C. A second gram-negative rod shaped bacterium that deposits $MnO_2$ inside the cells. Bar = 0.5 μm.
D. Same as C, but at higher magnification. Bar = 0.25 μm.
E. A third mode of $MnO_2$ deposition as seen in the bacillus in which the spores but not the cells become coated with $MnO_2$. Bar = 0.5 μm.
F. SG-1 spores similar to E. $MnO_2$ has been reduced. Bar = 0.5 μm.

Studies *in situ* have revealed the complex life cycles consisting of single cells, filaments, budlike terminal cells, aggregations, and others. Standard methods of culturing *Metallogenium* isolates yield variable results and it has not been possible to obtain accurate numbers of viable cells present. In fact, whether it is really possible to obtain pure cultures of *Metallogenium* is a matter of some debate (Schweisfurth *et al.,* 1978). No *Metallogenium*-like organisms or precipitates have been reported for marine sediments. Sorokin (1971) attempted to identify types using capillary methods, but was not successful. It may be that such organisms are not marine, or that sufficient effort has not yet been made to isolate them. Clearly, the situation of bacterial abundance and types in deep-sea samples is yet to be resolved, especially for the manganese bacteria, but the prognosis using the microstructural methods described above, is hopeful.

## MEASUREMENTS OF ACTIVITIES *IN SITU*

Even if many bacteria with manganese-oxidizing capabilities were isolated from a given environment, proof of biological mediation still requires measurement of activities *in situ*. This is particularly true for manganese oxidizers and reducers, since in no case has manganese oxidation or reduction been clearly shown to be required for the success of these bacteria in pure culture. Thus, while abundance may be used to support arguments of their importance in manganese transformation, it can in no way be used to prove it. However, when correlated with activity measurements, microbial abundances and structures form a convincing basis for establishing their biological involvement in precipitate formation. But actual measurements of *in situ* activities of manganese oxidizers and reducers are rare. In neutral soils, manganese oxidation was demonstrated to be due to biological factors by the comparison of nonsterilized and sterilized soil samples (Bromfield and Skerman, 1950; van Veen, 1973). In lake sediment samples, sterilization has been shown to inhibit stratification and microzone formation (Gabe and Gal'pirina, 1965). Harris *et al.* (1978) and Nealson *et al.* (1979) have demonstrated that manganese removal in Oneida Lake, New York is microbiologically mediated, a conclusion based on control experiments with bacterial-specific poison mixtures in which manganese oxidation was completely inhibited. Similar experiments in Saanich Inlet, British Columbia, have led to similar conclusions (Emerson *et al.,* 1982; Rosson *et al.,* 1979). Recent reports by Wollast *et al.* (1979), indicate that similar situations prevail in estuaries, where the removal of manganese is primarily mediated by biological oxidation.

In the above studies, it seems likely that manganese bacteria are geochemically significant, and as more data are assembled it should be possible to quantitate the microbial contributions. Such studies have not been done in the deep sea. As with other deep-sea processes, the biological oxidation of manganese is very likely quite slow compared with rates in the warmer surface waters. Furthermore, the Mn(II) concentration in seawater is generally quite low, making the overall rates of these processes difficult to measure. Structural analyses of deep-sea precipitates may supply valuable information as to the biogenic nature of the deposits; it may be possible to catalog typical structures that would indicate biological involvement in the precipitate formation. For the moment, however, it must be admitted that the role of the bacteria in deep-sea manganese chemistry is not well understood. Although laboratory studies have been used to suggest that bacteria are involved with both manganese oxidation and reduction, and even with manganese nodule formation, no convincing field data have yet been presented. More studies of the chemistry and microbiology of deep-sea water, sediments, and nodules will be needed to identify environments suitable for study of these processes and for assessment of the microbial role in deep-sea manganese chemistry.

# REFERENCES

Ali, S. H., and Stokes, J. L., 1972, Stimulation of heterotrophic and autotrophic growth of *Sphaerotilus discophorus* by manganous ions: *Antonie van Leeuwenhoek*, v. 37, p. 519–528.

Bromfield, S. M., 1974, Bacterial oxidation of manganous ions as affected by organic substrate concentration and composition: *Soil Biol. Biochem.*, v. 6, p. 383–392.

——, 1979, Manganous ion oxidation at pH values below 5.0 by cell free substances from *Streptomyces* sp. cultures: *Soil Biol. Biochem.*, v. 11, p. 115–118.

——, and David, D. J., 1976, Sorption and oxidation of manganous ions and reduction of manganese oxide by cell suspensions of a manganese oxidizing bacterium: *Soil Biol. Biochem.*, v. 8, p. 37–43.

——, and Skerman, V. B. D., 1950, Biological oxidation of manganese in soils: *Soil Sci.*, v. 69, p. 337–347.

Burns, R. G., and Burns, B. M., 1979, Manganese oxides, in Burns, R. G., ed., *Marine Minerals:* Washington, D.C., Mineral Society of America, p. 1–40.

Crerar, D. A., Cormick, R. K., and Barnes, H. L., 1979, Geochemistry of manganese—an overview, *in* Varentsov, I. M., ed., *Geology and Geochemistry of Manganese*, v. 1, Budapest, Hungarian Academy of Sciences, p. 293–334.

Edmond, J. M., Measures, C., Magnum, B., Grant, B., Sclater, F. R., Collier, R., Hudson, A., Gordon, L. I., and Corliss, J. B., 1979, On the formation of metal rich deposits at ridge crests: *Earth Planet. Sci. Lett.*, v. 46, p. 19–30.

Ehrlich, H. L., 1966, Reaction with manganese by bacteria from marine ferromanganese nodules: *Dev. Ind. Microbiol.*, v. 7, p. 279–286.

——, 1968, Bacteriology of manganese nodules, II. Manganese oxidation by cell-free extracts from a manganese nodule bacterium. *Appl. Microbiol.*, v. 16, p. 197–202.

——, 1974, The formation of ores in the sedimentary environment of the deep sea with microbial participation—the case for ferromanganese concretions: *Soil Sci.*, v. 119, p. 36–41.

——, 1976, Manganese as an energy source for bacteria, *in* Nriagu, J. O., ed., *Environmental Biochemistry*, Vol. 2: Mich., Ann Arbor Press, p. 633–644.

——, 1978a, Conditions for bacterial participation in the initiation of manganese deposition around sediment particles: *Env. Biogeochem. Geomicrobiol.*, v. 3, p. 839–845.

——, 1978b, Inorganic energy sources for chemolithotrophic and mixotrophic bacteria: *Geomicrobiol. J.*, v. 1, p. 65–83.

——, Ghiorse, W. C., and Johnson, G. L., 1972, Distribution of microbes in manganese nodules from the Atlantic and Pacific oceans: *Dev. Ind. Microbiol.*, v. 13, p. 57–65.

Emerson, S., Jacobs, L., Kalhorn, S., Rosson, R. A., Tebo, B. M., and Nealson, K. H., 1982, Environmental oxidation rate of manganese (II)—Bacterial catalysis: *Geochem. Cosmochem. Acta*, in press.

Gabe, D. R. and Gal'perina, A. M., 1965, The development of the microzonal mud profile in the absence of microflora, *in* Perfil'ev, B. V., Gabe, D. R., Gal'perina, A. M., Rabinovich, V. A., Sapotnitskii, A. A., Sherman, E. E., and Troshanov, E. P., *Applied Capillary Microscopy: The Role of Microorganisms in the Formation of Iron-Manganese Deposits:* New York, Consultants Bureau (transl. from Russian), p. 110–116.

——, Troshanov, E. P., and Sherman, E. E., 1965, The formation of manganese-iron layers in mud as a biogenic process, *in* Perfil'ev, B. V., Gabe, D. R., Gal'perina, A. M., Rabinovich, V. A., Sapotnitskii, A. A., Sherman, E. E., and Troshanov, E. P., *Applied Capillary Microscopy: The Role of Microorganisms in the Formation of Iron-Manganese Deposits:* New York, Consultants Bureau (transl. from Russian), p. 88–105.

Ghiorse, W. C., 1980, Electron microscopic analysis of metal depositing microorganisms in surface layers of Baltic Sea ferromanganese concretions, *in* Trudinger, P. A., Walter, M. R., and Ralph, B. J., eds., *Biogeochemistry of Ancient and Modern Environments,* Canberra City, Australian Academy of Science, and Berlin, Springer Verlag, p. 345–354.

——, and Ehrlich, H. L., 1974, Effects of seawater cations and temperature on maganese dioxide-reductase activity in a marine *Bacillus: Appl. Microbiol.,* v. 28, p. 785–792.

——, and Hirsch P., 1978, Iron and manganese deposition by budding bacteria, *in* Krumbein, W. E., ed., *Env. Biogeochem. Geomicrobiol.,* v. 3, Ann Arbor Science, Ann Arbor, Mich., p. 879–909.

——, and Hirsch, P., 1979, An ultrastructural study of iron and manganese deposition associated with extracellular polymers of *Pedomicrobium*-like budding bacteria: *Arch. Microbiol.,* v. 123, p. 213–226.

Hajj, H., and Makemson, J. 1976, Determination of growth of *Sphaerotilus discophorous* in the presence of manganese: *Appl. Environ. Microbiol.,* v. 32, p. 699–702.

Harris, T. G., Moore, W. S., and Nealson, K. H., 1978, The role of bacteria in the fixation of $Co^{++}$, $Fe^{++}$, and $Mn^{++}$ in Lake Oneida, N.Y.: *Proc. Spring Meeting Amer. Geophys. Un.,* N-28a.

Jung, W. K., and Schweisfurth, R., 1979, Manganese oxidation by an intracellular protein of a *Pseudomonas* species: *Zeit. fur Allg. Mikrobiol.,* v. 19, p. 107–115.

Krumbein, W. E., 1971, Manganese oxidizing fungi and bacteria in recent shelf sediments of the Bay of Biscay and the North Sea: *Die Naturwissenschaften,* v. 58, p. 56–57.

Kutzova, R. S., 1975, Study of iron-manganese microorganisms by the overgrowth-method: *Mikrobiologiya,* v. 44, p. 156–159.

LaRock, P. A., and Ehrlich, H. L., 1975, Observations of bacterial microcolonies on the surface of ferromanganese nodules from Blake Plateau by scanning electron microscopy: *Microbiol. Ecol.,* v. 2, p. 84–96.

Margulis, L., 1974, The classification and evolution of prokaryotes and eukaryotes *in, Handbook of Genetics, Vol. 1:* New York, Plenum Publishing Co., p. 1–41.

Marshall, K. C., 1980, Biogeochemistry of manganese minerals, *in* Trudinger, P. A., and Swaine, D., eds., *Biogeochemical Cycling of Mineral Forming Elements:* Elsevier, Amsterdam, 354p.

Morgan, J. J., and Stumm, W., 1965, Analytical chemistry of aqueous manganese: *J. Amer. Water Works Assoc.,* v. 57, p. 107–119.

Murray, J. W., 1974, The surface chemistry of hydrous manganese dioxide: *J. Colloid Interface Sci.,* v. 46, p. 357–371.

Nakhshina, Y. P., 1975, Manganese in fresh water: *Hydrobiological J.,* v. 11, p. 77–90.

Nealson, K. H., 1978, The isolation and characterization of marine bacteria which cata-lyze manganese oxidation, *in* Krumbein, W. E., ed., *Env. Biogeochem. Geomicro-biol.*, v. 3, Ann Arbor, Mich., Ann Arbor Science, p. 847–858.

——, and Ford, J., 1980, Surface enhancement of bacterial manganese oxidation—implications for aquatic environments: *Geomicrobiol. J.*, v. 2, p. 21-37

——, and Tebo, B. M., 1980, Structural features of manganese precipitating bacteria: *Origin of Life, Vol. 10:* p. 117–126.

——, Moore, W., and Chapnick, S., 1979, Distribution and activity estimates of manga-nese oxidizing bacteria in Oneida Lake: *EOS, Transactions Amer. Geophys. Un.*, v. 60, p. 851.

Novitsky, J. A., and Morita, R. Y., 1977, Survival of a psychrophilic marine vibrio under long-term nutrient starvation: *Appl. Environ. Microbiol.*, v. 33, no. 3, p. 635–641.

Perfil'ev, B. V. and Gabe, D. R. 1965, The use of the microbial-landscape method to investigate bacteria which concentrate manganese and iron in botton deposits, *in* Perfil'ev, B. V., Gabe, D. R., Gal'perina, A. M., Rabinovich, V. A., Sapotnitskii, A. A., Sherman, E. E., and Troshanov, E. P., eds., *Applied Capillary Microscopy: The Role of Microorganisms in the Formation of Iron-Manganese Deposits:* New York, Consultants Bureau (transl. from Russian), p. 9–54.

Rosson, R. A., and Nealson, K. H., 1980, Manganese binding and oxidation by spores of marine *Bacillus: Abstr. Ann. Meeting Amer. Soc. Microbiol.*, N105, p. 181.

——, Tebo, B. M., Nealson, K. H., and Emerson, S., 1979, Vertical distribution and potential oxidative activity of manganese bacteria in Saanich Inlet: Proc. Fall Mtg. Amer. Geophysical Union.

Schutt, C., and Ottow, J. C. G., 1977, Mesophilic and psychrophilic manganese-precipitat-ing bacteria in manganese nodules of the Pacific Ocean: *Zeitschr. fur Allgemeine Mikrobiologie*, v. 17, p. 611–616.

——, and Ottow, J. C. G., 1978, Distribution and identification of manganese-precipitating bacteria from noncontaminated ferromanganese nodules: *Env. Biogeochem. Geo-microbiol.*, v. 3, p. 869–878.

Schweisfurth, R., 1971, Manganoxidierende Pilze I Vorkommen, Isolierungen und mikro-skopische Untersuchungen: *Zeitschrift fur Allg. Mikrobiologie.*, v. 11, p. 415–430.

——, Eleftheriddis, E., Gunlach, H., Jacobs, M., and Jung, W., 1978, Microbiology of the precipitation of manganese: *Env. Biogeochem. Geomicrobiol.*, v. 3, p. 923–928.

Silverman, M. P., and Ehrlich, H. L., 1964, Microbial formation and degradation of minerals: *Adv. Appl. Micro.*, v. 6, p. 153–206.

Sorokin, Y. I., 1971, Microflora of iron-manganese concretions from the ocean floor: *Mikrobiologiya,* New York, Consultants Bureau (transl. from Russian), v. 40, no. 3, p. 563–566.

Trimble, R. B., and Ehrlich, H. L., 1968, Bacteriology of manganese nodules, III. Reduc-tion of $MnO_2$ by strains of nodule bacteria: *Appl. Microbiol.*, v. 16, p. 695–702.

——, and Ehrlich, H. L., 1970, Bacteriology of manganese nodules, IV. Induction of an $MnO_2$-reductase system in marine bacillus: *Appl. Microbiol.*, v. 19, p. 966–972.

Troshanov, E. P., 1965, Bacteria which reduce manganese and iron in bottom deposits, *in* Perfil'ev, B. V., Gabe, D. R., Gal'perina, A. M., Rabinovich, V. A., Sapotnitskii, A. A., Sherman, E. E., and Troshanov, E. P., *Applied Capillary Microscopy: The Role of Microorganisms in the Formation of Iron-Manganese Deposits:* New York, Consultants Bureau (transl. from Russian), p. 106–110.

Tyler, P. A. and Marshall, K. C., 1967, Hyphomicrobia—A significant factor in manganese problems, *J. Am. Waterworks Assoc.*, v. 59, p. 1043-1048.

van Veen, W. L., 1973, Biological oxidation of manganese in soils: *Antonie van Leeuwenhoek,* v. 39, p. 657–662.

——, Mulder, E. G., and Deinema, M. H., 1978, The *Sphaerotilus Leptothrix* group of bacteria: *Microbiol. Rev.,* v. 42, p. 329–356.

Wollast, R., Billen, G. and Duinker, J. C., 1979, Behavior of manganese in the Rhine and Scheldt Estuaries. I. Physico-chemical aspects: *Estuarine Coastal Mar. Sci.,* v. 9, p. 161–169.

Yayanos, A. A., Deitz, A. S., and Van Boxtel, R., 1979, Isolation of deep-sea barophilic bacterium and some of its growth characteristics: *Science,* v. 205, p. 808–810.

Zavarzin, G. A., 1962, Symbiotic oxidation of manganese by two species of *Pseudomonas: Microbiology,* v. 31, p. 481–482.

Peter A. Jumars and Eugene D. Gallagher
School of Oceanography
University of Washington
Seattle, Washington 98195

# 10

# DEEP-SEA COMMUNITY STRUCTURE: THREE PLAYS ON THE BENTHIC PROSCENIUM

# ABSTRACT

From the available literature in spring 1979, we examine the compatibility of data and theories of deep-sea benthic community structure. We begin with a brief review of data, showing that deep-sea infaunal species are small and sparsely distributed, that deposit feeders dominate the deep-sea infauna and frequently are very high in species diversity, that size-frequency distributions of individuals within deep-sea populations generally peak at the larger sizes, and that insufficient data exist to generalize about deep-sea population dynamics. Having thereby set the stage, we explore the respective abilities of theories at the individual, population, and community levels of ecological organization to explain the available observations and, more importantly, to provide predictions that can be tested with available technology in the foreseeable future.

At the individual level, foraging theory holds a great deal of promise for explaining the relative success of feeding guilds and for providing testable predictions. Excluding unusual environments such as hydrothermal vents and regions of high current activity, deep-sea suspensate levels apparently do not repay the costs of pumping water. Even passive suspension feeders show adaptations for intercepting enhanced particle fluxes by projecting their feeding appendages into the turbulent portion of the benthic boundary layer. The most extensive predictions, however, can be made for scavengers. Taking into account the diffusion patterns of scent trails away from carrion on the bottom, we would expect crawling scavengers to spend most of their time searching in cross-stream movements. Small, swimming scavengers should be generalists that take what living and non-living food items they encounter. Large, swimming scavengers, on the other hand, should spend more of their time at those greater heights above the bottom where (due to upward and lateral turbulent diffusion of scent trails) they can capture, by virtue of detection ability, the most carrion per unit time, and they should be specialists on carrion. Foraging theory further suggests that motile deposit feeders should move in such a fashion as to minimize recrossing of recently depleted deposits; such minimization has the potential for explaining such seemingly disparate phenomena as the coiled feeding traces of enteropneusta and the herding behavior of urchins. A major impediment to additional applications of foraging theory to the deposit feeders that dominate the deep sea, however, is the inability to identify clearly the resources they utilize. This problem is best attacked in shallow-water communities before a deep-sea answer is attempted.

Theories at the population level are disappointing in their predictive abilities. Stochastic models and $r$-$K$ theories, for example, provide conflicting predictions of expected life-history traits, and the meager life-history data as yet obtainable from the deep sea do not promise to provide any definitive tests of these and other models. Shallow-water, terrestrial, and laboratory testing of general population models seems prudent before attempting definitive deep-sea application.

At the community level, Connell and Slatyer's individual-by-individual successional models, especially as formalized through a Markovian approach, deserve further consideration for deep-sea application and manipulative testing. We demonstrate via some simple examples, however, that such applications and tests are premature until predator-prey and competitive relations have been established. The latter, in turn, are not likely to be discovered until the resources utilized by the ubiquitous deposit feeders are better identified.

On a more optimistic note, at each biological level of organization, the physics of the deep-sea environment is seen to provide potential explanations of phenomena for which other, biological explanations have been sought or used in the past. The physical structure of the deep-sea benthic boundary layer may allow unique foraging tactics among scavengers and certainly does limit viable suspension feeding methods. At the population level, the reduced incidence of physically mediated disturbances capable of causing size-independent mortality may cause size-selective predation to be a relatively more important phenomenon in the deep sea than it is in shallow water. At the community level, in turn, relatively weak bottom currents allow the persistence of biologically generated environmental heterogeneity (e.g., fecal mounds, tubes, burrows) that may facilitate the persistence of higher species diversity than generally is seen in shallow-water communities. Just as biological parameters can influence the survival and persistence of particular species in physically disturbed environments, so can physical parameters influence the relative success of deep-sea species.

> The belief that science proceeds from observation to theory is so widely and so firmly held that my denial of it is often met with incredulity. . . . But in fact the belief that we can start with pure observation alone, without anything in the nature of a theory, is absurd; as may be illustrated by the story of the man who dedicated his life to natural science, wrote down everything he could observe, and bequeathed his priceless observations to the Royal Society to be used as inductive evidence. This story should show us that though beetles may profitably be collected, observations may not.
>
> Sir Karl Popper (1965, p. 46)

## INTRODUCTION

Direct observations of biological processes in the deep sea have been rare. The few outstanding exceptions have merited their own chapters in the present volume. Our chapter, by contrast, serves two somewhat divergent purposes. First, it introduces an assortment of data and generalizations on deep-sea community structures as essential background for this and later chapters. The second, more central purpose is to explore an equally wide spectrum of theories which may serve to explain the observed patterns. Recent historians and philosophers of science, while failing to concur on a method for arriving at scientific explanations (see Feyerabend, 1975), do concur that the development of theories takes precedence over the collection of observations both in the recognition of successful research programs from the past (Lakatos, 1970; Kuhn, 1962; Platt, 1964), and in the prescription of how science ought logically to proceed in the future (Popper, 1959).

The extant observations of deep-sea community structure are, for the moment, the only available actors and props, and they inspire the theories which will serve as this season's repertoire. For the sake of brevity, but at the expense of smooth transitions, we will introduce the principal actors and props all at once. All will not be used in each of our subsequent three plays, and a few specialized extras will be introduced in the individual plays. The decisions we seek are what new actors and props should be hired or obtained and, given the actors who will likely be available, whether a new repertoire of plays should be attempted next season. What new data would be desirable, and what observations would allow rejection of existing theories?

## PATTERNS OBSERVED

The size of our theatrical company is small. Because it takes years to plan and obtain funds for deep-sea expeditions, hours to lower and retrieve a sampler, roughly a month of microscope work per typical sample just to separate the larger animals from sediments too large to sieve away, and months of many taxonomists' time to assign animals from a single locality to species categories, the data are few. Without the ability to observe live, naturally interacting animals, the data are also of few kinds. Much of what we think we know about ecology of deep-sea animals in fact comes from analogy with "similar" shallow-water species—an analogy as dangerous as likening poisonous mushrooms to similar edible species. Despite this paucity of data, but in keeping with our opening quotation, we will treat only enough of the available observations to allow us to delve into the pertinent theory. The data-hungry reader will soon have available a much more detailed review (Rowe, in press).

We will draw repeatedly upon examples from four localities (Table 10-1), each studied via samples taken with the same device, the 0.25-m² USNEL spade or box corer described by Hessler and Jumars (1974). It recovers a cube of minimally disturbed mud of about 0.5 m in each dimension, often with animals still swimming in the overlying water (until the temperature rises too high). Quantitative comparisons among regions sampled with different devices are virtually impossible. Hessler and Jumars (1974), for example, show the disparate results obtained in sampling the same locality (CNP) with the box corer versus a towed, fine-meshed (0.5-mm) trawl, the epibenthic sled. No device samples all the deep-sea fauna equally well, and those highly mobile animals (e.g., fishes) living on or just above the bottom (mobile *epi*fauna as opposed to *in*fauna, the latter living within the sediments) have not yet been sampled quantitatively with any device. It should be remembered, then, that the samples of Table 10-1 include only the infauna, together with the less mobile epifauna.

TABLE 10-1. Sampling Localities Treated Frequently in the Text[a]

| Name | Abbre-viation | Depth (m) | North Latitude | West Longitude |
|---|---|---|---|---|
| Santa Catalina Basin | SCB | 1130 | 39°58' | 118°22' |
| San Diego Trough | SDT | 1230 | 32°28' | 117°30' |
| Central North Pacific | CNP | 5500–5800 | 28°29' | 155°23' |
| Aleutian Trench | AT | 7298 | 50°58' | 171°38' |

[a]For additional data on these localities, see Hessler and Jumars (1974, CNP), Jumars (1976, SCB, SDT), and Jumars and Hessler (1976, AT).

DEEP-SEA COMMUNITY STRUCTURE

# Body Size and Areal Density of the Fauna

Perhaps the most obvious peculiarity of deep-sea infaunal samples is the small average body size and frailty of individuals. In shallow-water, soft-bottom sampling, the term "meiofauna" is usually employed for animals passing through a 0.50- or 1.00-mm mesh sieve but retained on a roughly 50-$\mu$m sieve, while the term "macrofauna" applies to animals collected on the 0.50- or 1.00-mm meshes. A 1.00-mm sieve would retrieve virtually no infauna from the CNP site and very few individuals from most other deep-sea localities. Furthermore, unless the samples are handled very delicately, few animals will be retained at all. There is little need of heavy exoskeletons or thick integuments in an environment of weak currents and little sediment motion; excessive agitation during sieving will leave few identifiable remains.

Some taxonomic groups (taxa) of animals in shallow water characteristically are meiofaunal in size—nematode worms and harpacticoid copepods, for example. Adults of other taxa (e.g., bivalves, annelid worms) characteristically are larger than 0.50 or 1.00 mm. Hessler and Jumars (1974) have suggested the terms "meiofaunal taxa" and "macrofaunal taxa" to allow deep-sea recognition of parallel taxomonic groups. At the SCB and SDT sites, a 0.42-mm sieve retains adults of the macrofaunal taxa, while the CNP and AT sites require 0.30-mm meshes. We prefer the taxonomic distinction to one purely of size; it is more reasonable to compare oranges of different sizes than apples with oranges.

One extreme example of deep-sea miniaturization, reminiscent of trends seen in the fauna living in the interstices of near-shore sands (Swedmark, 1964), is the proto-branch bivalve genus *Microgloma,* in which cell size and germ-cell numbers have been so reduced as to produce mature adults less than 1.00 mm long (Sanders and Allen, 1973). While such reduction in size of deep-sea macrofaunal taxa has been discussed at length (Gage, 1978; Thiel, 1975, 1979), its causes remain open to speculation concerning both physiological and ecological mechanisms. Depth alone is not the cause. Of the localities of Table 10-1, the smallest average body size is found at CNP, where food input from the surface is also likely to be lowest. Polloni *et al.* (1979) in fact find little consistent change in macrofaunal body size in the North Atlantic between 400-m and 4000-m depths, although few large individuals are seen below 360 m. They suggest that the greatest reduction in body size may occur between 0 m and 400 m. Although the precise mechanism is unclear, the miniaturization of macrofaunal taxa occurs primarily by species replacements rather than by decreases in average body size within species; Gage (1978), for example, finds no evidence of decreasing body size with depth within three annelid worm species having broad depth ranges (compared at 20 to 150 m versus 1800 to 2875 m). Correlations of depth and distance from shore with sizes of individuals belonging to meiofaunal taxa have not been documented as clearly. However, the ratio (on the basis of numbers of individuals) of members of meiofaunal taxa to members of macrofaunal taxa does appear to increase toward the abyss (Thiel, 1979).

Confusing the issue of size and the definitions of meiofauna and macrofauna still further are the foraminifera [a group of (usually) test-building, acellular Protozoa]. A large majority of shallow-water foraminifera are meiofaunal in size range. In the deep sea, however, they encompass the entire size ranges of other macrofaunal and meiofaunal

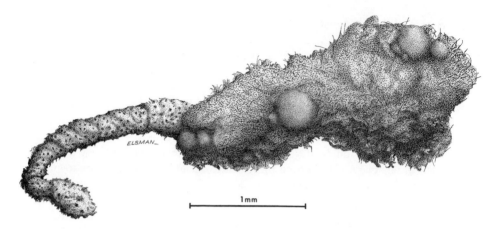

Fig. 10-1. *Baculella hirsuta*, a foraminiferan (suborder Textulariina) from the CNP locality (Table 10-1). It is much larger than individuals of macrofaunal taxa taken there. Its normal posture and ecological role are unknown. With permission, from Tendal and Hessler, 1977, Fig. 9.

taxa (e.g., Fig. 10-1). At the CNP locality, for example, the foraminifera larger than 0.30 mm outnumber the comparably sized metazoans (multicellular animals) by at least an order of magnitude (Bernstein *et al.*, 1978). Similar quantitative comparisons have not been made at the SCB, SDT, and AT sites, but large foraminifera are present in abundance at all these localities. Small animals are the rule among deep-sea infauna,

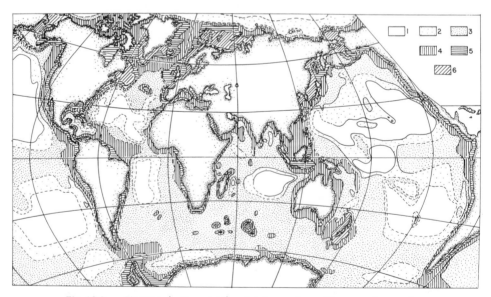

Fig. 10-2. Biomass (wet weight) of infauna from Okean grab samples (Belyayev *et al.*, 1973). Symbols (all in $g\,m^{-2}$): 1, <0.05; 2, 0.05–0.1; 3, 0.1–1.0; 4, 1.0–10.0; 5, 10.0–50.0; 6, 50.0– >1000.0 With permission, from Hessler and Jumars, 1977, Fig. 2.

DEEP-SEA COMMUNITY STRUCTURE

but foraminifera may well be the exception; a single pseudopod may extend up to 12 cm (Lemche *et al.*, 1976). Size trends among the epifauna, in general, also have not been documented reliably, but demersal (near-bottom dwelling) fishes *increase* in size with depth on the western Atlantic continental slope and rise (Haedrich and Rowe, 1977; Polloni *et al.*, 1979).

Accompanying the reduction of body size in macrofaunal taxa is an even more dramatic decline in standing stock (per unit area) with depth and distance from shore. The very lowest standing stocks in general occur in the food-poor central oceanic regions (Fig. 10-2). The surface waters here have very low rates of input of plant nutrients and consequently low production rates of food, most of which is consumed well before it can sink to the abyss. As a crude generalization, numbers of individuals per unit area decrease roughly exponentially with depth, but the same exponential function will not fit data from all regions (e.g., northwest Atlantic versus Gulf of Mexico, Fig. 10-3), and an exponential function will not fit at all in some regions (e.g., Gulf of Gascogne, Fig. 10-3). Rates and mechanisms (e.g., slumping and turbidity currents over steep topography) of food input to a given depth vary regionally and locally, being especially sensitive to the distance from productive, near-shore waters at which that particular depth is reached. Nor can one assume that deep-sea communities everywhere have equal conversion efficiencies of food into numbers or grams of individuals. In crude analogy with human populations, exceptionally dense assemblages can be expected to occur both where supply is plentiful and where utilization is efficient, or where (for any reason) birth and immigration rates greatly exceed death and emigration rates. In the deep sea, as elsewhere, the temptation to equate standing stocks (Figs. 10-2 and 10-3) with rates of production must scrupulously be avoided.

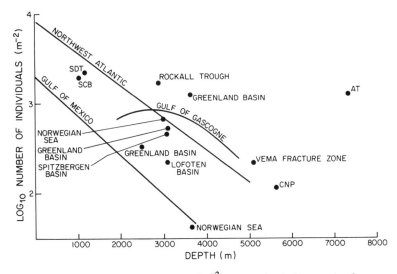

Fig. 10-3. Number of individuals (m$^{-2}$) versus depth for samples from a wide range of geographic localities. Curves summarize numerous data points from the Northwest Atlantic, Gulf of Mexico, and Gulf of Gascogne. Modified, with permission, from Khripounoff *et al.*, 1980, Fig. 6.

TABLE 10-2. Overall Macrofaunal Abundance and Identity of the Most Abundant Macrofaunal Species at the Sampled Localities[a]

| Locality: | SCB | SDT | CNP | AT |
|---|---|---|---|---|
| Total macrofaunal abundance (m$^{-2}$): | 1880 | 2251 | 115 | 1272 |
| Numbers of most abundant species (m$^{-2}$): | 350 | 142 | 15 | 184 |
| Species: | *Paraonis gracilis oculata* | *Tharyx* sp. | *Chaetozone* sp. | *Chaetozone* sp. |
| Family: | Paraonidae | Cirratulidae | Cirratulidae | Cirratulidae |

[a]All are polychaetes (bristle worms).

## Characteristic Taxonomic Composition and Trophic Structure

What kinds of albeit small and rare animals dominate the deep-sea bottom? Bristle worms (polychaetes) usually constitute half to three quarters of the individuals of macrofaunal taxa and usually contain the most abundant species at any soft-bottom deep-sea location (Table 10-2, Fig. 10-4). At even the grossest taxonomic levels (Table 10-3), however, deep-sea samples are often identifiable to broad depth zones—bathyal, abyssal, and

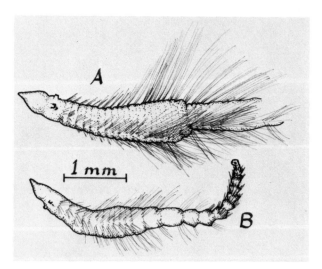

Fig. 10-4. *Chaetozone* sp., the most abundant macrofaunal species found at the AT locality. It is large by CNP standards. The paired feeding palps have been broken off both specimens, but their scars can be seen near the anterior (left) ends of the animals. A specialized construction apparent in the midsection of larger specimens (A), together with the relatively short tail (B) and head sections of some specimens, is suggestive of asexual reproduction.

hadal. Bathyal zones (e.g., SCB, SDT) have high proportions of polychaetes and amphipods; food-poor central oceanic (abyssal) regions (e.g., CNP) have smaller proportions of both polychaetes and amphipods, with correspondingly higher relative abundances of tanaids and isopods; hadal samples (e.g., AT) often have unusually high abundances of taxa which are poorly represented elsewhere (e.g., Aplacophora, Enteropneusta, Echiura). Shallow-water communities, while frequently consisting of fewer phyla, seem to be more highly variable in the proportions of the standing stocks comprised by those phyla (e.g., Friedrich, 1969).

An ecologically more meaningful comparison between deep-sea and shallow-water community compositions can, however, be made at the level of functional groups or guilds. Root (1975, p. 92) defines a guild as "a group of species that exploit the same class of resources in a similar way." Because there is very little direct evidence concerning

TABLE 10-3. Percentages (by number of individuals) of Macrofaunal Taxa Found in Box Cores (common names in parenthesis)

|  | SCB | SDT | CNP | AT |
|---|---|---|---|---|
| Annelid worms: |  |  |  |  |
| Polychaeta (bristle worms) | 76.6 | 75.5 | 55.1 | 49.0 |
| Oligochaeta | – | – | 2.1 | 1.4 |
| Hirudinea (leeches) | 0.1 | – | – | – |
| Crustaceans: |  |  |  |  |
| Cirripedia (barnacles) | – | <0.1 | – | – |
| Mysidacea (opossum shrimps) | 0.1 | 0.2 | – | – |
| Cumacea (lollipop shrimps) | 1.9 | 1.2 | – | – |
| Tanaidacea | 3.8 | 3.7 | 18.4 | 6.1 |
| Amphipoda (sand fleas) | 5.9 | 4.2 | – | 1.4 |
| Isopoda | 3.9 | 2.9 | 6.0 | 0.7 |
| Mollusks: |  |  |  |  |
| Gastropoda (snails) | 0.4 | 1.0 | 0.4 | 0.7 |
| Aplacophora | 0.4 | 0.5 | 0.4 | 10.5 |
| Bivalvia (clams) | 1.5 | 3.1 | 7.1 | 11.5 |
| Scaphopoda (tusk shells) | 1.2 | 0.3 | 2.5 | – |
| Other groups: |  |  |  |  |
| Porifera (sponges) | <0.1 | – | 1.1 | – |
| Cnidaria | 0.1 | 0.1 | 1.4 | – |
| Nemertinea (proboscis worms) | 0.6 | 1.6 | – | – |
| Pycnogonida (sea spiders) | – | – | – | 0.3 |
| Pogonophora | – | 0.3 | – | – |
| Sipuncula (peanut worms) | 0.4 | 0.5 | 0.4 | 0.3 |
| Echiura (gutter worms) | – | – | 0.4 | 3.0 |
| Priapulida | – | <0.1 | – | 0.7 |
| Bryozoa (moss animals) | 0.1 | 1.4 | 2.0 | 6.4 |
| Brachiopoda (lamp shells) | – | – | 0.7 | – |
| Enteropneusta (acorn worms) | 0.4 | 0.2 | – | 8.1 |
| Ophiuroidea (brittle stars) | 2.5 | 2.6 | 0.7 | – |
| Holothuroidea (sea cucumbers) | 0.1 | 0.5 | 0.4 | – |
| Pterobranchia | <0.1 | – | – | – |
| Ascidiacea (sea squirts) | <0.1 | 0.3 | 1.1 | – |

patterns of day-to-day resource utilization by deep-sea species (as noted, with cautions, in the introduction), deep-sea guilds are usually erected by analogy with better-known shallow-water relatives, in conjunction with examination of morphology and gut contents of deep-sea specimens. Very rarely, however, have investigators devised such classifications for all the taxa from a particular set of samples. In the following discussion, then, we often will be forced to jump from taxon to taxon in illustrating supposed generalizations.

One trend has nonetheless become clear across a wide spectrum of taxa: Suspension feeders show a marked decrease in abundance with increasing depth and distance from shore. No more than 7% (by numbers of individuals) of the total CNP macrofaunal taxa potentially are suspension feeders (Hessler and Jumars, 1974). A dramatic corroboration of the dearth of suspension feeders in the deep sea is the shift among guilds within major taxa as depth increases. Most deep-sea bivalves, for example, are deposit feeders (protobranchs) or even carnivores (septibranchs), while a much greater proportion of shallow-water bivalves are suspension feeders. Perhaps even more surprising is the tendency for deep-sea sea squirts (ascidians) to evolve carnivorous or deposit-feeding habits (Monniot and Monniot, 1978). Only in regions of anomalously high rates of food-supplying water flow (Lonsdale, 1977) or of high *in situ* chemoautotrophic (bacterial) production (Rau and Hedges, 1979), such as at the newly discovered hydrothermal vents near oceanic spreading centers (Ballard, 1977), are suspension feeders prevalent in the deep sea.

As they are in the classical Eltonian pyramid, carnivores too are relatively rare among animals captured in cores. Among the polychaetes, for example, we estimate (using the guilds defined by Fauchald and Jumars, 1979, together with our own observations) the following proportions (by numbers of individuals) of carnivores: SCB, 2% SDT, 13%; CNP, 12%; AT, 7%. Operationally, however, carnivores are difficult to distinguish from scavengers and omnivores; gut contents do not reveal whether food is taken alive. As pointed out above, the larger, more mobile members of the latter guilds have not yet been sampled quantitatively. Highly mobile scavengers are nonetheless known to be present at all deep-sea depths and to respond quickly (hours or less) and in large numbers to bait (animal flesh) lowered to the deep-sea floor (Isaacs and Schwartzlose, 1975). Scavenging amphipods (family Lysianassidae), in contrast to the general disappearance of other amphipod guilds at greater deep-sea depths, are attracted to bait at any oceanic depth. While fishes are quick to respond to bait at bathyal and abyssal localities, and fishes are known to occur at hadal depths, scavenging fishes appear to be missing from many trenches (Hessler *et al.*, 1978). The less motile (crawling as opposed to swimming) scavengers, such as brittle stars (ophiuroids) and quill worms [curious polychaetes which drag their tubes along with them (family Onuphidae)], by contrast, disappear more gradually with increasing depth and distance from shore and have not been observed to respond to bait at the CNP locality (Dayton and Hessler, 1972). One question left unanswered by these otherwise highly informative studies of species responding to bait, however, is what the animals do when bait is not present. Do they wait, or search for comparable windfalls, or do they hunt live prey?

Surprisingly, herbivores (or, more precisely, animals feeding on recognizable plant remains) are not wholly absent from the deep sea. The fauna colonizing wood has been the most thoroughly studied (Turner, 1977), but there are deep-sea species that seem to utilize more ephemeral plant structures such as sea grass blades (Jumars, 1976). Because of the proximity of most trenches to land, such vegetable matter and these sorts of vegetarians have been recorded relatively frequently there (Wolff, 1976).

The majority (usually 80% or more by numbers in macrofaunal taxa) of animals captured in deep-sea cores are deposit feeders, ingesting sediments and the smaller, less motile organisms they contain. Thus, while we tentatively call these animals deposit feeders, their exact trophic positions and impacts depend on the degrees to which they are incidental carnivores (Anderson, 1976; Dayton and Hessler, 1972; Feller *et al.*, 1979). Deposit feeders, then, are the animals to which the earlier remarks about size and areal density largely pertain. This guild, in turn, is divided between surface deposit feeders and species which feed on sediments below the sediment-water interface. In many deep-sea areas, deposit-feeding polychaetes, for example, are divided approximately equally (by numbers) between surface and subsurface deposit feeders (Fig. 10-5).

## Species Diversity

Perhaps more surprising than the above patterns, as judged by the wealth of speculation it has produced, is the finding of high species diversity in the deep sea. We will forgo reviewing the initial and now well-known documentation of this finding (Hessler and Sanders, 1967) and will instead make a few pertinent points by again relying on the

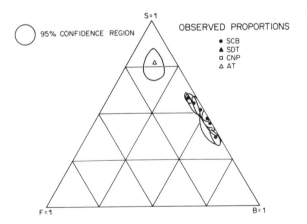

Fig. 10-5.   Relative proportions (by numbers of individuals) of suspension feeding (F), surface deposit feeding (S), and subsurface deposit feeding (B) polychaetes found at the localities of Table 10-1. In all but the Aleutian Trench, surface and subsurface deposit feeders share dominance approximately equally. Modified, with permission, from Jumars and Hessler, 1976.

internally consistent set of data from the North Pacific. The curves of Fig. 10-6 were generated from this data set via the Hurlbert (1971) rarefaction procedure. This statistically rigorous method uses data from field samples to provide unbiased (Smith and Grassle, 1977) estimates of the number of species that would most likely be seen in field samples of smaller size (fewer individuals). It provides an objective means of comparing species diversity in samples of varying sizes. A high slope near the origin in the curves produced by this method indicates great evenness (equitable distribution of individuals among species), while a high asymptote represents a substantial species richness (total number of species in the community).

Several facts are apparent from the figure. First, there is high variability in deep-sea species diversity from one basin to another; while species diversity is high in some deep-sea regions, it is not uniformly so. Second, high species diversity standardized to numbers of individuals does not necessarily imply that the number of species per unit area will

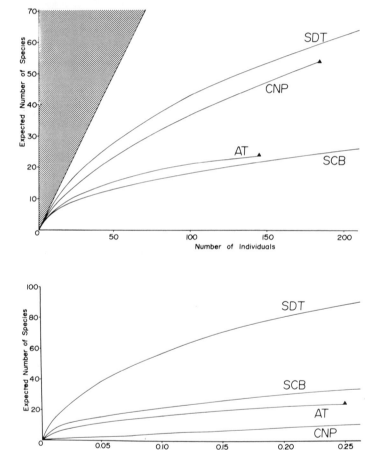

Fig. 10-6.   Polychaete numerical and areal diversity at the localities of Table 10-1. Triangles: total number of species and individuals collected; stippling: region cannot be occupied. With permission, from Jumars and Hessler, 1976, Fig. 2.

DEEP-SEA COMMUNITY STRUCTURE

be correspondingly high. Third, and not as obvious to those working in more accessible environments, is the fact that no one has fully enumerated the total species richness of any deep-sea macrofaunal community; none of the curves of Fig. 10-6, for example, closely approaches the asymptote where additional sampling would cease to provide additional species. Few authors have been foolish enough to estimate what such a total might be, but one guess (Jumars, 1975) is 177 to 212 polychaete species for the SDT, versus the 146 species (among 2125 individuals) actually sampled. The polychaetes comprise roughly half the macrofaunal species sampled there, so that one might expect the total macrofaunal community of the SDT to contain roughly $4 \times 10^2$ species.

It is not difficult to appreciate the intrinsic problem of obtaining basic ecological information such as microhabitat preferences, food resources, motilities, predator-prey relations, or generation times to accompany a species list of this length, even without the logistical problems of deep-sea sampling and observation. Consequently, the usual sorts of correlations between species diversity and habitat diversity, trophic position, or dietary specialization have been fragmentary. Two of these fragments are nonetheless germane to the theories we will discuss. First, high species diversity has not been demonstrated in all the deep-sea guilds. Most of the species of Fig. 10-6 are deposit feeders. Diversity in deep-sea scavengers (e.g., Shulenberger and Barnard, 1976), predators (e.g., Fig. 2 in Rex, 1976), and suspension feeders is rather low by contrast. Second, the greater part of deep-sea diversity is found among the more sedentary, infaunal species (Jumars, 1975, 1976). Successful explanations of high deep-sea diversity thus must hold for the small, relatively sedentary, largely infaunal deposit feeders that comprise the majority of this diversity.

## Size-Frequency Distribution and Population Dynamics

Among the other ecologically relevant parameters which can be measured or estimated from preserved samples are size-frequency distributions. Implicit in their estimation is the possession of a random sample of individuals with respect to body size. No piece of deep-sea sampling gear is ideal from this standpoint. Discrete samplers such as grabs and corers recover individuals which lived in close proximity to each other and may be of similar size [e.g., the maps of *Polyophthalmus* sp. (Polychaeta, Opheliidae) abundance and size in Jumars, 1978, Figs. 3 and 4]. Towed samplers such as dredges and trawls, on the other hand, have winnowing problems which are likely to eliminate selectively the smaller individuals.

Sampling problems aside, size-frequency distributions produced for deep-sea species typically have one of two forms. They may be bimodal or unimodal, but small individuals are almost invariably infrequent. While larger individuals will, on the average, be older, one must avoid the temptations to equate size increments with time increments. Figure 10-7 illustrates the difference and leads naturally to the question of population dynamics.

In many shallow-water environments, population growth and death rates may be followed by sampling a cohort, or group of organisms born at the same time, as they grow in size and dwindle in numbers. Such cohorts are tracked as moving (over time) peaks in size-frequency distributions. Most deep-sea size-frequency distributions produced for the same population over a number of seasons have, however, been monotonously

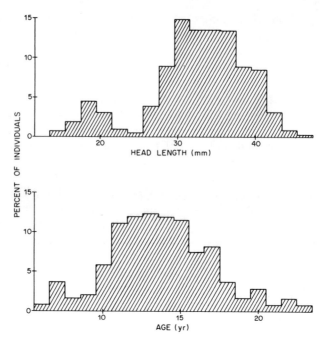

Fig. 10-7.   Size-frequency and age-frequency plots for *Nezumia sclerorhynchus*, a rattail fish found in the eastern Mediterranean Sea at bathyal (300 m to over 3000 m) depths. Based on data of Rannou, 1976.

similar, without apparent cohorts. Many deep-sea populations apparently reproduce year round, producing no large, synchronized cohorts (Rokop, 1974) and closing that convenient avenue to population dynamics information.

The other major way of gaining a knowledge of population dynamics is to estimate the ages of individuals. In shallow-water species, laboratory rearing often provides the necessary estimates of age (versus growth stage or size); similar methods have been unavailable to deep-sea biologists. Alternative aging techniques have been applied to deep-sea populations in only a few cases. One example is given in Fig. 10-7 and relies on the existence of annual growth rings in the otoliths of the fish. Turekian *et al.* (1975), by contrast, used an isotopic ($^{228}$Ra content of the shell) chronology to establish that the generation time of *Tindaria callistiformis* (a deposit-feeding, protobranch bivalve) is about 50 y, and its life span about 100 y. Unfortunately, the (95%) confidence intervals around these estimates are quite wide, being ±76 y for the latter estimate. Temptations to infer slow individual and population growth rates as special characteristics of the deep sea should further be tempered by long life spans documented for offshore, shallow-water species (e.g., Jones *et al.*, 1978), and by the more recent finding of Turekian *et al.*, (1979) which shows much more rapid growth rates in a (suspension-feeding) deep-sea bivalve from the Galapagos Rise. While fecundities of many deep-sea species do appear to be low, with few, relatively large ova being present at any one time (e.g., Sanders and Allen,

1973), the (unknown) frequencies of spawning must be compounded over the entire year for continuously reproducing species (Rokop, 1974) to allow comparison of reproductive rates with seasonally reproducing, shallow-water populations.

A third means of gaining information on population dynamics, namely on potential for reproduction and immigration, has been applied to opportunistic species. In these experiments, new resources (space, food) are provided and colonization is monitored. By placing wooden objects on the deep-sea floor at known times, Turner (1973) has established that some members of the genus *Xylophaga* (wood-boring bivalves) can have generation times as short as three months. In analogous experiments with azoic muds, Grassle (1978) demonstrated that *Nucula cancelleta*, a deposit-feeding, proto-branch bivalve, and *Polycarpa delta*, an ascidian, may reach maturity within two years. Seki *et al.* (1974) similarly showed that bacterial populations can grow rapidly in the deep sea, given the proper combination of bacterial inoculum and growth medium. While such demonstrations show how quickly deep-sea species *can* colonize and repro-duce, the degree to which they reflect natural rates depends critically upon the degree to which they mimic naturally occurring conditions.

## PROCESSES IMPLICATED

We present three plays. They explain the behaviors of the actors in very disparate ways. The first play suggests that the characters can best be understood as individuals, the second is a study of clans, and the third submits that the players are controlled by more varied interactions among groups of characters.

By treating, successively, the individual, population, and community levels of organization, we seek by example to show that processes operating at all these levels exert control over community structure. The lengths of the respective treatments have been set by the availability of compatible data and theories and do not reflect our relative degrees of faith in explanations at the respective levels. While we will cite a large number of past theoretical studies of the problems, we are not attempting a balanced review. The latter can be found elsewhere (Gage, 1978; Rowe, in press).

### Individuals

To gain insight into the lifestyles of the individuals that comprise deep-sea communities, we will explore applications of the theory of optimal foraging. The development of optimal foraging theory has recently been summarized by Pyke *et. al.* (1977). We will review a few of its basic tenets before we attempt deep-sea applications and will intro-duce additional complexities only as needed. The simplest and most frequently applied models assume that evolution has constrained the individual to maximize its net rate of energy gain. According to this model, the animal should engage in that foraging activity, and at that rate, which produces the optimal balance between the rate of energy gain and the rate of energy loss (Fig. 10-8).

We do not mean to imply that individual behaviors and species' ultimate capacities for survival are governed entirely by their net rates of energy gain. Just because the rate of food supply to the deep-sea benthos is relatively low does not mean that most individuals or populations there are food limited, as we shall discuss below. The surprising degree of success of optimal foraging theory as applied to a wide variety of environments (Pyke *et al.*, 1977), however, suggests that other constraints do not carry most animals far from their energetic optima. Guild by guild, then, we shall examine the ability of foraging theory to explain the behaviors and relative successes of deep-sea species.

Suspension feeders come in two not entirely distinct varieties: active and passive. Active suspension feeders use their own energy to pump water, while passive suspension feeders rely on externally produced flows, often accentuated by their morphologies or orientations (e.g., Vogel, 1978), to carry food through or onto their food-catching devices. To a first approximation, active filterers are independent of external flow speeds and directions, being influenced only by the concentration of suspended food particles. Passive filterers, on the other hand, depend on external flow velocity and often on flow direction (e.g., sea fans) as well as on suspended load. Active suspension feeders succeed only where particle concentrations repay the costs of pumping; passive suspension feeders succeed only where particle concentrations are high enough and flow conditions are predictable and fast enough to allow use of their particular passive collectors (Fig. 10-9). Suspended particle concentrations in general decrease with depth in the deep sea, but flow conditions become more predictable. Because the energetic costs of pumping water

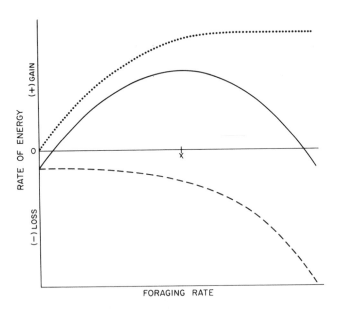

Fig. 10-8.    Gross rates of energy gain (dotted curve) and loss (dashed curve) of a hypothetical individual as a function of foraging rate. An optimal forager should forage at the rate (*x*) which maximizes its net rate of energy gain (solid curve).

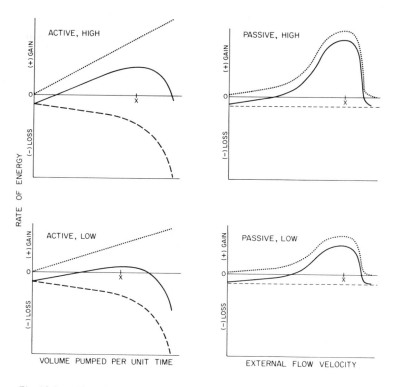

Fig. 10-9. Hypothetical optimization curves for both active and passive suspension feeders at high and low concentrations of suspended food. Symbols as in Fig. 10-8.

stay relatively constant with depth, active suspension feeders disappear as suspended loads dwindle (Jørgenson, 1966). The large suspension feeders that are most obvious in deep-sea photographs (e.g., Heezen and Hollister, 1971) all appear to be passive and to utilize the structure of the benthic boundary layer effectively.

Friction with the bottom causes physical mixing of deep-sea bottom water from ten to several hundred meters above the bottom, depending on the velocity of these bottom water masses and on bottom topography (e.g., Armi, 1978). This mixing maintains a higher suspended load near the bottom than is found in the clearer, overlying water. The near-bottom, slightly cloudier (or "nepheloid") layer typically contains 50 to 100 $\mu$g 1$^{-1}$ of solids, a very low suspended load by shallow-water standards. The friction, however, keeps water in the thin layer immediately over the bottom (the viscous layer of Table 10-4) moving very slowly and regularly, rather like molasses (in June). Many of the passive deep-sea suspension feeders are large (such as the passively feeding groups of sea squirts; Monniot, 1979), or have long stalks (such as the glass sponges, e.g., Rice *et al.,* 1979, Pl. 5), extending them into water moving fast enough to induce flow through their feeding structures and to bring by numerous particles (per unit time) for potential capture. The younger stages of such passive filterers are often found on objects such as manganese nodules or bits of wood which protrude through and disrupt the viscous sublayer, thereby providing a means of escaping the slowest flows during ontogeny.

TABLE 10–4. Typical Structure and Nomenclature of the (unstratified) Deep-Sea Benthic Boundary Layer[a]

| Layer Name | Distance Above Bottom (m) | | Characteristics |
| | Lower Boundary | Upper Boundary | |
| --- | --- | --- | --- |
| Viscous sublayer | 0 | $10^{-2}$ | Slowest, viscous flow |
| Logarithmic layer | $10^{-2}$ | $10^0$ to $10^1$ | Turbulent flow; velocity rapidly increasing with height above the bed (to $10^1$ cm $sec^{-1}$); most rapid mixing. |
| Bottom mixed layer | None usually given | $10^1$ to $10^2$ | Weakly turbulent flow; mixing of suspended and dissolved con- stituents. |

[a]For further details, see Wimbush (1976) and Armi (1978).

Close scrutiny of this application of foraging theory, however, leaves at least two problems unsolved. The resources utilized in the field by suspension feeders in general and by deep-sea suspension feeders in particular have not been identified clearly. This problem is not particularly severe with respect to application of the theory so long as food (now speaking only of material actually digested and absorbed) quality stays constant and its quantity remains strictly proportional to total suspended load. Secondly, if there are a few suspension feeders in the deep sea, why are they not more abundant? It is virtually inconceivable that the few existing suspension feeders could cause any appreciable decrease in the suspended resource. Either these suspension feeders select microhabitats with enhanced particle fluxes (i.e., all the good spots for larval settlement have been taken) or their abundance is controlled by factors other than foraging success (e.g., predation).

Resource identification at first seems no problem for the abundant scavengers attracted to meat bait in the deep sea. They are clearly generalists in terms of the variety of baits to which they will respond; in one experiment at 1200-m depth in the San Diego Trough[1] the ever-present lysianassid amphipods responded to dead fish, raw egg, and bologna with roughly equal avidity. Our observations on *Eogammarus confervicolus*, an intertidal and shallow subtidal gammarid amphipod in Puget Sound, however, suggest some caution in concluding from observations of bait that deep-sea, scavenging amphipods rely solely on such windfalls from the plankton and nekton (drifting and swimming organisms of the water column). *E. confervicolus* is attracted in large numbers to baits at least as varied as dead opossum and filamentous algae. Immunological techniques (Feller *et al.*, 1979), our visual field observations, and our laboratory experiments (unpublished), however, reveal that this species also feeds on a wide variety of smaller in-

[1]Jumars and Hessler, unpublished.

fauna and that many of these items are taken alive. Wolcott (1978) showed, analogously, that the beach-dwelling ghost crab *Ocypode quadrata,* which is generally regarded as a scavenger, obtains most of its food through active predation. Without similar observations on deep-sea scavengers, we cannot know what proportions of their diets come from the sinking of planktonic and nektonic remains versus dead or easily captured infauna and epifauna. Available information on deep-sea demersal (near-bottom living) fishes, however, suggests that even closely related species can differ markedly in the degree to which they depend on benthic versus nektonic and planktonic foods (e.g., Pearcy and Ambler, 1974), but strict specialization has not been seen.

One can argue on the basis of foraging theory that when additional experiments and observations become feasible for deep-sea scavenging amphipods that they will be found to be generalists as well, i.e., that their diets will be found to include easily captured, living fauna. The simplest optimal foraging model that leads to this conclusion is neatly outlined by MacArthur (1972). He divides foraging into four phases: deciding where to search; searching, looking out for palatable items; deciding whether to pursue a located food item; and pursuing (and possibly capturing and eating) the item. The animal should search where the expected yield is maximal; we will return to this consideration later. With respect to the third phase, "an animal should elect to pursue an item if and only if, during the time the pursuit would take, it could not expect both to locate and to catch a better item." The division of time (=energy) between search and pursuit in order to maximize energy intake is now easy to make and depends in a general way on the overall abundance of food. When food is abundant, such that food items of several kinds can be located at any time (are always in "sight"), and if more energy must consequently be spent in pursuit than in search, the animal should elect only those food items whose ease of capture and caloric content provide the greatest net gains, making it a specialist. If food is scarce, on the other hand, and most energy must be spent in searching, no easily caught and ingested food item, once located, should be passed over. The latter situation seems more likely for epibenthic scavengers in the deep sea (naturally occurring carrion rarely having been photographed or found in samples), arguing that they should be generalists.

The argument is complicated somewhat by the probably enhanced detection abilities of deep-sea predators via any one of a number of sensory modes. Deep-sea diffusion rates are relatively low, especially in the viscous sublayer, making chemical gradients more persistent in the deep sea than elsewhere. Background acoustic noise is low, making weak acoustic signals relatively easy to detect. Turbulence levels are low by comparison with shallow-water benthic boundaries, making the sorts of high-frequency pressure waves generated by prey (e.g., Ockelmann and Vahl, 1970) easier to detect. Even some visual cues may be easier to detect in the deep sea than in sunlit, near-surface waters; there is no background "noise" to hide even weak bioluminescence. While such enhanced sensory possibilities may allow more prey items to be detected in a given period than would otherwise be possible, the prey may also more easily detect and evade their predators. We will avoid such potential co-evolutionary interactions by restricting further discussion to carrion feeders. For the sake of brevity, and to tie our discussion to the preceding material, we will discuss only the chemosensory mode. We do not deny that any of the other modes may be important (e.g., Hawkins and Rasmussen, 1978; Reid and Reid, 1974).

As implied in the mention of the first phase of foraging (above), optimal foraging theory suggests not only what should be eaten, but also what foraging pathways should be useful in finding and harvesting food. Hydrodynamics of the benthic boundary layer dictate divergent strategies for two kinds of searchers capable of detecting "prey" at a distance. Because a chemical stimulus in the viscous sublayer will move much more rapidly downstream than cross-stream, animals restricted to the bed (e.g., brittle stars or quill worms) should spend most of their search time traveling perpendicular to the flow. Once a prey item is detected, such benthic scavengers should proceed upstream (oriented movement with respect to flow being called "rheotaxis"), turning only when they fail to detect the chemical stimulus. Time-lapse photographs of scavenging ophiuroid movement in response to bait should demonstrate such behavior. Scavengers capable of swimming or hovering off the bed (and out of the viscous layer) might adopt quite a different strategy, more akin to that used by mosquitoes for long-distance host location (e.g., Kennedy, 1939). Just as the passive suspension feeder can use the energy of the external flow to avoid energy expenditure in pumping, the hovering scavenger can use turbulent mixing within the bottom mixed layer to bring the chemical signal from a food parcel sitting on the bottom. The fact that this chemical plume is widened and extended vertically by turbulent mixing downstream of its point source (much the way the smoke from a smokestack broadens in both vertical and horizontal extent downwind) means that the scavenger (or mosquito) need not be directly downstream of (or at the same vertical elevation as) its respective food item in order to detect it. The optimal forager should hover if the improved detection ability provided by turbulent diffusion provides otherwise undetected food in excess of the additional energetic costs of hovering and pursuing the food item from the chosen hovering height. Such hovering costs can be reduced and amortized by using buoyancy mechanisms such as fat bodies or gas bladders.

What should this hovering height be? We assume that a piece of carrion emanates a chemical signal proportional in strength to its size. We further assume that a threshold concentration for detection exists below which the carrion cannot be sensed. Above this threshold concentration, we assume that the stimulus elicits a pursuit response which might be as simple as positive rheotaxis (moving upstream) in conjunction with a positive geotaxis (swimming downward), with more frequent turning when the signal is lost. Similar mechanisms, as alluded to above, have been demonstrated in insects (Shorey, 1977). Note that we do not invoke a chemical gradient as a necessary prerequisite of this location mechanism; such gradients are unlikely in the presence of turbulent mixing. Given the threshold concentration for response, however, larger pieces of carrion can be detected at greater distances from the seabed and at greater distances from the source of the stimulus. The optimal hovering height can then be deduced by analogy with the optimal foraging height in visually searching buzzards or with optimal perching height in kingfishers (MacArthur, 1972, pp. 67-68; refined and extended by Orians and Pearson, 1979). Just as a kingfisher should perch at the height from which "the greatest number of grams of fish per day can be captured," so should the scavenger hover. This height will depend upon, among other parameters, the effective vertical diffusion rates of the chemical stimulus and the size-frequency distribution of carrion. If competition for carrion is prevalent, then swimming speed of the scavenger (Hessler and Jumars, 1974) and its ability to sequester food once it has arrived would also be important. Whereas catches of scavengers above the bottom have been interpreted as being due to attraction of scavengers off the bed (e.g., Smith *et al.,* 1979), we suggest instead that these scavengers may spend most of their time hovering well above the bed. We would predict that larger scavengers, which could more effectively sequester large food parcels, would hover higher above the bottom and that larger pieces of bait would attract animals from higher in the water column. Animals hovering higher above the bed must, however, spend more time and energy pursuing a detected item than do scavengers hovering lower in the water column. Hence, based on the four-phase foraging model presented above, we would predict that scavengers further from the bed would be more specialized and less likely to prey incidentally on easily captured benthic individuals.

Geographic variation in the structure of the benthic boundary layer would also be expected to influence foraging strategies of hovering scavengers. Increased turbulence intensity could decrease optimal foraging height above the bed by decreasing the height above the bed at which threshold concentrations for detection would be reached and might increase the locomotory costs of hovering. Decreased turbulent mixing, on the other hand, might decrease the advantage of higher hovering in a rather different way. If upward diffusion were very slow, carrion could be detected and consumed by lower-hovering or epibenthic scavengers before it could be detected higher in the water column. Thus, it is tempting to speculate that the apparent absence of fishes and other large (presumably hovering) scavengers in trenches (Hessler *et al.,* 1978) is due to the weak currents throught to prevail there (Bishop and Hollister, 1974).

Geotaxis and rheotaxis toward upstream prey, however, is not the only conceivable pursuit strategy that could be used by a hovering scavenger. Hamner and Hamner (1977) have documented that neritic zooplankton can detect sinking carrion by following the scent trail left during settling through the water column. Despite the relatively low eddy diffusion rates in the deep sea, we do not feel that the flux of large particles arriving in the deep sea is sufficient to give a hovering scavenger a reasonable probability of detecting such an item during its fall. Even in the deep sea, a vertical scent trail would dissipate in a matter of minutes. Carrion on the bed, by contrast, would emanate a chemical plume for hours or even days, depending on its size and rate of consumption.

We can find fewer explicit predictions to apply to deep-sea deposit feeders than we have generated for scavengers for several intertwined reasons, the majority of which apply to shallow water as well as to the deep sea. First, foraging theory for deposit feeders is in its infancy (Levinton and Lopez, 1977; Taghon *et al.,* 1978). Second is the recurrent problem of identifying the resources used and the resource axes on which particle selection might occur (Self and Jumars, 1978). Third, infaunal deposit feeders are difficult to observe directly, even in shallow water. Consequently the applications we propose are even more tentative than those for suspension feeders and scavengers. Sessility in deposit feeders is feasible only if the morphologically available (e.g., via tentacles) foraging area provides resources at a sufficient rate to meet metabolic demands (Jumars and Fauchald, 1977), either by microbial renewal *in situ* (Levinton, 1972; Levinton and Lopez, 1977), or by advection. The prevalence of motility among the deposit feeders of food-poor deep-sea regions (Fig. 10–10) suggests that the requisite renewal rates are not met there in general.

Assuming that the resource (e.g., microbial coatings on sediment grains) is renewed slowly, what sorts of foraging strategies should be successful? Foraging theory suggests that the animals should move in such a fashion as to minimize ingestion of deposits from recently depleted foraging sites. One way of accomplishing this goal is to produce fecal pellets that either by virtue of location (e.g., vertical stratum in the sediments) or of physical-chemical characteristics (e.g., size) are unlikely to be ingested. Another, not necessarily mutually exclusive, method available to motile deposit feeders is to avoid returning to recently foraged areas. We know of foraging pathway evidence for only two deep-sea feeders, enteropneusts and herd-forming urchins. [See Kitchell (1979), however, for a stimulating discussion of fossil and deep-sea foraging trails irrespective of their producers.] Enteropneusts [acorn worms] appear to avoid crossing their feeding trails (e.g., Fig. 4 in Thiel, 1979). They may, however, turn more tightly in response to encountering food-rich deposits (e.g., Risk and Tunnicliffe, 1978), which may account for the interspersion of relatively straight tracks and tightly coiled ones. Herds of the urchin *Phormosoma placenta* on the continental slope, as described by Grassle *et al.* (1975), may form and move as an alternative solution to the recrossing problem (as Grassle *et al.* imply). Cody (1971, 1974) studied finch flocks in the Mohave Desert and hypothesized that they were "return time regulators," minimizing the variance of the time intervals between successive visits to a foraged area and adjusting the mean return time to a given area to match the rate of resource renewal in that area. The precise predictions of the flock turning frequency in Cody's model are heavily dependent upon the size of the habitat and the turning behavior of the flock at the boundary of the habitat (Pyke, 1978; Pyke *et al.,* 1977). We cannot even hazard a guess for either of these crucial parameters for *Phormosoma* and so are unable to make predictions in terms of precise turning frequency. In the absence of any boundary, there is nothing in the

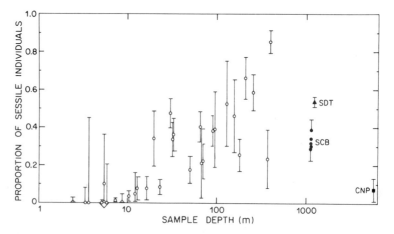

Fig. 10-10.   The proportion of sessile polychaete species as a function of depth offshore from southern California. Bars indicate 95% confidence intervals. Modified, with permission, from Jumars and Fauchald, 1977, Fig. 7.

DEEP-SEA COMMUNITY STRUCTURE

theory of optimal foraging pathways that would preclude the herd traveling in a straight line. If there is a boundary which necessitates *Phormosoma* returning to previously foraged areas, the pattern of turning and the rate of return to a given area should be coupled with the resource renewal rate. Herd size should also be proportional to the degree of resource limitation. With no resource limitation, herds should not form, while herds should increase in size with increasing resource limitation. We realize that there are a considerable number of explanations, other than optimal foraging, that could be invoked to explain the herding behavior of *Phormosoma,* including avoidance of predation and reproductive aggregation. We have opted for the present explanation to complete our survey of applications of foraging theory to the entire spectrum of major deep-sea guilds and to offer some predictions which might be testable in the deep sea.

## Populations

At the population level, there is a particularly wide variety of often conflicting theories from which to choose (Stearns, 1976). We will avoid the more elaborate treatments of age-specific reproductive behaviors in order to avoid a gross mismatch in detail between theory and available deep-sea observations. We will again utilize an optimization approach, but rather than maximizing net rate of energy gain by an individual, we will maximize either population growth rate or the probability of long-term persistence of a population.

Despite the controversy surrounding its application (e.g., Stearns, 1977), we will first introduce the concept of "*r*" versus "*K*" selection. These terms derive from the usual symbology in the (Verhulst-Pearl) logistic equation (Wilson and Bossert, 1971) (Fig. 10-11), which has been reasonably successful in describing the growth rates of many

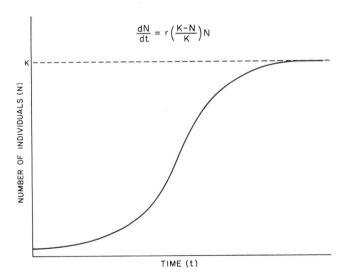

$$\frac{dN}{dt} = r\left(\frac{K-N}{K}\right)N$$

Fig. 10-11.    Graph of the logistic equation, which is the basis of *r-K* selection theory. *K* is the carrying capacity, while *r* is a measure of the potential rate of population growth.

laboratory and field populations. Populations which are disturbed so frequently that they rarely approach their carrying capacities (population levels supportable by the available resources) ensure their persistence by investing much of their energies into reproduction, (i.e., into maximizing $r$). A particularly lucid mathematical treatment of the critical frequency of disturbance is given by Southwood *et al.* (1974), but we will avoid such details, again due to the lack of applicable deep-sea data. The most effective methods for achieving high population growth rates have been recognized for some time (Table 10-5).

Optimal means for ensuring persistence of populations which frequently approach their respective environments' carrying capacities, on the other hand, are not agreed upon with comparable unanimity (Stearns, 1977). While agreement as to the specific life history tactics which achieve this end most effectively is lacking, it is clear that, if the carrying capacity stays constant, *(K)* selection will favor the population that keeps its abundance as high and as constant as possible from generation to generation (Table 10-6). Part of the problem in defining optimal life-history tactics of such a "*K* strategist" is that both intraspecific and interspecific competition are likely to occur as populations approach their carrying capacities, and similar population behaviors may not be optimal under both sorts of competition: The component problems of zero population growth are complex and may have competing solutions.

While it would thus be difficult to decide whether species are *K* selected, it may be easier to assess whether or not they are *r* selected. Grassle and Sanders (1973) undertook this sort of analysis for deep-sea benthos (a term including all organisms that live in the benthic environment) and concluded that deep-sea species in general have smaller clutch sizes than do comparable shallow-water species, even when corrected for the relative sizes of the animals. They pointed out, however, that the time markers needed to assess any of the other population growth enhancing methods of Table 10-5 are generally lacking. In addition, it is becoming ever clearer that there are some deep-sea environments that are more ephemeral or variable than others, e.g.: logs utilized by boring bivalves and their associates (Turner, 1977); hydrothermal vents utilized by suspension feeders and their associates (Ballard, 1977); catastrophic, avalanche-like ("turbidite") deposits which decimate local populations but whose organic contents subsequently can be mined by deeper burrowers (Griggs *et al.*, 1969); regions of at least occasionally erosive flow velocities (Gage, 1977; Greenwalt and Gordon, 1978); and areas of fluctuating oxygen concentration (e.g., some of the bathyal basins discussed by Emery, 1960). Hence, the

TABLE 10-5. Life-History Tactics Enhancing Population Growth Rate $(r)$, Listed in Generally Decreasing Order of Effectiveness (the first is by far the most effective)[a]

---

1. Decrease age at first reproduction.
2. Increase number of individuals per clutch.
3. Increase number of clutches born per unit time.
4. Increase reproductive lifespan.
5. Reduce prereproductive mortality rate.
6. Reduce mortality rate of reproductive stages.

---

[a]After Cole, 1954.

TABLE 10-6.  Example of $K$ Selection[a]

| Popu-lation | Initial Size (no.) | Calculation of Reproductive Success Over Time Interval | | |
|---|---|---|---|---|
| | | Time 0 to Time 1 | Time 1 to Time 2 | Time 2 to Time 3 |
| A | 1000 | $1.1 \times 1000 = 1100$ | $1.0 \times 1100 = 1100$ | $0.9 \times 1100 = 990$ |
| B | 1000 | $1.5 \times 1000 = 1500$ | $1.0 \times 1500 = 1500$ | $0.5 \times 1500 = 750$ |

[a]Note that although both populations on the average leave one individual for each individual starting a time interval, the population with less variation in reproductive success (Population $A$) leaves more progeny in the long run.

ability to make sweeping generalizations for the entire deep-sea fauna should not be expected.

In attempting to apply $r$-$K$ selection models, we thus again run headlong into two poorly answered questions which arise in each of our three theoretical sections, namely, "What are the resources utilized by each of the deep-sea guilds, and what sorts of disturbances (if any) depress deep-sea populations below their respective carrying capacities?" Levinton (1972) has presented some cogent arguments that deposit feeders will, by virtue of being buffered from environmental variability by the sediments, experience less environmental variability than will suspension feeders. How stable the resource supply for scavengers might be is unknown; it depends on the rate of carrion production (or arrival) within areas of the size in which individuals forage. The succeeding discussion, then, will be restricted implicitly to deposit feeders, the group for which small clutch size and relatively low abundance of juveniles hold most surely (Grassle and Sanders, 1973), and to the physically stable regimes typical of the abyss.

Grassle and Sanders (1973), in response to Dayton and Hessler's (1972) suggestion that predation regulates population size of the smaller deep-sea species, cite the aforementioned small clutch size of and the preponderance of larger (older) life stages as evidence that neither predation nor any other sort of disturbance is likely to exert a strong effect on life-history traits in the deep sea. We suggest, alternatively, that this size-frequency distribution can arise precisely because of the sort of size-selective predation that Dayton and Hessler (1972) hypothesize. Briefly, Dayton and Hessler suggest that the deep-sea benthos is comprised largely of generalist "croppers," animals that will eat any food item small enough to ingest. Deposit feeders dominate the benthos, and there is little reason to think that they would (or could) reject larvae and small juveniles from the items they select for ingestion (Isaacs, 1976; Self and Jumars, 1978).

We begin, in development of an alternative to $K$ selection for explaining the albeit meager data on deep-sea life-history traits, at the same deep-sea environmetal feature that provides a springboard for Dayton and Hessler (1972) as well as for Grassle and Sanders (1973)—the extreme physical constancy of the deep-sea milieu. We find no compelling reasons (and certainly no data) to suggest that predation is more intense, in an absolute sense, in the deep sea than elsewhere among benthic environments. Temperature, salinity, and oxygen stresses capable of killing adults as well as juveniles are all but absent, however, making predation *relatively* more important as a mortality source. Particularly because deposit feeders are generally selective of smaller particles (Taghon et al., 1978), we concur with Dayton and Hessler (1972) that small forms are exposed to more intense and more diverse predators (more mouths accommodating them) than are larger ones. Hence, larvae and juveniles of deep-sea species will be subjected to higher mortality rates, and likely also to more variable mortality rates, than will adults. To reiterate, the essential difference between shallow water and deep sea is that the relatively more frequent and severe environmental variations (capable of eliminating both adults and juveniles indiscriminately) in shallow water mitigate the predation-produced disparity between adult and juvenile mortalities.

The simplest model we can find that will evoke this behavior is that of Charnov and Shaffer (1973), as lucidly summarized by Schaffer and Gadgil (1975). We use the symbols (slightly modified) from the latter reference as follows (all per unit generation time):

$\lambda$ = the rate at which a population multiplies

$B$ = the number of larvae produced

$c$ = the probability that a larva survives to reproduce

$p$ = the probability that an adult survives (through one generation time)

For a population in which the adult dies upon shedding gametes (denoted by the subscript "$s$," for semelparity or one reproductive event per adult), then $\lambda_s = cB_s$. For a closely related species (i.e., one having no difference in $c$) surviving beyond first reproduction (subscript "$i$" for iteroparity or repeated reproduction), the reproductive rate is $\lambda_i = cB_i + p$. What condition, then, is necessary for $\lambda_i > \lambda_s$? It is $B_s < B_i + p/c$. By our previous arguments, $p/c$ is higher in the deep sea than in shallow water, making multiple reproductions (iteroparity) and relatively higher abundances of adults more likely in the deep sea.

If the greater diversity of predators to which juveniles are susceptible or some other biogenous variability (Jumars, 1975, 1976) makes juvenile mortality also more variable than adult mortality, then the relative fitness of the iteroparous form is increased further. Let $\overline{\lambda}$ = the mean (per generation time) rate of multiplication, and let the probability of larval survival be $c(1 + q)$ in "good" times and $c(1 - q)$ in "bad" (where $1 > q > 0$). Then

$$\overline{\lambda_s^2} = (1 + q)\, cB_s\, (1 - q)\, cB_s$$
$$= (1 - q^2)\, c^2 B_s^2$$

and

$$\overline{\lambda_i^2} = [(1 + q)\, cB_s + p]\, [(1 - q)\, cB_s + p]$$
$$= (cB_i + p)^2 - q^2 c^2 B_i^2$$

Then for $\overline{\lambda_i}$ to exceed $\overline{\lambda_s}$, it follows that

$$B_s^2 < B_i^2 + \frac{2\, cpB_i + p^2}{c^2 (1 - q)^2}$$

Hence, the larger variability ($q$) in survival of larvae, the more likely it is that this inequality will hold, and the iteroparous population will be favored.

Besides favoring iteroparity and consequently greater standing stocks of adults, more intense and more variable predation on larvae and juveniles will favor evolution of fewer, larger progeny (Schaffer and Gadgil, 1975) and will favor rapid growth past the predation-susceptible sizes (Connell, 1975; Lynch, 1977). Unfortunately, in terms of providing an unequivocal test of whether predation is an important cause of the evolution of low fecundity, with few large progeny per clutch in deep-sea species, the theory of $K$ selection (Gadgil and Solbrig, 1972) and other considerations of increased competitive demands on young stages (e.g., Harper *et al.*, 1970; Schaffer and Gadgil, 1975) suggest that these same life-history tactics will result from the evolutionary pressures of competition.

Whatever its ultimate and proximate causes, the existence of few large larvae, often without any obviously free-swimming stages, in deep-sea species (i.e., in the majority of deposit feeders of physically stable regions) suggests a limited requirement for long-distance dispersal. Suitable habitat for deposit feeders is more or less continuous and easily accessible, physically caused disasters are rare, and the environment may be as variable on small scales (10 cm) as on large scales (km) (Jumars, 1976, 1978). Both long-term and short-term selective forces for high dispersal ability (summarized by Strathmann, 1974) thus appear to be lacking. Exceptions to this generalization are again to be expected in the more ephermeral and unstable deep-sea habitats.

## Communities

So much theory has been written about deep-sea community structure—and with so little consensus—that we would be remiss to omit at least a brief review before entertaining the theory we regard as the best contender for providing new insights. Most of this past discussion has centered on species diversity, and so will ours. The application of theories to the empirical finding of high deep-sea diversity has been a meandering process and has treated phenomena on both evolutionary (Slobodkin and Sanders, 1969) and ecological time scales (Dayton and Hessler, 1972). We will limit our attention primarily to what we interpret to be the focus of recent discussions of factors potentially maintaining (rather than producing) high deep-sea diversity, (i.e., to ecological rather than evolutionary time scales). Initially we will address three questions: Are diverse deep-sea assemblages resource limited? If so, how is the resource partitioned? If not, how is resource limitation for these diverse assemblages prevented?

Early theoretical treatments (i.e., Sanders, 1968, 1969; Slobodkin and Sanders, 1969) assumed that competition is a strong structuring force in communities inhabiting stable environments and hence that resources are indeed limiting these deep-sea populations. Various resource partitioning mechanisms were invoked (Slobodkin and Sanders, 1969, p. 86): "We might expect stenotopy, complex behavior of rather specific and stereotyped kinds, and the possibility of specialization to specific foods, hiding places, hunting methods, and environmental periodicities—in short, to the details of the most significant parts of its environment." Dayton and Hessler (1972) challenged the assumption of resource limitation. They suggested that if resource partitioning occurred, it would be unlikely to occur on the basis of habitat (space) or time "because of the high physical homogeneity, both temporally and spatially, of the deep-sea environment" (*op. cit.*, pp. 199–200). Hence, they reasoned that measures of food-type specialization would provide a test of whether food resources were indeed limiting for most deep-sea populations—that the documentation of catholic taste would support the view that most deep-sea populations are limited by predation, whereas the finding of specialized diets would support the alternative position.

Grassle and Sanders (1973) rebutted the idea that most deep-sea populations are predator controlled, mainly by citing life-history data. They suggested that (p. 644), "In a community that is intensively cropped (or suffers a sustained mortality from any source), the population of prey species will be composed preponderantly of younger stages." We have discussed alternative interpretations of the general rarity of smaller life stages above. Size-frequency histograms cannot refute unequivocally the limitation of populations by predation.

Grassle and Sanders (1973) suggested, however, that food-type specialization is not the only likely resource partitioning mechanism. As they proposed, subsequent study (reviewed by Jumars and Eckman, in press) has revealed a great deal of spatial inhomogeneity and has implicated local successional series (Richerson *et al.*, 1970) in structuring the benthic community. Jumars (1975, 1976) has argued that a large part of the environmental heterogeneity is biogenous and that the deep-sea environment is unusual in permitting relatively fragile, small-scale structures (e.g., animal tubes and burrows) to persist without physical homogenization. Thistle (1979) tested this hypothesis and found that at least part of the diversity of harpacticoid copepods in one deep-sea community may be attributed to microhabitats provided by other animals. The existence of environmental heterogeneity vitiates the proposed critical tests of Dayton and Hessler (1972). If food is limiting, one would expect habitat specialization and not food-type specialization or resource partitioning by time of feeding (Schoener, 1974); an animal can obtain more food by feeding in a habitat or microhabitat which is not depleted by its competitors. In fact, population limitation by predation would probably permit more food-type specialization among the limited populations by leaving more resources and more kinds of resources available—exactly the reverse of Dayton and Hessler's predictions.

To demonstrate the complexity of the above issues alone, we have, however, omitted one critical question from our list: Whichever mechanism prevents competitive exclusion, how is a higher diversity maintained in the deep sea than in shallow water? We can envisage four possibilities: (1) Resources are more finely partitioned due to refined biological interactions (Slobodkin and Sanders, 1969) in the deep sea; (2) predators are somehow more effective at controlling populations in the deep sea (implicit in Dayton and Hessler, 1972); (3) some combination of these two explanations holds; or (4) neither predators nor competitors are appreciably different in shallow water versus deep sea, but the unique deep-sea environment somehow affects the interactions to maintain high species diversity.

We have already suggested that if finer resource partitioning occurs in the deep sea than elsewhere, it is probably accomplished on the basis of (micro)habitats. The arguments for increased efficacy of predation in the deep sea are homologous with our previous reasoning with respect to the increased importance of size-selective predation.

If predator populations are more rarely or never decimated by physical disturbances, predation becomes a relatively more important source of mortality, and prey become less likely to escape predator control and thereby to risk competitive exclusion (Dayton and Hessler, 1972; Connell, 1975). In support of such reasoning, Dayton and Hessler (1972) and Menge and Sutherland (1976) show that deep-sea species of lower trophic levels are more diverse, and Rex (1977) provides indirect evidence (via their weak patterns of zonation) that infaunal deposit feeders are kept below population levels at which strong competition occurs.

Huston (1979), in $r$ and $K$ terms, gives a more mechanistic explanation of deep-sea predator efficacy. He points out, assuming an abundant supply of species with an assortment of life histories, that dynamic equilibrium maintaining relatively high species diversity will be produced by a particular balance between population growth rates and (density-independent) disturbance rates. What would be special about deep-sea predator-prey relationships would be this balance between predation rate and prey population growth rate. If population growth rates of prey are slow, then the highest species diversity can be maintained by a relatively modest predation rate. Where growth rates are higher, as suggested for hydrothermal vents by Turekian *et al.*'s (1979) data on clam ages, a higher disturbance rate would be necessary to maximize species diversity via Huston's (1979) dynamic equilibrium process.

For lack of an appropriate theory, we will not discuss the possibility that both predation and resource specialization in concert are effective in promoting high deep-sea diversity. We will, however, discuss how physical habitat differences may make both predation and competition less effective at eliminating deep-sea species. This discussion is in the spirit of the "neutral model" of Caswell (1976). While in our chronological development of deep-sea diversity theory we have left the question until last, it should probably be asked first, i.e.: Can one develop a reasonable model explaining higher species diversity in the deep sea versus shallow water without invoking any differences either in the processes of resource partitioning via competition or in predation? The need for discrediting a neutral model that could account for this phenomenon—without invoking special biology—can be likened to the need for rejecting the null hypothesis in statistical hypothesis testing. Why invoke the special when the ordinary will do?

There is reason to suspect (Jumars and Eckman, in press) that the deep seabed may exhibit more environmental heterogeneity on small scales than do shallow, soft substrata. Increased environmental heterogeneity is perhaps the single most universally accepted correlate of higher species diversity. As Menge and Sutherland (1976) discuss in the context of marine benthos, such environmental heterogeneity buffers against competitive exclusion by making resource partitioning on the basis of habitat more likely and buffers against overexploitation by predators by providing structural (habitat) refuges for prey. If the deep sea is indeed a more heterogeneous environment than the physically homogenized shallow seabed, then there is good reason to expect both predation and competition to be less effective at eliminating species.

Osman and Whitlatch (1978) formalize such a patch model for species diversity. They also deal with processes over evolutionary time, but we will consider only their discussion of relatively short-term dynamics. Their model suggests that deep-sea species diversity will depend heavily on four parameters, i.e., areal extent of the community in question, patch sizes within it, disturbance (patch-clearance) rates, and animal motilities (dispersal rates among patches). The dearth of estimates for any of these parameters is sobering, but Jumars (1976) has argued on the basis of spatial sampling that the typical deep-sea "patch" may, for lack of physical disturbance, be close in size to that of the individuals which comprise the community; the relevant patch size or extent of a given microhabitat in the deep sea may be the area occupied or influenced by a single individual.

Connell and Slatyer (1977) recently have presented a model of succession that is uniquely suited to this spatial scale. They conceptualize succession as an individual-by-individual replacement process; given that a spot is occupied by species *X*, there is a finite probability that at the next time of observation it will be occupied by species *Y*. Their family of models subsumes the various possibilities discussed above and again points toward the kinds of data that desperately are needed for the deep sea. We find Markov representation a convenient means of formalizing their models, as can be illustrated with the classical ("facilitative" in Connell and Slatyer's terms; meaning that early colonists make the environment more suitable for later colonists and less suitable for themselves) picture of succession (Fig. 10-12).

We will not introduce the Markov approach in detail; little would be added to the conceptual content. The best such introductions we know of are the classic text by Kemeny and Snell (1960) and the considerably shorter but quite intriguing version by Roberts (1976, Chap. 5). A selection of past applications of Markov models to succession can be found in Horn (1975, 1976), Maynard Smith (1974, Chap. 6), and Kauppi *et al.*, (1978). Knowing the short-term ("transition") probabilities that one species will be replaced by another or that the space occupied by the individual will be vacated, one can easily (via elementary matrix algebra) calculate the long-term (dynamic) equilibrium composition of the community, as well as numerous other ecologically interesting parameters (e.g., the mean time between successive occupations of a spot by a given species).

To show what may be a more realistic Markov model of succession in the deep sea, we present Fig. 10-13, which bears a vague resemblance to Dayton and Hessler's (1972) "cropper" model. A location occupied by species *A* can be invaded and taken over by any other species because all the others can eat *A*. *B* can be eaten only by *C* and *D*, and *C* can be eaten only by *D*. *D* is the largest, and *A*, the smallest of the species. Con-

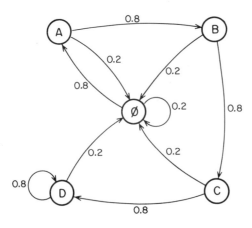

Fig. 10-12. Diagram of Markovian transition probabilities representing the classical (though suspect) picture of succession, wherein species *D* succeeds *C* succeeds *B* succeeds *A* unless some disturbance clears the space occupied by any of these individuals (P = 0.2 in this example), leading to an empty space (∅) and restarting the succession. See also Table 10-7.

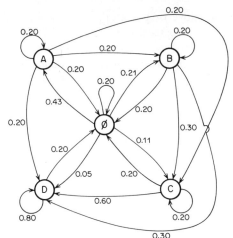

Fig. 10-13. Diagram of Markovian transition probabilities for a hypothetical community wherein species $A$ is the best colonist ($P = 0.43$) of cleared space ($\emptyset$), and $D$ is the worst ($P = 0.05$). Species $A$, however, is also the most easily displaced, while no other species can displace $D$. See also text and Table 10-7.

sequently (viz., Fenchel, 1974), we make $A$ the best colonist of unoccupied territory and $D$, the worst.

The importance of knowing the details of such transition probabilities is then illustrated by comparing the hypothetical communities of Figs. 10-12 and 10-13 under varying disturbance rates. While the communities are similar in equilibrium composition when the probability that any space will be cleared is 0.20, increasing the disturbance rate (transition probability to the vacant state) has decidedly different effects on the communities (Table 10-7); the change in the community of Fig. 10-12 is far more severe. Eliminating disturbance entirely, however, has exactly the same effect on the equilibrium compositions of both communities (Table 10-7).

We could present a wide variety of alternative Markov models to correspond with the various views of deep-sea community organization presented above, but the present examples (Figs. 10-12 and 10-13, Table 10-7) suffice to illustrate several salient facts. Neither these transition probabilities nor the biological processes determining them

TABLE 10-7. Equilibrium Composition of the Two Communities Depicted in Figs. 10-12 and 10-13 as the Probability of Disturbance Is Altered (and all other transition probabilities are changed proportionately)[a]

| Community of Fig. | Probability of Disturbance | Relative Abundance of Species | | | |
|---|---|---|---|---|---|
| | | $A$ | $B$ | $C$ | $D$ |
| 10-12 | 0.00 | 0.0 | 0.0 | 0.0 | 1.0 |
| 10-12 | 0.20 | 1.6 | 1.3 | 1.0 | 4.0 |
| 10-12 | 0.80 | 80.0 | 16.0 | 3.0 | 1.0 |
| 10-13 | 0.00 | 0.0 | 0.0 | 0.0 | 1.0 |
| 10-13 | 0.20 | 1.3 | 1.0 | 1.0 | 6.6 |
| 10-13 | 0.80 | 3.1 | 1.7 | 1.1 | 1.0 |

[a]The 0.20 probability level of disturbance is illustrated in the figures.

(e.g., competitive or predator-prey networks, other kinds of disturbances causing mortality, and the spatial extents and frequencies of such biotic and abiotic disturbances) are well known for any deep-sea community, and without them the results of manipulative experiments (e.g., caging to exclude certain predator-induced disturbances) will be difficult to interpret (cf. Table 10–7, and Paine, 1979).

## CONCLUSIONS AND FUTURE DIRECTIONS

In our opinion, the first play is the best, and the second is the worst. Optimal foraging theory appears to have the power for generating a large number of predictions which can be tested in the foreseeable future. The most detailed predictions are now possible for scavengers. The weakest predictions concern deposit feeders; more detailed information on their food resources and their foraging methods and on the way these characteristics vary among species is certainly required for more definitive applications of the theory. Such information, however, is also lacking for shallow-water animals in the same guild, and it would seem foolish for purely logistical reasons to attempt a deep-sea answer first.

In the case of optimal life-history tactics, both adequate theory and adequate data are lacking. As Stearns (1976, 1977) points out in his reviews, there is no generally accepted theory to explain life-history tactics. Tests of the $r$–$K$ models in accessible environments continue to be ambiguous; part of the results supports the theory, while other parts refute it, though the refutations have not been strong enough to discredit the approach entirely. Easily manipulated systems, i.e., populations outside the deep sea, should be used to test the multiplicity of models (Stearns, 1977). With the theoretical understanding of deep-sea populations blocked by the plethora of competing theories, the dearth of time (age) markers in deep-sea populations all but closes the door to an alternative, empirical approach.

At the community level, Connell and Slatyer's (1977) individual-by-individual successional models, especially as formalized through a Markovian approach, deserve further consideration for deep-sea application and manipulative testing. Such applications and tests are unlikely to be possible, however, without a knowledge of predator-prey and competitive relations. The latter, in turn, are not likely to be discovered until the resources utilized by the ubiquitous deep-sea deposit feeders are better identified. Again, the deep sea seems to be a poor place to try out new methods for determining these ecological unknowns, when they are nearly as poorly understood for shallow-water deposit feeders (e.g., Feller *et al.,* 1979; Self and Jumars, 1978).

At each biological level of organization, however, the physics of the deep-sea environment is seen to provide potential explanations of phenomena for which other, biological explanations have been used or sought. The physical structure of the deep-sea benthic boundary layer may allow unique foraging tactics among scavengers and certainly does limit viable suspension feeding methods. At the population level, the reduced incidence of physically mediated disturbances capable of causing size-independent mortality may cause size-selective predation to be a relatively more important phenomenon in the deep sea than it is in shallow water. At the community level, in turn, relatively weak bottom currents allow the persistence of biologically generated environmental heterogeneity (e.g., fecal mounds, tubes, burrows) which may facilitate the persistence of higher species diversity than generally is seen in shallow-water communities. Just as biology can influence the survival and persistence of particular species in physically disturbed environments (e.g., Menge and Sutherland, 1976), so can physics influence the survival of deep-sea species.

## ACKNOWLEDGMENTS

We gratefully acknowledge stimulating discussions with Drs. Arthur R. M. Nowell and J. Dungan Smith on the physics of the deep-sea benthic boundary layer. We also express thanks to Dr. James Morin and to his class on scientific writing for their comments on an earlier draft, and to the National Oceanic and Atmospheric Administration (Contract #03-78-B01-17) and the Office of Naval Research (Contract #N00014-75-C-0502) for their generous financial support. This chapter is contribution number 1242 from the School of Oceanography, University of Washington, Seattle.

## REFERENCES

Anderson, N. H., 1976, Carnivory by an aquatic detritivore *Clistoronia magnifica* (Trichoptera: Limnephilidae): *Ecol.,* v. 57, p. 1081–1085.

Armi, L., 1978, Mixing in the deep ocean—the importance of boundaries: *Oceanus,* v. 21, no. 1, p. 14–19.

Ballard, R. D., 1977, Notes on a major oceanographic find: *Oceanus,* v. 20, no. 3, p. 35–44.

Belyayev, G. M., Vinogradova, N. G., Levenshteyn, R. Ya., Pasternak, F. A., Sokolova, M. N., and Filatova, Z. A., 1973, Distribution patterns of deep-water bottom fauna related to the idea of the biological structure of the ocean: *Oceanology,* v. 13, p. 114–121.

Bernstein, B. B., Hessler, R. R., Smith, R., and Jumars, P. A., 1978, Spatial dispersion of benthic foraminifera in the abyssal central North Pacific: *Limnol. Oceanogr.,* v. 23, p. 401–416.

Bishop, W. P., and Hollister, C. D., 1974, Seabed disposal—where to look: *Nuclear Technol.,* v. 24, p. 425–443.

Caswell, H., 1976, Community structure—a neutral model analysis: *Ecol. Monogr.,* v. 46, p. 327–354.

Charnov, E. L., and Shaffer, W. M., 1973, Life history consequences of natural selection—Cole's result revisited: *Amer. Nat.,* v. 107, p. 791–793.

Cody, M. L., 1971, Finch flocks in the Mohave Desert: *Theor. Pop. Biol.,* v. 2, p. 142–158.

———, 1974, Optimization in ecology: *Science,* v. 183, p. 1156–1164.

Cole, L. C., 1954, The population consequences of life history phenomena: *Quart. Rev. Biol.,* v. 29, p. 103–137.

Connell, J. H., 1975, Some mechanisms producing structure in natural communities— a model and evidence from field experiments, *in* Cody, M. L., and Diamond, J. M., eds., *Ecology and Evolution of Communities:* Cambridge, Mass., Belknap Press, p. 460–490.

——, and Slatyer, R. O., 1977, Mechanisms of succession in natural communities and their role in community stability and organization: *Amer. Nat.,* v. 111, p. 1119–1144.

Dayton, P. K., and Hessler, R. R., 1972, The role of disturbance in the maintenance of deep-sea diversity: *Deep-Sea Res.,* v. 19, p. 199–208.

Emery, K. O., 1960, *The Sea Off Southern California:* New York, Wiley, 366 p.

Fauchald, K., and Jumars, P. A., 1979, The diet of worms—a study of polychaete feeding guilds: *Oceanogr. Mar. Biol. Ann. Rev.,* v. 17, p. 193–284.

Feller, R. J., Taghon, G. L., Gallagher, E. D., Kenny, G. E., and Jumars, P. A., 1979, Immunological methods for food web analysis in a soft-bottom benthic community: *Mar. Biol.,* v. 54, p. 61–74.

Fenchel, T., 1974, Intrinsic rate of natural increase—The relationship with body size: *Oecologia,* v. 14, p. 317–326.

Feyerabend, P., 1975, *Against Method:* London, Verso, 339 p.

Friedrich, H., 1969, *Marine Biology:* London, Sidgwick and Jackson, 474 p.

Gadgill, M., and Solbrig, O. T., 1972, The concept of *r-* and *K-*selection—evidence from wild flowers and some theoretical considerations, *Amer. Nat.,* v. 106, p. 14–31.

Gage, J. D., 1977, Structure of the abyssal macrobenthic community in the Rockall Trough, *in* Keegan, B. E., Ceidigh, P. O., and Boaden, P. J. S., eds., *Biology of Benthic Organisms:* Oxford, Pergamon, p. 247–260.

——, 1978, Animals in deep sea sediments: *Proc. Roy. Soc. Edinburgh,* v. 76B, p. 77–93.

Grassle, J. F., 1978, Diversity and population dynamics of benthic organisms: *Oceanus,* v. 21, no. 1, p. 42–49.

——, and Sanders, H. L., 1973, Life histories and the role of disturbance: *Deep-Sea Res.,* v. 20, p. 643–659.

——, Sanders, H. L., Hessler, R. R., Rowe, G. T., and McLellan, T., 1975, Pattern and zonation—a study of the bathyal megafauna using the research submersible *Alvin: Deep-Sea Res.,* v. 22, p. 457–481.

Greenwalt, D., and Gordon, C. M., 1978, Short-term variability in the bottom boundary layer of the deep ocean: *J. Geophys. Res.,* v. 83, p. 4713–4716.

Griggs, G. B., Carey, A. G., Jr., and Kulm, L. D., 1969, Deep-sea sedimentation and sediment-fauna interaction in Cascadia Channel and on Cascadia Abyssal Plain: *Deep-Sea Res.,* v. 16, p. 157–170.

Haedrich, R. L., and Rowe, G. T., 1977, Megafaunal biomass in the deep sea: *Nature,* v. 269, p. 141–142.

Hamner, P., and Hamner, W. M., 1977, Chemosensory tracking of scent trails by the planktonic shrimp *Acetes sibogae australis: Science,* v. 195, p. 886–888.

Harper, J. L., Lovell, P. H., and Moore, K. G., 1970, The shape and sizes of seeds: *Ann. Rev. Ecol. Syst.,* v. 1, p. 327–356.

Hawkins, A. D., and Rasmussen, K. J., 1978, The calls of gadoid fish: *J. Mar. Biol. Assoc. U.K.,* v. 58, p. 891–911.

Heezen, B. C., and Hollister, C. D., 1971, *The Face of the Deep:* New York, Oxford Univ. Press, 659 p.

Hessler, R. R., Ingram, C. L., Yayanos, A. A., and Burnett, B. R., 1978, Scavenging amphipods from the floor of the Philippine Trench: *Deep-Sea Res.*, v. 25, p. 1029-1047.

——, and Jumars, P. A., 1974, Abyssal community analysis from replicate box cores in the central North Pacific: *Deep-Sea Res.*, v. 21, p. 185–209.

——, and Jumars, P., 1977, Abyssal communities and radioactive waste disposal: *Oceanus*, v. 20, p. 41–46.

——, and Sanders, H. L., 1967, Faunal diversity in the deep sea: *Deep-Sea Res.*, v. 14, p. 65–78.

Horn, H. S., 1975, Markovian properties of forest succession, *in* Cody, M. L., and Diamond, J. M., eds., *Ecology and Evolution of Communities:* Cambridge, Mass., Belknap Press, p. 196–211.

——, 1976, Successions, *in* May, R. M., ed., *Theoretical Ecology; Principles and Applications:* Oxford, Blackwell Scientific Publ., p. 187-204.

Hurlbert, S. H., 1971, The nonconcept of species diversity—a critique and alternative parameters: *Ecol.*, v. 52, p. 577–586.

Huston, M., 1979, A general hypothesis of species diversity: *Amer. Nat.*, v. 113, p. 81-101.

Isaacs, J. D., 1976, Reproductive products in marine food webs: *Bull. Southern Calif. Acad. Sci.*, v. 75, p. 220–223.

——, D., and Schwartzlose, R. A., 1975, Active animals of the deep sea floor: *Scient. Amer.*, v. 233, p. 84–91.

Jones, D. S., Thompson, I., and Ambrose, W., 1978, Age and growth rate determinations for the Atlantic surf clam *Spisula solidissima* (Bivalvia: Mactracea), based on internal growth lines in shell cross-sections: *Mar. Biol.*, v. 47, p. 63–70.

Jørgenson, C. B., 1966, *Biology of Suspension Feeding:* Oxford, Pergamon Press, 357 p.

Jumars, P. A., 1975, Environmental grain and polychaete species' diversity in a bathyal benthic community: *Mar. Biol.*, v. 30, p. 253–266.

——, 1976, Deep-sea species diversity—Does it have a characteristic scale? *J. Mar. Res.*, v. 34, p. 217–246.

——, 1978, Spatial autocorrelation with RUM (Remote Underwater Manipulator)— vertical and horizontal structure of a bathyal benthic community: *Deep-Sea Res.*, v. 24, p. 589–604.

——, and Eckman, J. E., Spatial structure of deep-sea benthic communities, *in* Rowe, G. T., ed., *The Sea, Vol. 8, Deep-Sea Biology:* New York, Wiley Interscience, in press.

——, and Fauchald, K., 1977, Between-community contrasts in successful polychaete feeding strategies, *in* Coull, B. C., ed., *Ecology of Marine Benthos:* Univ. South Carolina Press, Columbia, p. 1-20.

——, and Hessler, R. R., 1976, Hadal community structure—implications from the Aleutian Trench: *J. Mar. Res.*, v. 34, p. 547–560.

Kauppi, P., Hari, P., and Kellomäki, 1978, A discrete time model for succession of ground cover communities after clear cutting: *Oikos*, v. 30, p. 100–105.

Kemeny, J. G., and Snell, J. L., 1960, *Finite Markov Chains:* D. Van Nostrand, Princeton, N.J., 210 p.

Kennedy, J. S., 1939, The visual responses of flying mosquitoes: *Proc. Zool. Soc. London*, A 109, p. 221–242.

Khripounoff, A., Desbruyères, et Chardy, P., 1980, Les peuplements benthiques de la faille Verma—Donées quantitatives et bilan d'énergie en milieu abyssal: *Oceanologica Acta*, v. 3, p. 187–198.

Kitchell, J. A., 1979, Deep-sea foraging pathways—an analysis of randomness and resource exploitation: *Paleobiol.*, v. 5, p. 107–125.

Kuhn, T. S., 1962, *The Structure of Scientific Revolutions*: Univ. Chicago Press, 210 p.

Lakatos, I., 1970, Falsification and the methodology of research programmes, *in* Lakatos, I., and Musgrave, A., eds., *Criticism and the Growth of Knowledge:* London, Cambridge Univ. Press, p. 91–196.

Lemche, H., Hansen, B., Madsen, F. J., Tendal, O. S., and Wolff, T., 1976, Hadal life as analyzed from photographs: *Vidensk. Meddr dansk naturh. Foren.*, v. 139, p. 263–336.

Levinton, J. S., 1972, Stability and trophic structure in deposit-feeding and suspension-feeding communities: *Amer. Nat.*, v. 106, p. 472–486.

——, and Lopez, G. R., 1977, A model of renewable resources and limitation of deposit-feeding benthic populations: *Oecologia*, v. 31, p. 177–190.

Lonsdale, P., 1977, Clustering of suspension-feeding macrobenthos near abyssal hydrothermal vents at oceanic spreading centers: *Deep-Sea Res.*, v. 24, p. 857–863.

Lynch, M., 1977, Fitness and optimal body size in zooplankton populations: *Ecology*, v. 58, p. 763–774.

MacArthur, R. H., 1972, *Geographical Ecology*: New York, Harper & Row, Pub., 269 p.

Maynard Smith, J., 1974, *Models in Ecology*: London, Cambridge Univ. Press, 146 p.

Menge, B. A., and Sutherland, J. P., 1976, Species diversity gradients—synthesis of the roles of predation, competition, and temporal heterogeneity: *Amer. Nat.*, v. 110, p. 351–369.

Monniot, C., 1979, Adaptations of benthic filtering animals to the scarcity of suspended particles in deep water: *Ambio Spec. Rept.*, v. 6, p. 73–74.

——, and Monniot, F., 1978, Recent work on the deep-sea tunicates: *Oceanogr. Mar. Biol. Ann. Rev.*, v. 16, p. 181–228.

Ockelmann, K. W., and Vahl, O., 1970, On the biology of the polychaete *Glycera alba*, especially its burrowing and feeding: *Ophelia*, v. 8, p. 275–294.

Orians, G. H., and Pearson, N. E., 1979, On the theory of central place foraging, *in* Horn, D. J., Stairs, G. R., and Mitchell, R. D., eds., *Analysis of Ecological Systems*: Ohio State Univ. Press, Columbus, p. 155–177.

Osman, R. W., and Whitlatch, R. B., 1978, Patterns of species diversity—fact or artifact: *Paleobiology*, v. 4, p. 41–54.

Paine, R. T., 1979, Disaster, catastrophe, and local persistence of the sea palm *Postelsia palmaeformis*: *Science*, v. 205, p. 685–687.

Pearcy, W. G., and Ambler, J. W., 1974, Food habits of deep-sea macrourid fishes off the Oregon coast: *Deep-Sea Res.*, v. 21, p. 745–759.

Platt, J. R., 1964, Strong inference: *Science,* v. 146, p. 347–353.

Polloni, P., Haedrich, R., Rowe, G., and Clifford, C. H., 1979, The size-depth relationship in deep ocean animals: *Int. Rev. ges. Hydrobiol.*, v. 64, p. 39–46.

Popper, K. R., 1959, *The Logic of Scientific Discovery*: London, Hutchinson & Co., 480 p.

——, 1965, Science—conjectures and refutations, *in* Popper, K. R., *Conjectures and Refutations—The Growth of Scientific Knowledge*: New York, Harper & Row, Publ., p. 33–65.

Pyke, G. H., 1978, Optimal foraging—movement patterns of bumblebees between in-florescences: *Theor. Pop. Biol.*, v. 13, p. 72–98.

——, Pulliam, H. R., and Charnov, E. L., 1977, Optimal foraging—a selective review of theory and tests: *Quart. Rev. Biol.*, v. 52, p. 137–154.

Rannou, M. R., 1976, Age et croissance d'un poisson bethyal: *Nezumia sclerorhynchus* (Macrouridae: Gadiforme) de la Mer d'Alboran: *Cahiers de Biologie Marine*, v. 27, p. 413–421.

Rau, G. H., and Hedges, J. I., 1979, Carbon-13 depletion in a hydrothermal vent mussel—suggestion of a chemosynthetic food source: *Science*, v. 203, p. 648–649.

Reid, R. G., and Reid, A. M., 1974, The carnivorous habit of members of the septibranch genus Cuspidaria (Mollusca: Bivalvia): *Sarsia*, v. 56, p. 47–56.

Rex, M. A., 1976, Biological accommodation in the deep-sea benthos—comparative evidence on the importance of predation and productivity: *Deep-Sea Res.*, v. 23, p. 975–987.

——, 1977, Zonation in deep-sea gastropods—the importance of biological interaction to rates of zonation, *in* Keegan, B. F., Ceidigh, P. O., and Boaden, P. J. S., eds., *Biology of Benthic Organisms*: New York, Pergamon, p. 521–530.

Rice, A. L., Aldred, R. G., Billet, D. S. M., and Thurston, M. H., 1979, The combined use of an epibenthic sledge and a deep-sea camera to give quantitative relevance to macro-benthos samples: *Ambio. Spec. Rpt.*, v. 6, p. 59–72.

Richerson, P., Armstrong, R., and Goldman, C. R., 1970, Contemporaneous disequilibrium, a new hypothesis to explain the "paradox of the plankton": *Proc. Natl. Acad. Sci. USA*, v. 67, p. 1710–1714.

Risk, M. J., and Tunnicliffe, V. J., 1978, Intertidal spiral burrows—*Paraonis fulgens* and *Spiophanes wigleyi* in the Minas Basin, Bay of Fundy: *J. Sed. Petrol.*, v. 48, p. 1287–1292.

Roberts, F. S., 1976, Discrete mathematical models: Englewood Cliffs, N.J., Prentice-Hall, 559 p.

Rokop, F. J., 1974, Reproductive patterns in the deep-sea benthos: *Science*, v. 186, p. 743–745.

Root, R. B., 1975, Some consequences of ecosystem texture, *in* Levin, S. A., ed., *Eco-system Analysis and Prediction*: Philadelphia, Pa., Society for Industrial and Applied Mathematics, p. 83–97.

Rowe, G. T., ed., *The Sea, Vol. 8, Deep-Sea Biology*: New York, Wiley-Interscience, in press.

Sanders, H. L., 1968, Marine benthic diversity—a comparative study: *Amer. Nat.*, v. 102, p. 243–282.

——, 1969, Benthic marine diversity and the stability-time hypothesis: *Brookhaven Symp. Biol.*, v. 22, p. 71–80.

——, and Allen, J. A., 1973, Studies on deep-sea Protobranchia; Prologue and the Pristi-glomidae: *Bull. Mus. Comp. Zool.*, v. 145, p. 237–261.

Schaffer, W. M., and Gadgil, M. D., 1975, Selection for optimal life histories in plants, *in* Cody, M. L., and Diamond, J. M., eds., *Ecology and Evolution of Communities*: Cambridge, Mass., Belknap Press, p. 142–157.

Schoener, T. W., 1974, Resource partitioning in ecological communities: *Science*, v. 185, p. 27–39.

Seki, H., Wada, E., Koike, I., and Hattori, A., 1974, Evidence of High organotrophic potentiality of bacteria in the deep ocean: *Mar. Biol.*, v. 26, p. 1–4.

Self, R. F. L., and Jumars, P. A., 1978, New resource axes for deposit feeders? *J. Mar. Res.*, v. 36, p. 627–641.

Shorey, H. H., 1977, Pheromones, *in* Sebeok, T. A., ed., *How Animals Communicate*: Indiana Univ. Press, Bloomington, p. 137–163.

Shulenberger, E., and Barnard, J. L., 1976, Amphipods from an abyssal trap set in the North Pacific gyre: *Crustaceana*, v. 31, p. 241–258.

Slobodkin, L. B., and Sanders, H. L., 1969, On the contribution of environmental predictability to species diversity: *Brookhaven Symp. Biol.*, v. 22, p. 82–92.

Smith, K. L., Jr., White, G. A., Laver, M. B., McConnaughey, R. R., and Meador, J. P., 1979, Free vehicle capture of abyssopelagic animals: *Deep-Sea Res.*, v. 26A, p. 57–64.

Smith, W., and Grassle, J. F., 1977, Sampling properties of a family of diversity measures: *Biometrics*, v. 33, p. 283–292.

Southwood, T. R. E., May, R. M., Hassell, M. P., and Conway, G. R., 1974, Ecological strategies and population parameters: *Amer. Nat.*, v. 108, p. 791–804.

Stearns, S. C., 1976, Life-history tactics—a review of the ideas: *Quart. Rev. Biol.*, v. 51, p. 3–47.

——, 1977, The evolution of life history traits—a critique of the theory and a review of the data: *Amer. Rev. Ecol. Syst.*, v. 8, p. 145–171.

Strathmann, R., 1974, The spread of sibling larvae of sedentary marine invertebrates: *Amer. Nat.*, v. 108, p. 29–44.

Swedmark, B., 1964, The interstitial fauna of marine sands: *Biol. Rev.*, v. 39, p. 1–42.

Taghon, G. L., Self, R. F. L., and Jumars, P. A., 1978, Predicting particle selection by deposit feeders—a model and its implications: *Limnol. Oceanogr.*, v. 23, p. 752–759.

Tendal, O. S., and Hessler, R. R., 1977, An introduction to the biology and systematics of Komokiacea: *Galathea Rept.*, v. 14, p. 165–194.

Thiel, Hj., 1975, The size structure of the deep-sea benthos: *Int. Rev. ges. Hydrobiol.*, v. 60, p. 575–606.

——, 1979, Structural aspects of the deep-sea benthos: *Ambio Spec. Rpt.*, v. 6, p. 25–31.

Thistle, D., 1979, Harpacticoid copepods and biogenic structures—implications for deep-sea diversity maintenance, *in* Livingston, R. J., ed., *Ecological Processes in Coastal and Marine Systems*: New York, Plenum, p. 217–231.

Turekian, K. K., Cochran, J. K., Kharkar, D. P., Cerrato, R. M., Vaišnys, J. R., Sanders, H. L., Grassle, J. F., and Allen, J. A., 1975, Slow growth rate of a deep-sea clam determined by $^{228}$Ra chronology: *Proc. Nat. Acad. Sci. USA*, v. 72, p. 2829–2832.

——, Cochran, J. K., and Nozaki, Y., 1979, Growth rate of a clam from the Galapagos Rise hot spring field using natural radionuclide ratios: *Nature*, v. 280, p. 385–387.

Turner, R. D., 1973, Wood-boring bivalves, opportunistic species in the deep-sea: *Science*, v. 180, p. 1377–1379.

——, 1977, Wood, mollusks, and deep-sea food chains: *Bull. Amer. Malacol. Un., Inc.*, 1977, p. 13–19.

Vogel, S., 1978, Organisms that capture currents: *Sci. Amer.*, v. 239, no. 2, p. 128–139.

Wilson, E. O., and Bossert, W. H., 1971. *A Primer of Population Biology*: Stamford, Conn., Sinauer Associates, Inc. Publishers, 192 p.

Wimbush, M., 1976, The physics of the benthic boundary layer, *in* McCave, I. N., ed., *The Benthic Boundary Layer*: New York, Plenum, p. 3–10.

Wolcott, T. G., 1978, Ecological role of ghost crabs, *Ocypode quadrata* (Fabricius), on an ocean beach—Scavengers or predators? *J. Exp. Mar. Biol. Ecol.*, v. 31, p. 67–82.

Wolff, T., 1976, Utilization of seagrass in the deep-sea: *Aq. Bot.*, v. 2, p. 161–174.

George N. Somero
Marine Biology Research Division
Scripps Institution of Oceanography
University of California
La Jolla, California 92093

# 11

# PHYSIOLOGICAL AND BIOCHEMICAL ADAPTATIONS OF DEEP-SEA FISHES: ADAPTIVE RESPONSES TO THE PHYSICAL AND BIOLOGICAL CHARACTERISTICS OF THE ABYSS

The biochemical characteristics of deep-sea fishes, notably the properties of enzyme systems, reflect adaptations to the physical (low temperature and high pressures) and biological (quantity and distribution of food) features of the abyss. Adaptations to high pressure are reflected in a high degree of insensitivity to pressure in the kinetic properties of enzymes. Studies of skeletal muscle lactate dehydrogenases (LDHs) of shallow- and deep-living fishes have revealed that the interactions between the enzyme and substrates are strongly perturbed by pressure in the case of LDHs of shallow forms, but markedly insensitive to pressure for LDHs of deep-sea fishes. Pressures of only 50–100 atm, corresponding to depths of approximately 500–1000 m, appear adequate to favor selection for pressure-insensitive enzymes. Differences in pressure sensitivity are found between enzymes of congeneric species (genus *Sebastolobus*) having different distribution patterns in the water column. This type of enzymic adaptation may be instrumental in establishing species zonation patterns in the sea. While pressure-insensitive, the LDHs of deep-living fishes have lower catalytic efficiencies, as measured by the rates of conversion of substrate to product per unit time per enzyme molecule. Pressure-adapted enzymes appear to compete unfavorably with enzymes of shallow-living, cold-adapted species in terms of catalytic power. This type of enzymic difference may be a partial contributor to setting the upper distribution limits of deep-living fishes.

The low rates of metabolism characteristic of deep-sea fishes appear to be due to a combination of inefficient enzymes and low intracellular enzyme concentrations. Enzymic activities per gram wet weight of tissue in skeletal muscle differ by three orders of magnitude between highly active surface fishes, like tunas, and sluggish deep-sea fishes. Muscle enzymic activities correlate strongly with whole organism respiration rates. At any given depth of occurrence, several-fold variation in muscle enzymic activity is found among species. Most of this interspecific difference can be accounted for in terms of locomotory and feeding habits. "Float and wait" predators in the deep sea have the lowest enzyme contents among deep-living fishes, while more active, cruising fishes like rattails have the highest enzymic activities. Unlike muscle tissue, brain exhibits no depth-related differences in enzymic activity among species. This observation suggests that the muscle differences are specifically related to feeding/locomotory strategies in the deep-sea, and that brain function is not reduced in deep-living forms. Enzyme contents in muscle vary much more than total muscle protein concentration, a result which suggests that the enzymes of energy metabolism are decreased in deep-sea fishes much more than structural and contractile proteins. The utility of using enzyme data to make predictions about hard-to-observe deep-sea fishes is discussed.

## INTRODUCTION: THE NATURE OF THE DEEP-SEA ENVIRONMENT

Before one can understand the physiological and biochemical "design" of deep-sea animals, it is necessary to become familiar with the physical, chemical, and biological features of the deep-sea environment that serve as major shaping influences on the characteristics of the deep-sea fauna. All organisms, whether terrestrial or aquatic, plant or

GEORGE N. SOMERO

animal, unicellular or multicellular, display vast numbers of specific adaptations to enable successful existence and reproduction in the particular niches the organisms occupy. Deep-sea animals are no exception, and because they experience a set of environmental conditions which it seems fair, if a bit terrestrially chauvinistic, to term "stressful," we will see that the study of deep-sea animals' physiologies and biochemistries offers a very good route to the investigation of adaptational processes in organisms. Indeed, what is unique about deep-sea organisms in terms of the stresses in their lives is the dual stress imposed by extremes of temperature and hydrostatic pressure. Both of these environmental factors have pronounced influences on physiological and biochemical structures and functions, and both temperature and pressure stresses have led to marked changes in the physiologies and chemistries of deep-sea species. Study of these changes—these adaptations—will allow us new insights into evolutionary processes in general and, in the special case of temperature and pressure adaptations, will reveal how adaptive processes enable organisms to cope with two physical stresses that at times seem to place conflicting demands on their biochemical systems.

In considering the characteristics of the deep-sea environment, it is important to emphasize that deep oceanic habitats are not rare, atypical marine environments; rather, the abyssal environment (Fig. 11-1) includes approximately three quarters of the benthic regions of the seas. The average depth of the sea floor is about 3800 m. Thus, in terms of sea-bottom area, the environment of deep-sea animals is more of an "average" type of environment than our own terrestrial realm. What is *not* average, at least from our terrestrial perspective, is the suite of chemical, physical, and biological characteristics of this deep-sea world. As indicated on Fig. 11-1, the physical features of the deep sea are such as to suggest a difficult life for animals. There is no light other than bioluminescence generated by the organisms themselves. Temperatures in the deep sea are now very low, averaging only 2-4°C. Some 100 million years ago the deep sea was a warmer place, with temperatures of 10-15°C higher. These paleotemperature data give us some idea of how much time modern deep-sea animals have had to erect the biochemical and physiological adaptations necessary for coping with contemporary deep-sea temperatures.

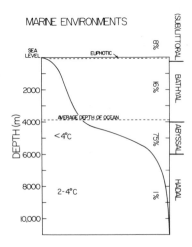

Fig. 11-1. Classification of marine environments. The percentages refer to the area of the ocean floor included in each depth zone.

Hydrostatic pressure increases by 1 atm for each 10-m increase in depth through the water column. Thus, in the deepest regions of the ocean, organisms are forced to deal with approximately 1000 atm pressure, and at average abyssal depth the pressure is near 400 atm. As we will see, pressures of this magnitude and, in fact, pressures of only several tens of atmospheres, are adequate to have severe effects on the biochemical structures and functions of organisms. This holds true even for animals lacking gas-filled spaces. Most of the key components of organisms' structures and metabolic apparati are dependent on weak bonding interactions, which high pressure and extremes of temperature seriously perturb.

In addition to stressful physical conditions, deep-sea environments are noted for their extremes of biological or ecological characteristics. In particular, the supply of food reaching the deep oceans appears to be very small. As Fig. 11-1 shows, light-driven primary productivity cannot occur in the deep sea. Some chemoautotrophic productivity does occur in the deep sea in the spreading center regions, where geothermally heated water, rich in mineral nutrients like sulfide, allows the growth of sulfide-oxidizing bacteria, which are found free-living in the vent water (Karl et al., 1980) and in symbiotic associations with the vent tubeworms (Cavanaugh et al., 1981; Felbeck, 1981) and bivalves (Felbeck et al., 1981). The contribution of the hydrothermal vent ecosystems to the overall food-energy budget of the deep sea remains to be clarified, however, and in the following discussion we assume that most of the nutrients obtained by deep-sea animals (excluding the vent species) derive from photosynthetic productivity in shallow waters (Fig. 11-1). The fraction of surface primary productivity reaching the deep sea is not known, but deep-sea biologists are in general agreement that the flux of food to the deep sea represents but a minor part of the total (net) surface productivity. Moreover, the manners in which the food reaching the deep sea is "packaged" also create problems for deep-sea animals. For example, much of the food reaching the deep-sea fauna may arrive packaged in the form of copepod fecal pellets (Honjo and Roman, 1978). This fact suggests that benthic species which eke out a living by scavenging for food on, or in, the deep-sea sediments must usually rely on minute food packets. There are, however, large fish in the deep sea, and many of these species are found near the bottom. The food sources, and the frequency with which meals are found, are not well understood for these species. Certain of these fishes, for example the rattails, may be highly mobile scavengers which cruise over the bottom in search of carcasses that have descended from the surface. Certain of these fishes may swim up and down in the water column searching for food. This feeding strategy is suggested by two different types of observations. Baited cameras dropped into deep water have recorded rapid food localization behavior by rattails and certain other deep-living fishes. And, as discussed below, some of the biochemical and physiological properties of these active searchers distinguish these species from other deep-sea fishes which employ a "float-and-wait" predatory strategy.

In summary, the deep-sea environment must be regarded as a habitat possessing extreme physical (temperature, pressure, and light) and biological (food quantity and distribution) stresses. The properties of deep-sea animals clearly reflect adaptations to cope with these different stresses, and these adaptations can be appropriately categorized according to whether they (1) facilitate life under stressful pressure and temperature conditions, or (2) adjust the rate of life (metabolism) to be consistent with the limiting energy resources available in the deep sea. We will consider the first of these two classes of adaptations, those making life possible, before we examine the second class, those which determine how much life is possible.

## ADAPTATIONS TO TEMPERATURE AND PRESSURE: EVOLUTION OF ENZYMES IN DEEP-SEA ANIMALS

When considering the manners by which extremes of temperature and pressure may upset the functionings of organisms, we are immediately led to the investigation of how these physical stresses influence the activities and structures of enzymes, the protein catalysts that allow metabolic reactions to occur at rates supportive of life. By lowering the "energy barriers" to chemical reactions, as discussed below, enzymes permit organisms to process energy and materials at rates several orders of magnitude higher than those characteristic of noncatalyzed processes. As the mechanisms of enzymic catalysis become more fully understood, we are learning why these key molecules are so sensitive to temperature and pressure. One of the most important features of enzyme design is an inherent flexibility in their structures. This flexibility allows enzymes to undergo rapid and reversible changes in shape ("conformation") during the catalytic sequence. In a certain sense enzymes are like other machines which convert a starting material into one or more final products. All such machines have moving parts, and if the machine is to be used more than once, the movements of these parts must occur in a reversible manner. Also, it is fair to say that a machine capable of undergoing the requisite changes in shape in a very rapid and energetically inexpensive manner will be the most productive machine. Enzymes, in this sense, are exemplary, since they may convert hundreds or thousands of substrate molecules to product molecules each second. For the purposes of this discussion of deep-sea animals' enzymes, the inherent flexibility of enzymes must be viewed as both a plus and a minus. Flexibility suggests structural lability, i.e., the very features of enzyme structure which underlie their capacities to work so rapidly and with such high specificity also render enzymes vulnerable to environmental perturbation. Distortion of enzyme structure by pressure and temperature thus represents a major evolutionary hurdle facing species in the process of colonizing deep-sea regions.

To appreciate the problems associated with maintaining controlled enzymic function in the deep-sea, one must first understand precisely what properties of enzymes must be conserved under all conditions. Three traits seem especially crucial (see Somero, 1978). First, as indicated above, enzymes must maintain the correct structures. Second, enzymes must possess adequate catalytic capacity to support the appropriate metabolic rate. Third, enzymes must not only be able to catalyse metabolic reactions at high rates, but they must also be capable of regulating their rates of catalysis according to the demands of the cell for particular metabolic products. The last of these three essential features, regulatory responsiveness, is particularly sensitive to temperature and pressure perturbation and thus serves as an appropriate initial focus in our analysis of enzymic adaptations in deep-sea animals.

An exceedingly important feature of enzyme systems is their capacity to respond to signals calling for increases or decreases in the rates at which specific enzymes are to function. We are all aware of our bodies' capacities to change the overall rate of metabolism, for example, during transitions from sedentary behavior to vigorous exercise. These transitions in metabolic activity are due in large measure to the opening of enzymic "gates" or "switches" that control the rates at which substrates, such as glycogen and glucose, are chemically degraded by enzymes to generate biologically useful energy in the form of compounds like adenosine triphosphate (ATP). If our metabolic machine always ran at full velocity, even under basal conditions, we would not have the capacity to alter our rates of locomotion, digestion, growth, etc. Needless to say, an organism lacking the ability to speed-up its metabolism to outrun a predator, or to capture a prey organism, is unlikely to survive and reproduce. Enzyme systems thus have been under high selective pressure to provide organisms with a large and tightly controllable capacity to vary their rates of metabolic function. An important cornerstone of this regulatory ability is the maintenance of a subsaturating relationship between enzymes and their substrates. By this we refer to the fact that for virtually all enzymes, the concentrations of substrates present in the cell are below the concentrations necessary to cause the enzyme to work at its maximal velocity ($V_{max}$) (Fersht, 1977). When studying an enzymic reaction in a test tube, one finds that for most enzymes the addition of increasing amounts of substrate to a given amount of enzyme leads to a roughly hyperbolic increase in reaction rate. Thus, as substrate concentrations reach higher and higher values, the rate of increase in enzymic activity slows down until at substrate concentrations yielding the highest velocity ($V_{max}$) the rate of the reaction is independent of substrate concentration changes. It should be apparent that cellular metabolism would be ill-served by an enzyme battery in which substrate concentrations were so high that enzymes always worked at $V_{max}$. Whereas such a design would allow the cell to extract optimal amounts of metabolic turnover from a given number of enzyme molecules, demands for increases in the rates of enzymic activity needed, e.g., to avoid a predator, would fall on deaf ears. Thus it would seem to behoove enzyme systems to hold part of their impressive catalytic potential in reserve. This reserve capacity is achieved in large measure by keeping substrate concentrations below values yielding $V_{max}$, as discussed above. The key parameter that is useful in gauging the regulatory status of an enzyme is not, therefore, the $V_{max}$ of the reaction, but rather a measure of the point where the reaction stands relative to $V_{max}$. This key regulatory parameter is the apparent Michaelis constant ($K_m$), defined as the concentration of substrate at which the velocity of the reaction is one-half of $V_{max}$. From what has been said above we can conclude that a consistent relationship between intracellular substrate concentrations and $K_m$ values should be maintained for regulatory purposes. In fact, substrate concentrations, which are highly conserved among different animal species, normally are near or below $K_m$ values (Fersht, 1977).

While the importance of maintaining this relationship between $K_m$ values and substrate concentrations is clear on the basis of the cell's regulatory requirements, achieving this relationship in deep-sea animals is not apt to be a simple evolutionary feat due to the fact that $K_m$ values are strongly perturbed by pressure and temperature (Hochachka and Somero, 1973; Somero et al., 1982). The weak bonds holding protein conformations in substrate-recognizing shapes and holding the substrate to the active site are readily broken by increases in pressure and extremes of temperature. Figure 11-2 illustrates how temperature changes affect the $K_m$ of pyruvate for the enzyme lactate dehydrogenase (LDH), an important enzyme for the function of skeletal muscle under conditions of high rates of muscular activity. LDH must be capable of converting pyruvate to lactic acid in order to maintain vigorous levels of muscular activity in the face of limited oxygen supplies. LDH is a particularly important component of the biochemical machinery of fish white muscle, which permits rapid bursts of swimming activity during prey capture and predator avoidance (Driedzic and Hochachka, 1978; Somero and Childress, 1980). To function effectively, an LDH system must have a $K_m$ of pyruvate in a range which at once permits a high rate of LDH activity, yet which still allows the enzyme to increase its rate of function as pyruvate concentrations increase during exercise. The conservation of pyruvate $K_m$ values (and pyruvate concentrations;

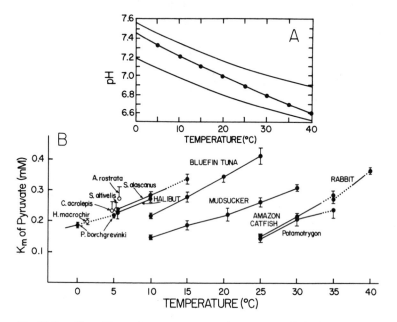

Fig. 11-2. The effect of temperature on the apparent Michaelis constant ($K_m$) of pyruvate for purified muscle-type ($M_4$) lactate dehydrogenases from several vertebrate species adapted to different temperatures and pressures. The inset to the figure shows the relationship between temperature and the pH of the intracellular fluids of a number of ectothermic vertebrates. The observed pH ranges are bracketed by thin lines; the points and the darker line show the effect of temperature on the pH of the imidazole buffer used in these studies (see Yancey and Somero, 1978). On the $K_m$-versus-temperature plots, the solid lines indicate the physiological temperature ranges of the species. Vertical bars show the 95% confidence intervals around the $K_m$ values.

Species studied: Deep-sea fishes (see Fig. 11-3 for depth distribution ranges): *Halosauropsis macrochir, Coryphaenoides acrolepis, Antimora rostrata,* and *Sebastolobus altivelis.* Shallow-living species: *Pagothenia (Trematomus) borchgrevinki,* an Antarctic teleost; the halibut species is *Hippoglossus stenolepis;* the mudsucker is *Gillichthys mirabilis,* an estuarine fish; the amazon catfish is *Hypostomus plecostomus;* and *Potamotrygon* sp. is a freshwater ray. From Somero *et al.,* 1982.

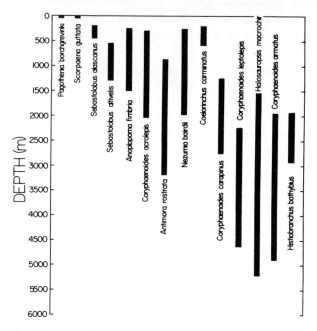

Fig. 11-3.   Depth distributions for 14 of the marine teleost fishes used in the enzyme studies discussed in this chapter. For the congeners *Sebastolobus alascanus* and *S. altivelis*, depths of maximal abundance are given. The full depth range of these two species is found in Siebenaller and Somero (1978). From Somero *et al.*, 1982.

see Yancey and Somero, 1978) among different fish species is extremely high, especially in view of the fact that temperature markedly affects the $K_m$ of all LDH's (Fig. 11-2). Thus, for any single LDH homologue, the $K_m$ of pyruvate rises with an increase in experimental—or body—temperature. If only a single version of LDH existed in all species, there would be an extremely large variation in the $K_m$ of pyruvate, at appropriate physiological temperatures, among different species. What we find, in fact, is that each interspecific homologue of LDH reflects the temperature range in which the enzyme functions, such that enzymes of fishes with low body temperatures (polar and deep-sea fishes, for example) have higher $K_m$'s of pyruvate, at any given temperature, than the LDHs of warmer water species or mammals (Fig. 11-2). These evolutionary adjustments in $K_m$ values are such as to lead to a high degree of conservation in $K_m$ of pyruvate at the normal body temperatures of the different species. Note that the deep-sea fishes (depth ranges given in Fig. 11-3) reflect adaptation to low, deep-sea temperatures in terms of where the $K_m$ of pyruvate values for their LDHs fit on the plot shown in Fig. 11-2. However, in view of the fact that $K_m$ values may be perturbed by pressure as well as by temperature, we must withhold judgment about the "success" of enzyme adaptation by deep-sea animals until we determine if the appropriate $K_m$ values seen at abyssal temperatures, but at 1 atm (Fig. 11-2), also are maintained in the face of *in situ* pressures of up to several hundred atmospheres.

The effects of hydrostatic pressure on the $K_m$ values of pyruvate and the cofactor or cosubstrate of the LDH reaction, nicotinamide adenine dinucleotide (NADH), at a measurement temperature of 5°C, are shown in Fig. 11-4. We see that, in spite of similarities between cold-adapted shallow water and deep-sea fishes' LDHs at 1 atm (Fig. 11-2), pronounced differences between the enzymes of the shallow- and deep-living fishes are evident at elevated pressures. For all of the shallow-living fishes' LDHs, increases in hydrostatic pressure are accompanied by increases in $K_m$ for both pyruvate and NADH. The $K_m$ of NADH is particularly pressure-sensitive, and even for the deep-sea fishes' LDHs a slight increase is noted. However, this increase is small and occurs only between approximately 1 and 68 atm, whereas for the LDHs of all of the shallow-living fishes, the $K_m$ of NADH rises much more with increasing pressure and continues to increase at higher and higher pressures in most cases. We view these increases in $K_m$ with rising pressure as highly maladaptive for LDH function in the face of either high or variable pressures (Siebenaller and Somero, 1978, 1979) and regard the similar, essentially pressure-insensitive $K_m$'s of LDHs from deep-sea species as a striking example

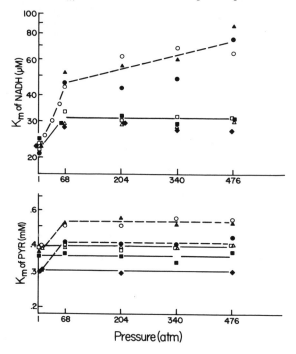

Fig. 11-4. The effects of hydrostatic pressure on the apparent $K_m$ of NADH (upper panel) and pyruvate (lower panel) for purified $M_4$-lactate dehydrogenases of several deep- and shallow-living marine teleost fishes. Deep-living species: *Sebastolobus altivelis* (□), *Antimora rostrata* (△), *Coryphaenoides acrolepis* (■), and *Halosauropsis macrochir* (◆). Shallow-living species: *Pagothenia* (*Trematomus*) *borchgrevinki* (●), *Scorpaena guttata* (▲), and *Sebastolobus alascanus* (○). All measurements were made at 5°C. The 95% confidence intervals around the $K_m$'s of NADH are approximately ±14% of the $K_m$ values, and for pyruvate $K_m$'s, are approximately ±10% of the $K_m$ values. From Siebenaller and Somero, 1979.

of convergent evolution. Particularly notable are the differences found between the LDHs of the two *Sebastolobus* congeners, *S. altivelis* (deep-living) and *S. alascanus* (shallow-living; Fig. 11-3). These congeners are extremely similar morphologically, and are separated by a relatively small genetic distance, as determined electrophoretically (Siebenaller, 1978). Also, the LDHs of the two species could not be differentiated by a number of sensitive electrophoretic procedures, a result which suggests very little difference in amino acid sequence between the two LDH homologues. Nevertheless, the two LDHs display significantly different responses to pressure, and these differences indicate that pressures of only 50–100 atms may be perturbing enough of LDH function to favor selection of pressure-resistant enzymes. Study of such closely related species, which differ in their depths of distribution but which experience similar thermal conditions, seems an especially good approach for distinguishing the effects of pressure adaptation from those associated with temperature adaptation.

Enzymic differences of the types found with the *Sebastolobus* congeners also provide a basis for framing conjectures about speciation processes and depth distribution mechanisms in the water column. It seems reasonable to hypothesize that in the case of the *Sebastolobus* species an ancestral population existed in which some individuals having a mutant, pressure-insensitive LDH (and, perhaps, many other enzymes) were present. These individuals having reduced pressure sensitivity of metabolism could have migrated to deeper regions in the water column and colonized habitats not previously exploited. This scenario must not be interpreted to argue that only enzymic, or only LDH, adaptations are required for successful penetration into the deep-sea world, for a wide variety of behavioral, physiological, anatomical, and biochemical adaptations are no doubt necessary to make a species a true deep-sea species. Nonetheless, we feel that initial colonization of deeper regions of the oceans may have been attendant on the evolutionary "discovery" of ways to fabricate pressure-insensitive enzyme systems.

What the foregoing scenario does not tell us, however, is why a pressure-insensitive species has not "taken over" at all of the depths at which the particular genus is found. Thus, while the acquisition of reduced enzyme pressure sensitivity may provide a partial basis for allowing an organism to live in deeper waters, pressure insensitivity *per se* does not seem to provide us with any basis for understanding why pressure-tolerant animals need to be restricted to deeper waters. To approach the latter question, we must broaden our focus and consider other important aspects of pressure effects on enzymes, notably the influences of pressure on the integrity of the enzyme-substrate-cofactor complex and on rates of catalytic function.

We stated earlier that enzymes must be flexible machines to accomplish their prodigious feats of catalysis. We also indicated that highly precise interactions between enzymes and their substrates and cofactors are necessary to allow the specificity that is a hallmark of enzymic activity. The complex formed between an enzyme and its substrate(s) and cofactor(s) thus must be viewed as a somewhat fragile entity, which exists for but a short period of time and possesses a high degree of precision in the alignment between the interacting groups that stabilize the complex. These interactions which maintain the integrity of the enzyme-substrate-cofactor complex are termed *weak bonds* to indicate that their energies are very low compared to the covalent bonds that link the atoms of a molecule together (for example, carbon-oxygen bonds). Weak bonds, which include hydrogen bonds, ionic bonds, and hydrophobic interactions, are metastable at biological temperatures and pressures, and continuously break and reform. As long as the rates of formation and rupture are not too different, and as long as changes in temperature and pressure do not lead to a significant reduction in the total number of weak bonds stabilizing enzyme-substrate-cofactor complexes, the positive contributions of weak bonds to biological function vastly outweigh the negative features, i.e., the sensitivity to perturbation by temperature and pressure, of weak bonds. Indeed, the reversible changes in enzyme structure to which we have alluded repeatedly are dependent on the weakness of weak bonds. If enzymes were stabilized entirely by bonds which could not be reversibly formed and broken at physiological temperatures, enzymes might be vastly stronger and have longer lifespans, but they would also function poorly if at all.

To link the importance of weak bonds in enzyme systems to the phenomenon of adaptation in the deep sea, and to develop a biochemical explanation that may account, in part, for the finding that deep-sea fishes tend to occur only in deep waters, we must examine the relationship that exists between enzyme flexibility and catalytic efficiency, and the manner in which this important relationship is affected by temperature and pressure. In the context of this discussion, catalytic efficiency is considered solely in terms of how effectively an enzyme can lower the activation energy "barrier" to its metabolic reaction. The degree to which this barrier is lowered determines how fast the enzyme will convert substrate to product, other things (e.g., substrate and cofactor concentrations) being equal. Since the high catalytic activities of enzymes derive from their abilities to lower activation free energy ($\Delta G^{\ddagger}$) barriers to reactions, it has been proposed that enzymes of organisms having low body temperatures would be able to offset the retarding effects of low temperature on metabolic rates if these enzymes had especially good abilities to reduce $\Delta G^{\ddagger}$. In other words, if this hypothesis is correct $\Delta G^{\ddagger}$ values should be proportional to organisms' body temperatures. In comparisons of several sets of homologous enzymes, this relationship has indeed been found (Somero, 1978). Not only do "cold-blooded" species have more efficient enzymes than "warm-blooded" species like birds and mammals, but within ectothermic ("cold-blooded") species the most cold-adapted animals have the most efficient enzymes, while tropical, warm-water species have the least efficient enzymes (see, for example, Johnston and Walesby, 1977). Taken at face value, this relationship makes eminent sense. By having enzymes that are especially efficient, a fish living in cold waters may be able to overcome the rate-retarding effects of low temperature, and its capacity to metabolize, grow, and reproduce will not be as low as might be predicted on the basis of temperature effects on a temperate or tropical fish. However, there is a seemingly paradoxical aspect to this relationship between body temperature and enzymic efficiency; this is the fact that, in viewing a broad historical perspective, we must account for a loss of catalytic efficiency in birds and mammals (and, perhaps, in many warm-adapted ectotherms as well). Thus, if we assume that the ancestor of a modern bird or mammal was a fish that had a body temperature well below 37°C, and if we further assume that the enzymes of this distant ancestor were as efficient as those of a modern-day fish living at temperatures below those of birds and mammals, then we are faced with having to explain the selective advantage of reduced catalytic efficiencies in birds and mammals (Low and Somero, 1974). One possible solution to this paradox is based on considerations of a compromise that may exist between catalytic efficiency and structural stability (Somero, 1978). In view of the role of weak bonds in stabilizing enzyme-substrate-cofactor complexes and in permitting reversible conformational changes in enzymes, it seems reasonable to conjecture that enzymes of warm-adapted species may require one or more additional weak bonds to confer adequate stability on their structures in the face of higher operating temperatures. This requirement can be appreciated as coming into conflict with catalytic efficiency, for if more weak bonds are formed to stabilize the enzyme-substrate-cofactor complex, then more weak bonds must later be broken to complete the catalytic cycle. Indeed, the studies of Borgmann, Laidler, and Moon (1975; Borgmann and Moon, 1975) have shown that, for LDHs of a mammal and a cold-adapted fish, one or more additional weak bonds are present in the mammalian enzyme-substrate-cofactor complex, and the need to break these bonds during the catalytic cycle accounts for the higher $\Delta G^{\ddagger}$ of the mammalian LDH reaction. In summary, the evolution of enzymes in different thermal regimes involves a compromise wherein a balance is reached between catalytic efficiency and structural stability. $\Delta G^{\ddagger}$ barriers may be low for reactions of low temperature species because their enzymes require less thermal stability, while enzymes of warm-adapted species sacrifice some efficiency for enhanced tolerance of higher temperatures.

How do the enzymes of deep-sea animals fit into this scheme? We might predict that, due to the low temperatures of the deep-sea environment (Fig. 11-1), enzymes of deep-sea animals would possess high catalytic efficiencies like those of of cold-adapted, shallow-living fishes. However, as the data of Table 11-1 indicate, this expectation is not realized. The LDHs of all of the deep-sea species examined in this study (Somero and Siebenaller, 1979) are inefficient relative to LDHs of shallow-water, cold-adapted fishes.

TABLE 11-1. Activation Free Energies ($\Delta G^{\ddagger}$) and Relative Catalytic Velocities of Skeletal Muscle Type ($M_4$) Lactate Dehydrogenases from a Mammal and Fishes Adapted to Different Temperatures and Hydrostatic Pressures[a]

| Species (body temperature) | $\Delta G^{\ddagger}$ (kcal/M) | Relative Velocity |
|---|---|---|
| Pagothenia borchgrevinki (-2°C) | 14,000 | 1.00 |
| Sebastolobus alascanus (4 – 12°C) | 14,009 | 0.98 |
| Sebastolobus altivelis (4 – 12°C) | 14,249 | 0.64 |
| Coryphaenoides acrolepis (2 – 50°C) | 14,222 | 0.67 |
| Halosauropsis macrochir (2 – 5°C) | 14,227 | 0.66 |
| Antimora rostrata (2 – 5°C) | 14,343 | 0.54 |
| Thunnus thynnus (bluefin tuna) (15 – 30°C) | 14,152 | 0.76 |
| Rabbit (37°C) | 14,342 | 0.54 |

[a]The velocity of the LDH reaction of *Pagothenia borchgrevinki* was the highest measured for any LDH and is set equal to 1.00 for comparative purposes (velocities measured at 5°C). Data from Somero and Siebenaller, 1979.

At 5°C a single LDH molecule of a deep-sea fish functions at only about 60% the rate noted for shallow-water, cold-adapted fishes' LDHs, at least when measurements are made at 1 atm. This difference in catalytic efficiency is noted in comparisons of the LDHs of the two *Sebastolobus* congeners, showing another way in which these electrophoretically indistinguishable LDH homologues from fishes that are highly similar morphologically and genetically differ in kinetic properties.

From the results given in Table 11-1, it appears that the acquisition of pressure-insensitivity may carry as its price a loss in catalytic efficiency. If this is the case, then some interesting implications arise for the mechanisms of molecular evolution in the deep sea and for the factors contributing to depth distribution patterns. In the first case, the reductions in catalytic efficiency in high-pressure-adapted enzymes may be a reflection of selection for a more stable enzyme-substrate-cofactor complex. Much as a bird or a mammal may have to pay the cost of reduced catalytic efficiency to ensure adequate enzyme thermal stability, deep-living animals may have to tolerate a similar loss in enzyme performance in order to acquire enzymes with pressure-resistant structures.

These correlated decreases in pressure sensitivity and catalytic efficiency for LDHs also have implications for the possible roles of biochemical factors in establishing depth distribution patterns in the ocean. If, in shallow waters, an enzyme molecule of a deep-sea fish is able to function at only about 60% of the rate noted for the homologous enzyme of a shallow-living species, then the metabolic processes of the deep-sea fish would seem at a competitive disadvantage. To compensate for its poorer enzyme activity, the deep-sea fish might maintain a higher concentration of enzyme molecules in its tissues. However, this strategy too would put the deep-sea fish at a disadvantage in view of the fact that this species would have to channel increased energy into protein synthesis. This strategy also is inconsistent with what we know about the compositions of deep-sea fishes, for these species tend to be dilute and low in protein content (see below). It is suggested, therefore, that the observed inefficiencies of high-pressure-adapted LDHs provide a model for explaining how biochemical adaptations contribute to depth distribution patterns, specifically the restriction of pressure-adapted species to deeper regions of the water column. Certainly, additional studies of other enzyme systems are necessary before broad generalizations about enzymic contributions to depth zonation patterns are allowable. Biological interactions and adaptations at other levels of biological organization must also play key roles in setting distribution limits. Nonetheless, the trade-off between catalytic efficiency and pressure-sensitivity noted in comparative studies of LDHs supports the view that biochemical factors may contribute in important ways to the species zonation patterns found in the water column.

Before leaving the topic of enzyme activity/structure adaptations, it is pertinent to consider how increases in hydrostatic pressure affect the velocities of the LDH reactions of shallow- and deep-living fishes (Table 11-2). To this point in our discussion we have considered the relative efficiencies of the different LDH homologues only at 1–atm pressure. However, to provide a truly fair comparison of these differently pressure-adapted LDHs we must measure rates of activity under *in situ* pressures. When this experiment is done, we find that the inefficient enzymes of deep-sea fishes fare better than the preceding discussion might have led us to expect (Table 11-2). As shown by the inhibition data, a pressure of, for example, 340 atm inhibits the velocity of the LDH reaction of a deep-sea fish by an average of only 5%, while for the shallow-living fishes inhibition

TABLE 11-2. Apparent Volume Changes Associated with Pressure Inhibition of Optimal Velocities of Deep- and Shallow-living Fishes' $M_4$-Lactate Dehydrogenase Reactions[a]

| Species | $\Delta V \pm$ S.E. (cm$^3$/$M$) | % inhibition of 1 atm rate at | | |
|---------|------------------------------|---------|----------|----------|
| | | 68 atm | 204 atm | 340 atm |
| Deep-living fishes | | | | |
| Sebastolobus altivelis | 8.1 ± 0.22 | 2 | 7 | 11 |
| Antimora rostrata | 0.3 ± 1.45 | 0 | 0 | 0 |
| Coryphaenoides acrolepis | 2.6 ± 0.83 | 1 | 2 | 4 |
| Halosauropsis macrochir | 4.2 ± 0.40 | 1 | 3 | 5 |
| Shallow-living fishes | | | | |
| Sebastolobus alascanus | 12.8 ± 0.53 | 4 | 11 | 17 |
| Scorpaena guttata | 12.8 ± 0.94 | 4 | 11 | 17 |
| Pagothenia borchgrevinki | 5.2 ± 1.53 | 2 | 5 | 7 |

[a]Measurements were made at 10°C and at the indicated pressures. Data from Siebenaller and Somero, 1979.

is much greater, averaging 17%. At typical depths for the deep-sea fishes (Fig. 11-3), the loss of catalytic efficiency noted at 1 atm is at least partially counterbalanced by the abilities of these enzymes to retain catalytic ability in the face of high pressure. At the greatest depths frequented by deep-sea fishes, their LDHs may be able to function at essentially the same rates that would characterize the LDHs of shallow-living species subjected to the same high pressures. It must also be emphasized that the comparisons of Table 11-2 are based on saturating substrate concentrations, so the pronounced differences in $K_m$ sensitivity to pressure between shallow- and deep-living species' LDHs (Fig. 11-4) are not reflected in these rate data. At nonsaturating substrate and cofactor concentrations, the LDHs of deep-sea fishes would operate at a clear rate advantage at high pressures, since the LDHs of shallow-living species would have their rates reduced by both the optimal velocity and $K_m$ perturbations of pressure.

## METABOLIC AND BODY COMPOSITIONAL ADAPTATIONS OF DEEP-SEA ANIMALS: THE REQUISITES FOR LIFE IN AN ENERGY-POOR ENVIRONMENT

The enzymic adaptations discussed in the previous section of this chapter are adaptations which confer a fundamental tolerance of high-pressure and low-temperature effects on enzymes of deep-sea animals. To only a limited extent, however, do these kinetic and structural adaptations determine how much biological activity can occur in the deep sea. Thus, having found that enzymes (and other macromolecular systems; see Somero *et al.,* 1982) can acquire the capacities to function under the physical stresses characteristics of the deep sea, we must shift our attention to the factors that determine the quantity of metabolism that occurs in the dark, food-poor deep-sea environment.

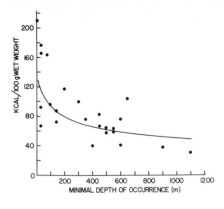

Fig. 11-5. The relationship between minimal depth of occurrence and energy content (kcal/100 g wet w) for a number of midwater teleost fishes. Modified after Childress and Nygaard, 1973; from Somero et al., 1982.

To begin this analysis it is appropriate to consider what is perhaps the most striking depth-related change in marine life, the exponential decrease in biomass with depth, where biomass is measured both as number of individuals and as total mass (Banse, 1964). Moreover, in stating that mass of organisms decreases with depth, we are really understating the situation, for not only does the total mass of living organisms fall off rapidly with depth, but the "living stuff" represented in each unit mass may also decrease with depth. As shown by the representative data of Childress and Nygaard (1973) in Fig. 11-5, the caloric content of fishes falls with increasing minimal depth of occurrence of the species. Deep-living fishes tend to be watery and low in protein content (also see Table 11-3). From a nutritional standpoint this means that not only are there fewer individuals and less biomass to support life, but a given mouthful of food is apt to be relatively poor in calories (at least if a deep-sea animal is feeding on its deep-dwelling neighbors rather than on more energy-rich material descending from the surface or obtained from the hydrothermal vent regions). The low input of food into the deep sea from surface waters, and the low caloric content of much of the constituent fauna of the deep sea, have had pronounced effects on all aspects of the "design" of deep-sea animals, ranging from predatory strategies to buoyancy mechanisms to body chemical composition (Agassiz, 1888; Dayton and Hessler, 1972; Grassle and Sanders, 1973; Hessler et al., 1978; Marshall, 1979). In this chapter it is only possible to discuss a few of these relationships between the biological and ecological aspects of the deep-sea environment and organismal design; we shall concentrate particularly on the factors that establish the rates at which energy must flow through an organism (i.e., on rates of metabolic activity).

The data of Figs. 11-6 and 11-7 illustrate the characteristic changes in respiratory

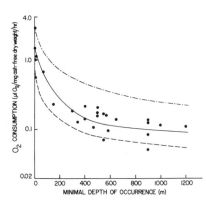

Fig. 11-6. The relationship between maximal respiration (_. _. _), respiration at 30-70 mm $H_g$ $O_2$ (————), and minimal respiration (------------) and species' minimal depths of occurrence for a number of midwater marine crustaceans. Modified after Childress, 1975; from Somero et al., 1982.

**PHYSIOLOGICAL AND BIOCHEMICAL ADAPTATIONS OF DEEP-SEA FISHES**

**TABLE 11-3.** Activities of Lactate Dehydrogenase (LDH), Pyruvate Kinase (PK), Malate Dehydrogenase (MDH), and Citrate Synthase (CS), and the Water and Protein Contents of White Epaxial Skeletal Muscle of Marine Fishes Having Different Depths of Occurrence and Locomotory and Feeding Habits[a]

| Species (n) | Enzyme Activity[b] LDH | PK | MDH | CS | Percent Water | Protein[c] Content | Depth Range |
|---|---|---|---|---|---|---|---|
| **Shallow species** | | | | | | | |
| *Auxis thazard*[d] (2) | 1186 | 192 | nd | 5.45 | nd | nd | surface–200 m |
| *Medialuna californiensis* (2) | 981 | 125 | 23 | 0.70 | 75.6 | 198 | surface–40 m |
| *Engraulis mordax* (16) | 540 | 60 | 61 | 1.52 | 80.8 | nd | surface–100 m |
| *Phanerodon furcatus* (1) | 414 | 90 | 77 | 0.71 | 77.3 | 251 | surface–45 m |
| *Atherinops affinis* (8) | 412 | 107 | 20 | nd | 77.7 | 222 | surface–40 m |
| *Paralabrax nebulifer* (6) | 397 | 71 | 21 | 0.52 | 75.6 | 211 | surface–185 m |
| *Paralabrax clathratus* (7) | 389 | 75 | 23 | 0.79 | 77.4 | 200 | surface–50 m |
| *Chromis puntipinnis* (1) | 388 | 92 | 51 | 0.85 | 77.8 | 230 | surface–50 m |
| *Rhacochilus toxotes* (1) | 351 | 42 | 63 | 1.15 | 78.5 | 211 | shallow–50 m |
| *Gillichthys mirabilis* (7) | 321 | 28 | 26 | 0.90 | nd | nd | shallow–10 m |
| *Paralabrax maculatofaciatus* (3) | 287 | 41 | nd | 0.50 | nd | nd | shallow–60 m |
| *Genyonemus lineatus* (6) | 267 | 41 | nd | 0.67 | nd | nd | shallow–100 m |
| *Caulolatilus princeps* (1) | 209 | 32 | 32 | 0.50 | 80.8 | 213 | shallow–95 m |
| *Squalus acanthias*[e] (2) | 182 | 58 | 18 | 0.56 | 69.5 | 166 | shallow–374 m |
| *Sebastes mystinus* (1) | 116 | 72 | nd | 0.74 | nd | nd | shallow–95 m |
| *Scorpaena guttata* (2) | 77 | 15 | nd | 0.25 | nd | nd | shallow–95 m |
| *Sebastolobus alascanus* (8) | 58 | 7 | nd | 0.35 | nd | nd | 25–1000 m |
| $\overline{X}$ | 337 | 63 | 38 | 0.71 | 77.9 | 211 | |
| S.D. (n) | 216(16) | 34(16) | 21(11) | 0.31(15) | 1.9(9) | 24(9) | |
| C.V.[f] | 65% | 53% | 56% | 44% | 2.4% | 11% | |
| **Deep-living species** | | | | | | | |
| *Histobranchus bathybius* (1) | 156 | 19 | 7 | 0.35 | nd | nd | 1885–2830 m |
| *Coryphaenoides acrolepis* (2) | 154 | 10 | 9 | 0.31 | 79.8 | 103 | 500–2300 m |
| *Coryphaenoides armatus* (2) | 153 | 21 | 16 | 0.97 | 80.3 | 151 | 1885–4800 m |
| *Anoplopoma fimbria* (5) | 107 | 16 | 16 | 0.48 | 71.5 | 112 | 200–1550 m |
| *Paraliparis melanurus* (1) | 66 | 10 | nd | 0.10 | 93.9 | nd | 1800–2500 m |
| *Antimora rostrata* (1) | 36 | 10 | 7 | 0.37 | 82.4 | 102 | 825–2500 m |
| *Sebastolobus altivelis* (8) | 25 | 4 | nd | 0.19 | nd | nd | 350–1500 m |
| *Coryphaenoides carapinus* (1) | 15 | 9 | 7 | 1.10 | 83.8 | 147 | 1250–2740 m |
| *Halosauropsis macrochir* (1) | 12 | 4 | 3 | 0.41 | 83.4 | 124 | 1500–5180 m |
| *Nezumia bairdii* (7) | 9 | 5 | 9 | 0.82 | 80.5 | 165 | 260–1965 m |
| *Coryphaenoides leptolepis* (1) | 5 | 4 | 4 | 0.40 | 82.5 | 139 | 2288–4640 m |
| $\overline{X}$ | 67 | 10 | 9 | 0.49 | 82.0 | 130 | |
| S.D. (n) | 63(11) | 6(11) | 5(9) | 0.32(11) | 5.8(9) | 24(9) | |
| C.V. | 94% | 60% | 53% | 65% | 7% | 18% | |

[a]From Sullivan and Somero, 1980.
[b]Activities are international units per gram wet weight of muscle at 10°C.
[c]Protein content is expressed as mg protein per gram wet weight of muscle.
[d]The data for the warm-bodied frigate mackerel (*Auxis thazard*) are not included in the statistical analyses.
[e]The water content of *Squalus acanthias* was not included in the average for the shallow-living species.
[f]C.V. (coefficient of variation) equals the standard deviation (S.D.) divided by the mean. For LDH, PK, and MDH activities and protein content, the means for the shallow-living species differ significantly from the means of the deep-living species ($P < 0.01$ in all cases; Student's t-test).

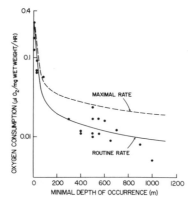

Fig. 11-7. The relationship between routine respiration (————), maximal respiration (------------), and minimal depth of occurrence for a number of midwater marine teleost fishes. For species occurring at depths of less than 100 m, measurements were made at 10°C; deeper-occurring species were measured at 5°C. Modified after Torres *et al.*, 1979; from Somero *et al.*, 1982.

activity (oxygen consumption) as a function of species' minimal depths of occurrence for crustaceans and fishes, respectively. These respiratory rates exhibit a depth-related pattern similar to that found for caloric content (Fig. 11-5). Thus, compositional and metabolic changes go hand in hand, as one would expect. Also note, in the fish respiratory data of Fig. 11-7 from Torres *et al.* (1979), that both routine and maximal respiration decrease with increasing minimal depth of occurrence. A similar pattern is found for crustaceans (Fig. 11-6). The paired decreases in routine and active metabolism with increasing minimal depth of occurrence suggest that both the so-called "basal" metabolic rates and the capacities for vigorous activity are greatly reduced in deep-living animals (note that the $Y$–axis in each graph is a logarithmic scale).

To understand more fully the bases of the decreases in metabolic rate with increasing minimal depth of occurrence, we have examined the levels of enzymic activity in various tissues of shallow- and deep-living fishes (Childress and Somero, 1979; Sullivan and Somero, 1980). Certain of these data are presented in Figs. 11-8, 11-9, and 11-10 and Table 11-3. For lactate dehydrogenase, pyruvate kinase (PK), and malate dehydrogenase (MDH), significant depth-related decreases in enzymic activity/gram of white skeletal muscle are noted. For example, LDH activity varies by almost three orders of magnitude between a highly active, shallow-living fish like *Medialuna californiensis* or *Auxis thazard* and certain sluggish, deep-living species (Table 11-3). Note, however, that a large amount of interspecific variation in enzymic activity is present within each group.

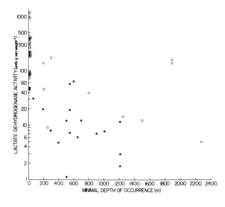

Fig. 11-8. The relationship between minimal depth of occurrence and lactate dehydrogenase activity of white skeletal muscle for a number of marine fishes. Units of activity are $\mu$moles substrate converted to product per minute per gram wet weight of muscle tissue. The assay temperature was 10°C. Open symbols are data from Table 11-3; closed symbols are data from Childress and Somero, 1979, which were obtained using identical experimental procedures. From Sullivan and Somero, 1980.

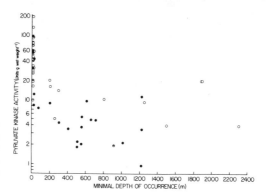

Fig. 11-9. The relationship between minimal depth of occurrence and pyruvate kinase activity of white skeletal muscle for a number of marine fishes. Units and symbols are defined as in Fig. 11-8. From Sullivan and Somero, 1980.

We have attributed this non-depth-related variation in enzymic activity of skeletal muscle to interspecific differences in feeding and locomotory habits. Highly active pelagic swimmers like *Medialuna* and *Auxis* have muscle enzymic activities reflecting the abilities of these fishes to maintain high swimming velocities and, in particular, to undergo bursts of high-speed acceleration. In contrast, sit-and-wait demersal predators (e.g., *Scorpaena guttata*) lack the enzymic power associated with the more active pelagic swimmers. The observed variation in muscle enzymic activity among different deep-sea fishes may derive in large measure from similar differences in locomotory ability and feeding strategy. The deep-sea fishes possessing the highest levels of muscle enzymic activity, e.g., the two rattail species, *Coryphaenoides armatus* and *C. acrolepis,* are species that cruise the deeper regions of the water column in search of food. These two rattails and the sablefish, *Anoplopoma fimbria,* are frequently observed in baited camera-drop studies. Much lower enzymic activities are characteristic of the deep-sea fishes that have been observed merely to drift over the bottom in a more passive feeding strategy, e.g., *Antimora rostrata* and *Halosauropsis macrochir* (see Sullivan and Somero, 1980). We conclude, therefore, that the observed interspecific differences in muscle enzymic activity derive from at least two important selective considerations: locomotory/feeding strategy and depth of occurrence. These two determinants are not, of course, cleanly separable. In the case of deep-sea fishes, the amount and distribution of food packets will strongly influence the frequency

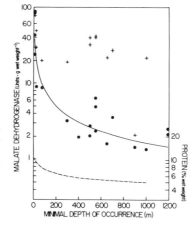

Fig. 11-10. The relationship between minimal depth of occurrence and malate dehydrogenase activity of white skeletal muscle (●) and brain (+) for a number of marine fishes. Changes in protein content with depth of occurrence (dashed line) are also shown. From Childress and Somero, 1979.

with which species adopt relatively passive feeding strategies. To a large extent, the depth-related decreases in enzymic activity in muscle noted in particular over the first 200 m of the water column may be a reflection of a shift toward more passive feeding/locomotory habits. Also, it is important to remember that the requirements for active swimming may derive not only from considerations of food gathering, but also from needs to avoid predation. In the deep sea, where the only light present is due to bioluminescence, the need for a powerful swimming ability to avoid visual predators may be greatly reduced. A number of different factors, then, may permit a reduction in the metabolic potential of certain tissues of deep-sea animals, a reduction which is in tune with the low energy input into the abyss.

In view of the greatly reduced activities of enzymes in skeletal muscle tissue of deep-sea fishes, we can inquire about the compositional basis of this relationship. Are the low enzymic activities simply a reflection of dilute cellular fluids, i.e., of high water contents and low protein concentrations? The data in Table 11-3 suggest that this simple mechanism cannot adequately account for the enzymic activity changes shown in this table and in Figs. 11-8, 11-9, and 11-10. Muscle water content did not differ significantly between the shallow- and deep-living fishes, and the observed decrease in muscle protein content in the deep-sea fishes was too small (of the order of 40%) to explain much of the enzymic change with depth. A similar conclusion was reached in the study of Childress and Somero (1979; Fig. 11-10). Thus it appears that the concentrations of specific enzymic proteins are reduced much more than total muscle protein, and that differential changes in the structural protein and catalytic protein components of muscle tissue characterize adaptation to depth.

The activity of citrate synthase (CS), an indicator enzyme of citric acid cycle activity, did not display any strong relationship to minimal depth of occurrence or feeding/locomotory strategy (Table 11-3). This finding can be reconciled with what is known about the metabolic poise of fish white muscle (Driedzic and Hochachka, 1978; Sullivan and Somero, 1980). The function of white muscle is essentially glycolytic in nature, with the degradation of glycogen stores via the activities of LDH, PK, and the remainder of the glycolytic enzymes supplying the energy for white-muscle-driven swimming. MDH may also play a role in the anaerobic functioning of white muscle (see Sullivan and Somero, 1980). The activities of aerobically-poised enzymes like CS are extremely low in white muscle, and seem apt to be unrelated to selective pressures arising from feeding and locomotory strategy considerations.

In view of the anaerobic poise of white skeletal muscle of fishes, it is important to understand how this anaerobic muscle function is integrated into an overall aerobic metabolic nature of fish metabolism. This point is particularly important if we are to explain the relationship between decreases in respiratory activity *and* anaerobic enzymic activity in deep-living fishes. The link between white skeletal muscle metabolism, which generates the anaerobic end-product lactic acid, and the overall aerobic metabolism of the whole fish is one involving temporal and spatial features of lactic acid metabolism. The lactic acid generated in white muscle during bouts of swimming activity is largely channeled into the circulatory system and carried to aerobic sites for further metabolism (oxidation to $CO_2$ and water or resynthesis into glucose and glycogen). Burning of lactate in aerobically-poised tissues like heart, brain, and gill is thus an important contributor to a fish's total oxygen consumption rate (Driedzic and Hochachka, 1978). Thus, high levels of enzymic activity for lactate generation must be paired with high aerobic potentials elsewhere in the fish. In other words, one predicts a correlation between white muscle anaerobic metabolic potential and the capacity of the whole fish for respiratory metabolism. Data presented in Figure 11-11 illustrate this correlation for a number of shallow-living and midwater fishes (Childress and Somero, 1979). This relationship between muscle enzymic activity and whole fish respiration rate has more than theoretical significance. In view of the difficulties entailed in capturing and retrieving healthy deep-living animals, the availability of a simple enzymic test for metabolic (respiratory) potential of such species may allow deep-sea biologists to estimate the metabolic requirements of many organisms that cannot be used in respirometer experiments. LDH, PK, and MDH are extremely stable enzymes that maintain activity during prolonged storage in frozen muscle. Thus, even tissues from specimens badly damaged during capture may yield useful insights into whole-organism metabolic rates.

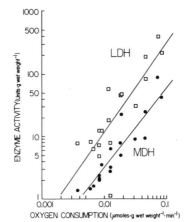

Fig. 11-11. The relationship between routine oxygen consumption rates of midwater fishes and their muscle LDH and MDH activities. Lactate dehydrogenase: □; malate dehydrogenase: ●. Modified after Childress and Somero, 1979; from Somero *et al.*, 1982.

The depth- and feeding/locomotory strategy-related changes in enzymic activity are not noted in comparisons of brain tissue (Fig. 11-10; Childress and Somero, 1979; Sullivan and Somero, 1980). This observation would seem to make intuitive sense. Whereas the amount and the vigor of swimming may differ between shallow- and deep-living species, it does not seem likely that requirements for neural coordination and sensing mechanisms would exhibit a corresponding change. Demands for olfactory capacities, sensory coordination, etc., may vary little among fishes having widely different habitats and life-styles. Also, in view of the miniscule fraction of total body mass represented by a fish's brain, there is apt to be little, if any, selective advantage in reducing brain density as a buoyancy regulating mechanism. Most of a fish's body mass consists of white skeletal muscle and the skeletal system, so that buoyancy-related changes in body composition would be expected to center on these two dominant tissues, as in fact has been observed (Blaxter *et al.*, 1971; Denton and Marshall, 1958; Marshall, 1979).

# CONCLUDING COMMENTS AND CAVEATS

The study of deep-sea physiology and biochemistry is yet in its infancy. This state of affairs accounts for much of the lure of such studies and for the limited extent to which generalizations can be made about the adaptations of deep-sea animals. The focus of this chapter has been largely on enzyme phenomena, and the range of organisms considered has been limited largely to fishes. While thus limited in scope, our discussion has nonetheless shown that the physical and biological characteristics of the deep-sea environment have dramatic effects on the biochemical and physiological features of deep-sea animals. Because temperature and pressure effects on enzymes will be similar regardless of whether the enzyme is contained within a fish, an invertebrate, a fungus, or a bacterium, many of the conclusions drawn about enzymic adaptation strategies may apply for all types of deep-sea organisms. Again, since flexibility and conformational changes appear to be general features of enzyme design, the conclusions reached from an examination of fish lactate dehydrogenases may have general applicability to many other enzyme systems. As in the case of the deep sea's physical features, the biological properties of the abyss, notably its low and unpredictable food supply, will affect many different types of organisms in similar ways. Thus, the compositional and enzymic activity differences noted in comparisons of fishes having different depths of occurrence and life-styles may also pertain for many invertebrate groups. However, in view of the fact that the species discussed in this chapter are largely found in the water column, not on the bottom, the adaptive strategies of benthic species might be markedly different from those of pelagic forms. Much further study of benthic animals is needed before we will understand how these species are affected by the biological features of their habitat.

Lastly, the enzymic emphasis of this chapter is largely the result of the author's specific interests, and should not be taken as an indication that other types of biochemical and physiological adaptations in deep-sea animals are insignificant. Whereas we know very little about, for example, neurophysiological adaptations and changes in structural elements like microtubules facilitating structural integrity at high pressures and low temperatures, the fact that so many different features of organisms are reliant on the weak bonds discussed in the context of temperature and pressure effects on enzymes makes it certain that widespread adaptations throughout the physiological and biochemical features of deep-sea animals remain to be discovered.

# ACKNOWLEDGMENTS

Discussions of the ideas in this chapter with Drs. James J. Childress, Joseph F. Siebenaller, Kenneth L. Smith and Robert R. Hessler and Ms. Kathleen M. Sullivan are gratefully acknowledged. Research was supported by National Science Foundation grant PCM-7804321.

Agassiz, A., 1888, Three cruises of the United States Coast and Geodetic Survey Steamer "Blake": *Bull. Harvard Mus. Comp. Zool.,* 14, p. 1–314.

Banse, K., 1964, On the vertical distribution of zooplankton in the sea, *in* Sears, M., ed., *Progress in Oceanography, Vol. II:* Oxford, Pergamon, p. 53–125.

Blaxter, J. H. S., Wardle, C. S., and Roberts, B. L., 1971, Aspects of the circulatory physiology of deep-sea fish: *J. Mar. Biol. Assoc. U.K.,* v. 51, p. 991–1006.

Borgmann, U., Laidler, K. J., and Moon, T. W., 1975, Kinetics and thermodynamics of lactate dehydrogenases from beef heart, beef muscle, and flounder muscle: *Can. J. Biochem.,* v. 53, p. 1196–1206.

——, and Moon, T. W., 1975, A comparison of lactate dehydrogenases from an ectothermic and an endothermic animal: *Can. J. Biochem.,* v. 53, p. 998–1004.

Cavanaugh, C. M., Gardiner, S., Jones, M. K., Jannasch, H. W., and Waterbury, J. B., 1981, Prokaryotic cells in the hydrothermal vent tube worm *Riftia pachyptila* Jones: possible chemoautotrophic symbionts: *Science,* v. 213, p. 340–342.

Childress, J. J., 1975, The respiratory rates of midwater crustaceans as a function of depth of occurrence and relation to the oxygen minimum layer off southern California: *Comp. Biochem. Physiol.,* v. 50A, p. 787–799.

——, and Nygaard, M. H., 1973, The chemical composition of midwater fishes as a function of depth of occurrence off Southern California. *Deep-Sea Res. 20:* 1093–1109.

——, and Somero, G. N., 1979, Depth-related enzymic activities in muscle, brain, and heart of deep-living pelagic marine teleosts: *Mar. Biol.,* v. 52, p. 273–283.

Dayton, P. K., and Hessler, R. R., 1972, Role of biological disturbance in maintaining diversity in the deep-sea: *Deep-Sea Res.,* v. 19, p. 199–208.

Denton, E. J., and Marshall, N. B., 1958, The buoyancy of bathypelagic fishes without a gas-filled swimbladder: *J. Mar. Biol. Assoc. U.K.,* v. 37, p. 753–767.

Driedzic, W. R., and Hochachka, P. W., 1978, Metabolism in fish during exercise, *in* Hoar, W. S., and Randall, D. J., eds., *Fish Physiology, Vol. VII:* New York, Academic, p. 503–544.

Felbeck, H., 1981, Chemoautotrophic potential of the hydrothermal vent tube worm, *Riftia pachyptila* Jones (Vestimentifera): *Science,* v. 213, p. 336–338.

——, H., Childress, J. J., and Somero, G. N., 1981, Calvin-Benson cycle and sulphide oxidation enzymes in animals from sulphide-rich habitats: *Nature,* v. 293, p. 291–293.

Fersht, A., 1977, *Enzyme Structure and Mechanism:* San Francisco, W. H. Freeman & Co., Publ., 371 p.

Grassle, J. F., and Sanders, H. L., 1973, Life histories and the role of disturbance: *Deep-Sea Res.,* v. 20, p. 643–659.

Hessler, R. R., Ingram, C. L., Yayanos, A. A., and Burnett, B. R., 1978, Scavenging amphipods from the floor of the Philippine Trench: *Deep-Sea Res.,* v. 25, p. 1029–1047.

Hochachka, P. W., and Somero, G. N., 1973, *Strategies of Biochemical Adaptation:* Philadelphia, W. B. Saunders and Co., 358 p.

Honjo, S., and Roman, M. R., 1978, Marine copepod fecal pellets—production, preservation and sedimentation: *J. Mar. Res.,* v. 36, p. 45–57.

Johnston, I. A., and Walesby, N. J., 1977, Molecular mechanisms of temperature adaptation in fish myofibrillar adenosine triphosphatases: *J. Comp. Physiol.*, v. 119, p. 195–206.

Karl, D. M., Wirsen, C. O., and Jannasch, H. W., 1980, Deep-sea primary production at the Galapagos hydrothermal vents: *Science,* v. 207, p. 1345–1347.

Low, P. S., and Somero, G. N., 1974, Temperature adaptation of enzymes—a proposed molecular basis for the different catalytic efficiencies of enzymes from ectotherms and endotherms: *Comp. Biochem. Physiol.*, v. 49B, p. 307–312.

Marshall, N. B., 1979, *Developments in Deep-Sea Biology:* Poole, England, Blandford Press, 566 p.

Siebenaller, J. F., 1978, Genetic variability in deep-sea fishes of the genus *Sebastolobus, in:* Battaglia, B., and Beardmore, J., eds., *Marine Organisms:* New York, Plenum, p. 95–122.

——, and Somero, G. N., 1978, Pressure-adaptive differences in lactate dehydrogenases of congeneric fishes living at different depths: *Science,* v. 201, p. 255–257.

——, and Somero, G. N., 1979, Pressure-adaptive differences in the binding and catalytic properties of muscle-type ($M_4$) lactate dehydrogenases of shallow- and deep-living marine fishes: *J. Comp. Physiol.*, v. 129, p. 295–300.

Somero, G. N., 1978, Temperature adaptation of enzymes—biological optimization through structure-function compromises: *Ann. Rev. Ecol. Syst.,* v. 9, p. 1–29.

——, and Childress, J. J., 1980, A violation of the metabolism-size scaling paradigm— activities of glycolytic enzymes in muscle increase in larger-size fish: *Physiol. Zool.,* v. 53, p. 322–337.

——, and Siebenaller, J. F., 1979, Inefficient lactate dehydrogenases of deep-sea fishes: *Nature,* v. 282, p. 100–102.

——, Siebenaller, J. F., and Hochachka, P. W., 1982, Biochemical and physiological adaptations of deep-sea animals, *in* Rowe, G. T., ed., *The Sea, Vol. VII:* New York, Wiley-Interscience, in press.

Sullivan, K. M., and Somero, G. N., 1980, Enzymic activities of fish skeletal muscle and brain—influences of depth of occurrence and feeding/locomotory habits: *Mar. Biol.,* v. 60, p. 91–99.

Torres, J. J., Belman, B. W., and Childress, J. J., 1979, Oxygen consumption rates of midwater fishes off California: *Deep-Sea Res.,* v. 26A, p. 185–197.

Yancey, P. H., and Somero, G. N., 1978, Temperature dependence of intracellular pH— its role in the conservation of pyruvate apparent $K_m$ values of vertebrate lactate dehydrogenases: *J. Comp. Physiol.*, v. 125, p. 129–134.

K. L. Smith, Jr. and G. A. White
Scripps Institution of Oceanography
University of California
La Jolla, California 92093

# 12

# ECOLOGICAL ENERGETIC STUDIES IN THE DEEP-SEA BENTHIC BOUNDARY LAYER: *IN SITU* RESPIRATION STUDIES

# ABSTRACT

The consumer populations of the soft-bottom benthic boundary layer are arbitrarily divided into four components: the sediment community, the epibenthic megafauna, the benthopelagic animals, and the plankton. The study of carbon flow through this system, as estimated from respiration rates, is currently underway using described *in situ* instrumentation including the free vehicle grab respirometer (sediment community respiration), the fish trap respirometer (benthopelagic animal respiration), and the slurp gun respirometer (plankton respiration).

The published respiration data (through 1979) on these groups using this instrumentation is presented and an initial attempt is made to compare rates for each component at a western North Atlantic station on the continental rise (DOS-2, depth 3650 m). Crude estimates of the respiration rates for three components were as follows: sediment community ($0.96$ g C $m^{-2}$ $y^{-1}$), benthopelagic animals ($0.04$ g C $m^{-2}$ $y^{-1}$), and plankton ($0.18$ g C $m^{-2}$ $y^{-1}$). The epibenthic megafauna were not considered because of unavailable data. Combined respiration is equivalent to 28% of the particulate organic flux to the benthic boundary layer as measured by sedimentation traps.

# INTRODUCTION

The open ocean ecosystem is bounded by an air-sea interface at the surface and a sediment-water interface on the bottom. The organisms which comprise this ecosystem can be grouped into functional units consisting of primary producers, primary consumers, secondary consumers, and tertiary consumers. Primary producers generally occupy the surface or euphotic zone and fix solar energy into organic carbon. This food energy is utilized by primary consumers which inhabit the upper layers of the water column. Primary consumers are fed on by secondary consumers which in turn are fed on by tertiary consumers. The secondary and tertiary consumers extend throughout the water column and include organisms inhabiting the bottom sediments. All three groups of consumers include populations which undergo ontogenetic and diurnal vertical migrations. Food energy is thus disseminated throughout the water column by either active dispersion mediated by organisms, or passive sinking of material such as fecal pellets, dead organisms, and exuvea. With each transfer of food energy from one organism or group of organisms to the next, there is subsequent loss of energy, primarily as heat.

The benthic boundary layer is an arbitrary delineation of the bottom portion of the water column (0–50 m) and the sediments. Secondary and tertiary consumers which inhabit this region can be divided into more descriptive components either by the niche and habitat they occupy or according to the methodologies used in sampling or measuring them. These components include the sediment community, the epibenthic megafauna, the benthopelagic animals, and the plankton. The sediment community includes all the organisms which inhabit the sediments, ranging in size from the macrofauna to the microbiota. The epibenthic megafauna consists of those animals that inhabit the sediment surface, such as ophiuroids, asteroids, holothurians, pycnogonids, and others. Both the epibenthic megafauna and the sediment community biota can undergo limited vertical migrations. The benthopelagic animals include such organisms as rattails, squid, and crustaceans which can undergo vertical migrations within and above the benthic boundary layer. Plankton includes the animals and bacteria which inhabit the water above the sediment and are passive swimmers using water currents as the main means of dispersion. These organisms can also undergo vertical migrations but are probably more restricted to the confines of the benthic boundary layer.

The energetics of these four groups within the benthic boundary layer system are unknown. Our aim is to determine activity rates of the various components in an attempt to understand the ecological energetics of this system. We have used oxygen consumption rates as the primary measurement to assess the energetic importance of benthic boundary layer organisms either singly or as a community. Oxygen consumption or respiration rate provides an instantaneous estimate of the food energy requirements of organisms, assuming a steady state situation, i.e., no growth, only maintenance processes. Since growth rates of these deep-sea organisms are largely unknown, such an assumption is necessary at this stage of analysis.

The purpose of this review is threefold: (1) to present the techniques which have been developed to measure the respiration of the benthic boundary layer organisms *in situ,* including the sediment community, the benthopelagic animals, and the plankton, (2) to review the results obtained with these various techniques, and (3) to synthesize the available results.

# METHODOLOGY

Equipment and techniques which have been developed for measuring the *in situ* respiration of the benthic boundary layer organisms are diverse. The majority of descriptions given here are of the equipment and procedures used in our research on the biotic components of this system, i.e., the sediment community, the benthopelagic animals, and the plankton. The electronic and mechanical systems for oxygen measurement and water sampling are, with a few variations, the same for all experimental enclosures used. The dimensions and configuration of the enclosures vary depending on the organism (e.g., fish, crustacean) or community (e.g., sediment) being studied and the means of equipment deployment (submersible or free vehicle). Descriptions of additional methodology used by other researchers are also given to cover fully the types of devices and techniques used in *in situ* respiration measurements.

*In situ* respiration is measured by one of two techniques: determination of oxygen consumption using a polarographic oxygen sensor or Winkler titration method, or incorporation of radiolabeled organic compounds to measure $CO_2$ evolution (Jannasch and Wirsen, 1973). In oxygen consumption measurements, a polarographic oxygen sensor (Kanwisher, 1959) is inserted into the experimental enclosure. The current output from the sensor is run through an autoscaling amplifier and recorded on a chart recorder (Smith *et al.,* 1976). Before each deployment these sensors are calibrated using air-saturated and nitrogen-purged seawater. Corrections for drift, dissolved oxygen consumption by the sensor, and pressure effects are made using laboratory generated curves for each individual sensor (Smith *et al.,* 1979a).

An electromagnetic stirrer (stirring motor) is also inserted into the respiration chambers to maintain circulation within the enclosure and over the oxygen sensor. It consists of a magnetic impeller rotated by the synchronous energizing of three electromagnets (Smith *et al.,* 1979a).

A general purpose electrolytic release capable of handling loads up to 30 pounds is used to activate timed events such as grab closure, control water sampling, or chemical injection into the respirometry systems. A nichrome wire loop, which secures the release, serves as the cathode in the electrolytic cell; the anode, a bronze rod mounted close to the nichrome wire, and seawater electrolyte complete the cell. A precision event timer, set to preselected firing times, applies an eight-volt potential across the cell. The resulting current causes the oxidation and breakage of the wire loop in approximately thirty seconds, releasing the load.

## Sediment Community Respirometry

We have developed four respirometer units capable of *in situ* oxygen uptake measurements in the deep-sea sediment. Two of these units, the bell jar respirometer and grab respirometer, are placed by a deep-sea submersible. The other two devices are free-vehicle versions of the bell jar and grab. An additional free-vehicle respirometer has been developed which uses oxygen end-point determinations rather than an oxygen sensor to measure uptake in deep sea sediment.

The bell jar respirometer consists of two side-by-side plexiglass cylinders which have been capped on one end. The open ends are inserted into the sediment, leaving 5–10 cm of overlying water enclosed. Oxygen uptake is measured with a polarographic oxygen sensor. A pressure sphere mounted on the end plate houses the electronics, recorder, and power source to drive the sensors and the stirring motors (Smith and Teal, 1973).

The need for an experimental package that did not require excessive amounts of valuable ship or submersible time resulted in the development of the free vehicle respirometer (FVR) (Smith *et al.,* 1976). This unit is capable of running replicate respiration experiments over periods of up to six days independent of support vessels. At the end of the desired incubation period, the FVR is recalled to the surface by acoustic command.

An aluminum tripod 2.2 m on a side and 2.3 m high serves as the frame for the FVR (Fig. 12-1). The primary instrument package consists of four acrylic bell jars cemented to a square polycarbonate plate. An electronic pressure cylinder containing two chart recorders, auto-scaling amplifiers and batteries, is located centrally on the plate. Associated with each bell jar unit is an oxygen sensor and stirring motor. The entire package is suspended from a central acoustic release and held in place by vertical guide rods which go through holes at the corners of the polycarbonate plate. Upon command from the surface, the release frees the tray which slowly drops along the guide rods into the sediment. After the desired experimental duration, a second acoustic command actuates the remote release which drops the descent weight. The FVR is brought to the surface by

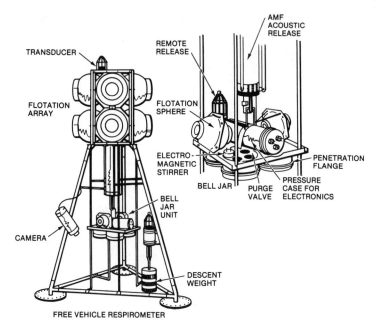

**FREE VEHICLE RESPIROMETER**

Fig. 12-1. Free-vehicle respirometer (FVR) in deployment position with a detailed sketch of the centrally located bell jar unit. Adapted from Smith *et al.*, 1976.

an array of ten flotation spheres which are attached to the top of the tripod. The 250-kg positive buoyancy provided by the spheres is sufficient to break the bell jar and pod suction with the sediment and bring the 164 kg FVR up at a rate of about 52 m min$^{-1}$. On the surface, a transmitter and flasher attached to the top of the flotation array aid in the rapid location and recovery of the FVR.

The other two units, the submersible-placed grab respirometer (DSGR) and the free-vehicle grab respirometer (FVGR), are able to retrieve the enclosed sediment sample at the end of the experiment. This allows the sediment to be analyzed for faunal biomass, community composition, ATP concentration, sediment organic carbon content, pore-water nutrient concentration, and grain size.

A stainless steel Ekman grab (21 × 21 × 30 cm) with a negatively buoyant free-floating top comprises the basic unit of the DSGR. The top contains a stirring motor and an oxygen sensor. The grab consists of two spring-loaded jaws which can penetrate to a depth of 30 cm, enclosing 441 cm$^2$ of sediment surface. After deployment, the top plate drops on guide rods to a 1-cm ledge on the top rim of the chamber; two latches secure the top plate and a silicone gasket on the ledge assures a tight seal (Smith *et al.*, 1978).

Two DSGR units and a pressure case, housing their associated electronics, are deployed on a free-vehicle elevator. On the bottom, the grabs are removed from the elevator by the submersible manipulator and positioned in the sediment. A 2-m long conducting cable to each grab connects the grab closure release mechanism, the oxygen sensor, and the stirring motors to the electronics cylinder, which remains on the elevator.

The free-vehicle grab respirometer (Fig. 12-2) represents a hybrid of the FVR and the DSGR. The FVGR instrument package consists of four replicate grabs which are similar in design and identical in function to the DSGR units. A top tray contains the accessory electronics for the respiration studies. Each respiration chamber has an oxygen sensor and a stirring motor. A separate battery cylinder to power the electronics systems is mounted next to the central acoustic release above the instrument tray.

As in the DSGR, the top tray is separated from the grab assembly until the grabs are in place, permitting flushing through the chambers. The top tray is held in place by an acoustic release. Two hydrostatic plungers slowly push the grab assembly into the sediment to a depth of about 15 cm. After a period which allows for the penetration of the grabs, the top tray is released on command. Both the grab assembly and electronics tray slide on four guide rods; the top tray seals to the grabs in a manner similar to the DSGR, and four spring-loaded latches secure the two assemblies together.

After a period of one to five days of continuous monitoring the grabs close and the FVGR is recalled. Upon release of the descent weights, the tripod achieves a positive buoyancy resulting in an ascent rate of 50-60 m min$^{-1}$. Surface recovery is facilitated by

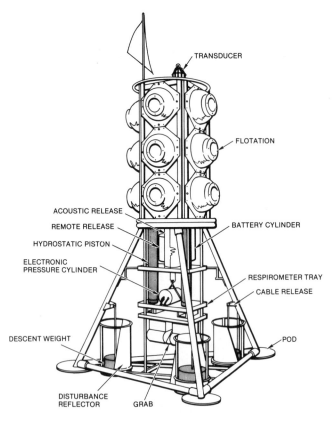

Fig. 12-2.    Free-vehicle grab respirometer (FVGR) in deployment position with the flotation and instrument packages. The respirometer is a modified version of that described by Smith *et al.,* (1979a).

ECOLOGICAL ENERGETIC STUDIES

a transmitter and flasher, and a 1-m-square daylight fluorescent flag on a 2-m mast. Recovery is usually within 30 minutes of surfacing.

Another free vehicle benthic respirometer has been designed to measure oxygen, carbon dioxide, silicate, nitrate, nitrite, phosphate, and ammonia fluxes across the sediment-water interface. This device, although similar in function to the FVR, requires no electronic event control other than that for the release of the descent weight (Hinga *et al.,* 1979).

Four Teflon-coated aluminum chambers, capped on one end, are suspended by two guide rods from a rectangular aluminum frame. Each chamber is held in position for deployment by corrosive links which have been machined to specific diameters for precise dissolving times. When the link breaks, the chamber slides down the guide rods, sealing the open end in the sediment. A Savonius rotor mounted on top of each chamber magnetically drives a stir bar within the enclosure, preventing stratification of dissolved substances.

To collect water samples from inside the chambers, a syringe system has been developed to pull a sample into a stainless steel tube which is formed into a coil. The intakes for the two coils associated with each chamber are attached 3 cm above the sediment surface on the enclosure. The valved outlet on each coil is connected to a hose which runs to a syringe. The syringe plunger is weighted and an assembly of eight syringes is held upright in a bracket. When tripped immediately prior to recall, the array of syringes is reversed and the plunger weight drops, withdrawing a 14-ml water sample from inside the enclosure into the holding coils. Upon recovery, the outlet valves are shut and the coils are removed and refrigerated for later analysis. A Niskin bottle, attached to the frame, is tripped at the same time as the syringes, providing a reference sample from outside the enclosure.

## Benthopelagic Animal Respirometry

The fish trap respirometer (FTR) is designed for trapping and measuring respiration rates of fish and other midwater scavengers at various altitudes above the ocean floor with either free vehicle or submersible deployment techniques (Fig. 12-3). The unit consists of dual acrylic chambers with trapezoidal side walls enclosing a total volume of 53.7 l. A spring-loaded double door system with separate locking latches is situated at the sloped end of each chamber. One trap is used as a control and the other as an experimental trap. The chambers have independent oxygen sensors and stirring motors; retractable bait cannisters are located on the top of each trap.

At the time of deployment, the FTR descends with the control trap screen door closed, the experimental trap screen door open and both bait cannisters in the down position (Fig. 12-3, left trap). The deployed traps are observed by personnel aboard a submersible or with the use of remote television cameras. When a fish enters the trap the manipulator of the submersible or remote vehicle closes the screen door. After a selected interval of time, the acrylic doors on both the experimental trap and the control trap close and the bait containers move into the up position. Changes in dissolved oxygen inside the trap are monitored continuously and recorded.

The combination screen and acrylic doors with separate release mechanisms allow

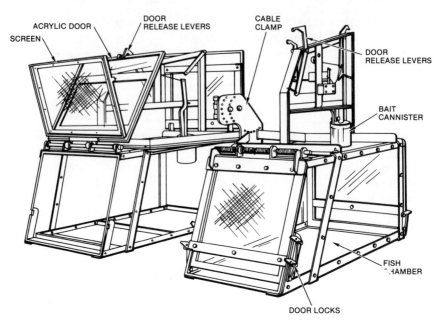

Fig. 12-3. Two fish trap respirometers (FTR) illustrated in a tandem free-vehicle configuration. The left chamber is in an open "fishing" mode while the right chamber is in a closed "measurement" mode.

starvation studies to take place *in situ*. With the screen closed, the fish is separated from food sources while water, flowing through the traps, maintains an otherwise normal environment. The delayed acrylic door closure allows the fish to adjust to the trap environment before the experiment begins; initial respiration rates for trapped fish have been noted to increase 10-15% for the first 20 to 95 minutes after capture (Smith, 1978b). An earlier modification of this trap respirometer was successfully used with the Remote Underwater Manipulator (RUM) (Smith and Hessler, 1974) and *Alvin* (Smith, 1978b).

## Plankton Community Respirometry

The animals comprising the plankton community are diverse, ranging in size from microorganisms to macroscopic free-floating individuals and colonies. To augment our research on the activity rates of planktonic organisms, we have built an instrument to measure the respiration rates of individual animals as well as the planktonic community in general. In addition, Jannasch and Wirsen (1973) have developed equipment for the study of microbial respiration.

The slurp gun respirometer (SGR) is a four-chambered unit used for gently collecting and measuring the oxygen consumption rates of small planktonic organisms. This instrument was designed to be used from a submersible of the DSRV *Alvin* class. Four main functional components of the SGR are: (1) the respirometer, (2) the pumping system, (3) the electronic circuitry, and (4) the mounting frame.

The respirometer consists of four identical modules (Fig. 12-4; only three modules are shown). Each module has an 8.9 cm × 40.5 cm plexiglass chamber equipped with two ports for the placement of 10-ml withdrawal syringes and another port for the placement of an oxygen sensor. Each end of this chamber fits into a valve with spring-loaded plungers. The intake valve has a moveable plexiglass piston for adjusting the volume of the chamber while the outlet valve contains a stirring motor and a screen which prevents the animal from being drawn into the pump. Intake and pump hoses are attached to a manifold which rides over the valves on a track. Two negator spring motors provide the tension to advance the manifold against pull-pins which restrain it sequentially over each chamber. When the SGR is ready for deployment, the manifold is positioned over the first chamber and held in place by the pull-pin. The valve plungers are also restrained in the open position by pull-pins.

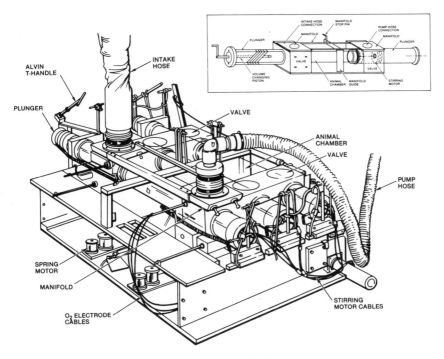

Fig. 12-4. Slurp gun respirometer (SGR) illustrated with three animal chamber modules and the common manifold positioned over the front chamber. Detailed sketch of one module shown to right.

The pumping system consists of a large funnel-shaped intake section, a section of intake hose, the manifold, a length of pump hose, and an electric pump housed in a separate cylinder on the mounting frame. The power necessary to run the pump comes from the submersible's power distribution system. To operate, the intake funnel is positioned over an organism such as a small crustacean, medusa, or fish by the submersible manipulator. The pump is then activated and the animal slurped into the intake hose and down through the intake valve to the transparent respiration chamber. At this point the intake and outlet valves are closed and the organism is thus sealed in the chamber. Depending on the size of the organism, the chamber can be constricted with the piston. This sequence is repeated for each chamber by positioning the manifold over each consecutive chamber valve series.

The associated electronics circuitry is housed in an aluminum pressure cylinder secured to the mounting frame. The systems used with the SGR include oxygen sensors, stirrers, and syringe systems. Power to run these systems is provided by a battery pack secured to the frame in a separate housing.

The mounting frame for the SGR (not illustrated) is adapted from the design used for the sampling basket frame on DSRV *Alvin*. The pump, which is mounted to this frame, is wired into an oil-filled patch panel and receives power from the submersible. The cable interconnections pull free when the SGR is released from the submersible.

After the respiration chambers are filled, a tether attached to the SGR is secured to a free-vehicle mooring system and the submersible then releases the entire SGR system at the depth at which the animals were collected. The SGR then incubates for varying intervals of time from 15–72 hours. At the termination of these measurements, the free vehicle is brought to the surface and the SGR is recovered.

Another piece of equipment designed to be deployed from a deep-sea submersible is used to study the response of microorganisms to *in situ* nutrient enrichment (Jannasch and Wirsen, 1973). Five groups of four bottles containing known quantities of radio-labeled growth media are arranged in a rack and secured to a T-handle. The rack fits into an aluminum pressure housing with a high-pressure valve fitting at one end. The unit is evacuated after closing at the surface. Once on the bottom the valve on the pressure cylinder is opened, allowing the cylinder and bottles to fill with water and suspended particles from the sediment surface. The bottles have punctured serum caps which allow for self-inoculation. After the pressure has equalized in the cylinder, the rack of samples is removed and allowed to incubate on the sediment surface. Upon retrieval the degree of media conversion into cell material and $CO_2$ is measured by liquid scintillation spectrophotometry.


## RESULTS


Available data concerning the functional aspects of the deep-sea benthic boundary layer are limited. Respiration measurements are certainly no exception. Using the methodology just described, an effort is being made to determine the energy flow through the benthic boundary layer. Of the four components of the benthic boundary layer, the sediment community has received the most attention, with measurements being made using both the *in situ* and shipboard coring methods (Hinga *et al.,* 1979; Pamatmat, 1973; Smith, 1978a). Benthopelagic and planktonic organisms have received limited attention, which consists primarily of respiration measurements on rattails (Smith, 1978b; Smith and Hessler 1974), and amphipods (Smith and Baldwin, in press), and [14]C uptake rate measurements on bacteria (Jannasch and Wirsen, 1973). Although the activity rates of the epibenthic megafauna are currently unknown, the functional role of the group in the system is now under investigation.

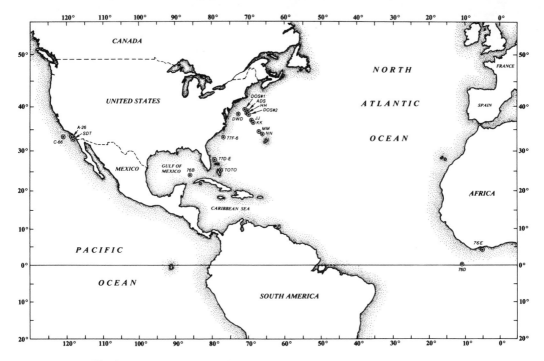

Fig. 12-5. Stations at which *in situ* sediment respiration has been measured in the North Atlantic, the Gulf of Mexico, and the eastern North Pacific.

## Sediment Community Respiration

Sediment community respiration has been measured using *in situ* methods at a total of 18 stations in the World Ocean (Fig. 12-5). Primary productivity in the surface waters overlying these stations ranges from the oligotrophic area of the western North Atlantic to the eutrophic areas of the eastern North Pacific.

These stations occur in a wide variety of soft-sediment environments spanning depths of from 278 to 5200 m and distance from shore up to 900 km (Tables 12-1 and 12-2). Bottom-water temperature is relatively constant between 2 and 4°C except at the shallow-slope stations where it rises to 11°C. Dissolved oxygen in the bottom water is relatively high at the Atlantic stations but declines at the Pacific stations to a low of 0.61 ml/l in the San Diego Trough, a nearshore basin in the Southern California Continental Borderland (Table 12-2).

Sediments are generally pelagic oozes and clays, with coarser constituents nearer shore. Organic carbon and nitrogen content of the sediments generally decreases with distance from shore except in eutrophic areas where such a pattern is not found, as in the eastern North Pacific stations.

The abundance of macrofauna in the sediments varies by two orders of magnitude. The higher densities are found closer to shore or in more eutrophic areas such as beneath the California Current upwelling region off southern California. Biomass of the macro-

TABLE 12-1. Sediment Community Respiration from 10 Stations in the Western North Atlantic with Associated Environmental Parameters[a]

| Station | Sediment Community Respiration ml $O_2$ m$^{-2}$ h$^{-1}$ | Depth (m) | Distance from Shore (km) | Annual Primary Productivity g C m$^{-2}$ y$^{-1}$ | Bottom Water Temp. °C | Dissolved Oxygen Bottom ml/l | Benthic Abundance #/m$^2$ | Benthic Biomass mg wet wt/m$^2$ | Sediment Organic Carbon mg/g dry wt | Sediment Organic Nitrogen mg/g dry wt | Particulate Organic Carbon Flux g m$^{-2}$ y$^{-1}$ |
|---|---|---|---|---|---|---|---|---|---|---|---|
| DOS-1 | 0.50 | 1,850 | 176 | 120 | 4 | 7.05 | 3,218 | 9,450 | 10.0 | 1.1 | — |
| DWD | 0.46 | 2,200 | 172 | 100 | 3 | 6.34 | 22,988 | 556 | 12.1 | 1.5 | 6.3 |
| ADS | 0.35 | 2,750 | 259 | 160 | 3 | 6.52 | 8,764 | 2,143 | 13.3 | 1.6 | 2.3 |
| HH | 0.20 | 3,000 | 291 | 160 | 3 | 6.15 | 2,146 | 653 | 9.1 | 1.1 | 2.3 |
| DOS-2 | 0.21 | 3,650 | 352 | 100 | 3 | 6.54 | 1,632 | 771 | 13.0 | 0.9 | 4.2 |
| JJ | 0.09 | 4,670 | 497 | 68 | 3 | 6.43 | 753 | 220 | 0.8 | 0.1 | — |
| KK | 0.04 | 4,830 | 612 | 68 | 3 | 6.04 | 285 | 180 | 6.9 | 0.7 | — |
| NN | 0.07 | 5,080 | 880 | 72 | 3 | 6.25 | 117 | 78 | 6.4 | 0.9 | 0.7 |
| MM | 0.02 | 5,200 | 806 | 72 | 3 | 6.15 | 259 | 142 | 6.4 | 0.9 | 0.7 |
| 77DE | 1.31 | 1,345 | 148 | 85 | 4 | 5.65 | — | — | 15.6 | — | 5.4 |

[a]Adapted from Hinga et al., 1979; and Smith, 1978a.

TABLE 12-2. Sediment Community Respiration from a Total of 8 Stations in the Gulf of Mexico–Straits of Florida, Eastern North Atlantic, and Eastern North Pacific with Associated Environmental Parameters[a]

| Station | Sediment Community Respiration ml O$_2$ m$^{-2}$ h$^{-1}$ | Depth (m) | Distance from Shore (km) | Annual Primary Productivity g C m$^{-2}$ y$^{-1}$ | Bottom Water Temp. °C | Dissolved Oxygen Bottom ml/l | Benthic Abundance #/m² | Benthic Biomass mg wet wt/m² | Sediment Organic Carbon mg/g dry wt | Particulate Organic Carbon Flux g m$^{-2}$ y$^{-1}$ |
|---|---|---|---|---|---|---|---|---|---|---|
| | | | | Gulf of Mexico–Straits of Florida | | | | | | |
| 77FG | 2.95 | 675 | 74 | 72 | 10 | 3.25 | — | — | 0.8 | 2.6 |
| TOTO | 3.10 | 2,000 | 18 | 72 | 4 | 5.56 | 102 | — | 4.4 | 2.1 |
| 76B | 0.69 | 3,450 | 220 | 50 | 5 | 4.16 | — | — | 6.0 | — |
| | | | | Eastern North Atlantic | | | | | | |
| 76E | 3.93 | 278 | 40 | 159 | 11 | 1.01 | — | — | 23.9 | — |
| 76D | 0.65 | 4,000 | 540 | 107 | 2 | 5.64 | — | — | 4.8 | — |
| | | | | Eastern North Pacific | | | | | | |
| SDT | 2.40 | 1,230 | 26 | 273 | 4 | 0.71 | 7,285 | 8,050 | 30.5 | 9.8 |
| A-26 | 2.22 | 1,193 | 28 | 273 | 4 | 0.62 | 7,285 | 8,050 | 30.5 | 9.8 |
| C-66 | 2.28 | 3,815 | 315 | 307 | 2 | 3.86 | 5,274 | 24,190 | 12.2 | 9.8 |

[a]Adapted from Hinga et al., 1979; Smith, 1974; and Smith et al., 1979a.

291

fauna follows the same trend as macrofauna abundance at the stations where it was measured.

Particulate organic carbon flux to the benthic boundary layer ranged from 0.7 g C $m^{-2}y^{-1}$ at the oligotrophic western North Atlantic stations to 9.8 g C $m^{-2}y^{-1}$ at the eutrophic eastern North Pacific stations (Tables 12-1 and 12-2).

Respiration of the sediment community varies by two orders of magnitude among the 18 stations measured, from 0.02 ml $O_2 m^{-2}h^{-1}$ at Station MM in the western North Atlantic (depth 5200 m) to 3.93 ml $O_2 m^{-2}h^{-1}$ at Station 76E in the eastern North Atlantic (depth 278 m). This seems to indicate that there is a strong correlation of sediment community respiration with depth. Such a relationship was reported by Smith (1978a) for the stations in the western North Atlantic along the Gay Head–Bermuda transect (DOS-1 to MM in Table 12-1). He found that depth accounted for 83.1% of the variation in sediment community respiration. Sediment organic carbon:nitrogen and sediment organic nitrogen contributed an additional 3.2 and 1.8%, respectively.

Further analyses of the combined *in situ* data for the World Ocean were performed. A predictive equation for sediment community respiration was generated using the ten environmental parameters in Tables 12-1 and 12-2 in addition to sediment organic carbon:organic nitrogen, and macrofauna abundance:biomass. This equation is

$$Y = 0.820 + 0.007(PP) - 0.192(DO) - 0.065(SON)$$

where $Y$ is sediment community respiration. The three independent variables are primary productivity (PP), dissolved oxygen (DO), and sediment organic nitrogen (SON). This equation accounts for 98% of the variation in sediment community respiration.

Depth ceases to be an important variable when all the stations are evaluated. This is in large part attributable to the variability in the overall productivity of the areas such as upwelling regions off southern California, where the rates of sediment community respiration do not change significantly over a distance of 287 km and a depth range of 2500 m (Stas. A-26 to C-66). A similar example is provided by two stations in the Straits of Florida area (Stas. 77FG and TOTO) which are close to land and receive terrigenous and shallow-water organic enrichments such as wood and grass (Table 12-2).

## Benthopelagic Animal Respiration

This group of animals generally represents the mobile scavengers of the benthic boundary layer and consists of fishes such as rattails, sablefish, halosaurs, hagfish and rockfish, and invertebrates such as amphipods and decapods. Respiration measurements have been made on a few members of this scavenging group such as the rattails and the hagfish.

Respiration measurements of five benthopelagic fishes have been made *in situ* in the western North Atlantic and eastern North Pacific. Three rattail specimens of *Coryphaenoides armatus* were trapped in baited respirometers similar to the FTR at depths of 2753 to 3650 m (Sta. DOS-2) along the Gay Head–Bermuda transect (Table 12-3). *Coryphaenoides armatus* is a cosmopolitan macrourid which commonly inhabits depths between 2000 and 4000 m (Iwamoto and Stein, 1974; Marshall and Iwamoto, 1973)

TABLE 12-3. Respiration Rates of Benthopelagic Fishes in the Western North Atlantic and Eastern North Pacific[a]

| Species | Depth (m) | Total Length (cm) | Wet Weight (kg) | Sex | Respiration ml $O_2$/h Total | Per kg wet wt |
|---|---|---|---|---|---|---|
| Western North Atlantic | | | | | | |
| *Coryphaenoides armatus* (1) | 3650 | 48.5 | 0.5 | M | 1.9 | 3.7 |
| *Coryphaenoides armatus* (2) | 2753 | 53.5 | 0.7 | M | 2.2 | 3.1 |
| *Coryphaenoides armatus* (3) | 3650 | 68.5 | 1.2 | F | 3.3 | 2.7 |
| Eastern North Pacific | | | | | | |
| *Coryphaenoides acrolepis* | 1230 | 68.0 | 1.8 | – | 4.4 | 2.4 |
| *Eptatretus deani* | 1230 | 51.0 | 0.1 | – | 0.2 | 2.2 |

[a]Smith, 1978b; Smith and Hessler, 1974.

and has been reported to contribute as much as 80% of the benthopelagic fish biomass on the continental rise (Haedrich and Rowe, 1977). This species when small feeds predominantly on benthic animals but when larger feeds on pelagic animals (Haedrich and Henderson, 1974; Pearcy and Ambler, 1974) and has been caught as far as 685 m above the bottom in the central North Pacific (Smith *et al.*, 1979b).

Respiration rates of *C. armatus* ranged from 1.9 to 3.3 ml $O_2$ $hr^{-1}$ while the weight specific rates decreased with increased size of the fish. This supports the basic physiological axiom that respiration increases as a fractional power of body weight (Smith, 1978b). This relationship between respiration and body weight is best described by the power function $Y = 0.03W^{0.65}$ where $Y$ is the oxygen consumption (ml $h^{-1}$) and $W$ is the wet weight (g). These rates were significantly lower than those of the phylogenetically related shallow-water Atlantic Cod (*Gadus morhua*) (Saunders, 1963; Smith and Hessler, 1974).

In the eastern North Pacific, the respiration of a congener, *Coryphaenoides acrolepis*, was measured at a station in the San Diego Trough (Table 12-3). This rate was not significantly different (P > 0.01) from that obtained for *C. armatus*.

## Plankton Respiration

The plankton of the benthic boundary layer include such organisms as hydromedusae, siphonophores, chaetognaths, copepods, amphipods, decapods, polychaetes, protozoans, and bacteria. The analysis of the available data for this component yields measurements only for the amphipods and the bacteria. The amphipods we are interested in range in size from 3 mm to 18 mm in length[1] and occupy both the sediment and overlying water up to 400 m off the bottom (Smith *et al.*, 1979b). The size of individuals and the habitat they are occupying at the time of capture are the criteria used to determine whether they should be grouped with the plankton, the benthopelagic animals, or the sediment com-

[1]R. Hessler and C. Ingram, unpublished data.

munity. The amphipods used in the following analysis were small in size and caught up to 2 m off the bottom. They are here considered to be plankton.

The slurp gun respirometer (SGR) was used in the western North Atlantic (Sta. ADS) to collect the amphipod, *Paralicella caperesca,* and measure its oxygen consumption (Smith and Baldwin, in press). These amphipods appear normally benthic, but when bait fish is available, they will swim into the overlying water in search of this food source. At this point, they become predominant members of the plankton. *P. caperesca* was attracted into the water column and two individuals were slurped up, each into a single chamber of the SGR. Respiration rates were measured for a period of 21 h on the bottom.

When first captured, these amphipods exhibited a rapid swimming behavior similar to that of the amphipods which had also been attracted to the bait but were outside the respirometer. This activity continued for the entire two-hour visual observation period. The dissolved oxygen recordings showed that respiration of the amphipods in the SGR was elevated for the first 6 to 7.25 h and then declined precipitously by one to two orders of magnitude to a constant resting rate for the duration of the measurements (Table 12-4). The initial respiration rate of these amphipods is similar in magnitude to that of shallow-water rates of other species of amphipods measured at comparable temperatures. The resting rate of respiration in *P. caperesca* is similar to that for a midwater amphipod measured at atmospheric pressure but similar temperature (Childress, 1975). Smith and Baldwin (in press) interpret these results as suggesting an adaptation to a low food-energy environment where these animals are capable of rapid response when food is available but can return to a state of dormancy when such fortuitous food sources are not available.

The other components of the plankton which have been studied, bacteria, were collected from immediately above the sediment-water interface at 1830 m (Sta. DOS-1, western North Atlantic) and incubated *in situ* in sterile sample bottles containing $^{14}$C-labeled glutamate and casamino acid substrates (Jannasch and Wirsen, 1973). Respiration rates were determined from the difference between total incorporated substrate (particulate) and metabolized substrate ($CO_2$ plus particulate); rates varied with type of substrate (Table 12-5). The $CO_2$ respired from casamino acids was more than twice that for glutamate.

TABLE 12-4. Respiration Rates of the Amphipod *Paralicella caperesca* from Sta. ADS in the Western North Atlantic[a]

| *P. caperesca* | Wet Weight (mg) | Overall Length (mm) | Sex | Time Interval (h) | Respiration ($\mu$l $O_2$/h) | |
|---|---|---|---|---|---|---|
| | | | | | Total | Per mg wet wt |
| No. 1 | 30.4 | 11.6 | F | 0–6 | 10.6 | 0.35 |
| | | | | 6–21 | 0.5 | 0.02 |
| No. 2 | 59.1 | 13.2 | M | 0–7.25 | 39.6 | 0.67 |
| | | | | 7.25–21 | 2.0 | 0.03 |

[a]Smith and Baldwin, in press.

294

**TABLE 12-5.** *In Situ* Respiration Rates of Enriched Bottom-Water Samples Taken at 1830 m in the Western North Atlantic (Sta. DOS-1)[a]

| Substrate | Original Conc. $\mu g/ml$ | Incubation Time (Days) | Total % $CO_2$ Respired | $CO_2$ Respired mg m$^{-3}$ y$^{-1}$ | $CO_2$ Respired mg C m$^{-3}$ y$^{-1}$ |
|---|---|---|---|---|---|
| Glutamate | 5 | 98 | 0.24 | 44.7 | 14.30 |
| Casamino acids | 2 | 98 | 1.38 | 102.0 | 32.6 |

[a]Jannasch and Wirsen, 1973. (Conversion factor of 1 mg $CO_2$ = 0.32 mg C.)

## SYNTHESIS OF RESULTS

To synthesize the limited data on the functioning of the benthic boundary layer, let us first develop a simplistic food web from which to work. This food web can be built from the four basic descriptive compartments we have been discussing: (1) the sediment community, (2) the benthopelagic animals, (3) the plankton, and (4) the epibenthic megafauna (Fig. 12-6).

Each component is represented in pyramidal form to reflect the relative vertical distribution of the included organisms. The four faunal components have organisms which undergo vertical migration, represented by the thickened vertical arrows. Benthopelagic animals and plankton are particularly capable of transcending the benthic bound-

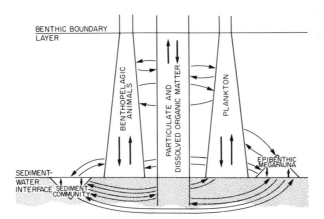

Fig. 12-6. Simplified food web diagram of the deep-sea benthic boundary layer showing the possible food energy exchange routes between components. (Bold arrows within compartments indicate the direction of movement of organisms or particles constituting each component.) The pyramidal form of some compartments reflects the relative vertical distribution of the included organisms. The dimensions of each compartment do not reflect quantity (standing crop) for intercompartmental comparisons).

ary layer into the midwater column. Examples of such migrations are provided by the benthopelagic rattail *Coryphaenoides armatus* which has been caught in baited traps up to 685 m off the bottom (Smith *et al.*, 1979b). To a more limited extent, organisms of the sediment community and epibenthic megafauna undergo excursions into the water column. For example, the amphipod *Paralicella caperesca* (see Results section) can be classified either as a member of the sediment community or as a component of the plankton. Holothurians and polychaetes, both generally considered to be benthic animals (epibenthic megafauna), have likewise been observed in the water column (Barnes *et al.*, 1976; Polloni *et al.*, 1975).[2]

The possible food energy exchanges between each component are shown by the intercomponent arrows. Food energy for each component can be derived from essentially three sources within the benthic boundary layer: (1) intracompartment sources, (2) intercompartment sources, and (3) particulate and dissolved organic matter. Most of these pathways have not yet been identified or quantified; this should be kept in mind during the following analysis.

Working from this framework, we can apply the limited data so far compiled. The broadest data base we have is from the sediment community in the western North Atlantic. Station DOS-2 is at a depth of 3650 m along the Gay Head–Bermuda transect (Table 12-1, Fig. 12-5). Annual sediment community respiration there is estimated at 2.10 1 $O_2 m^{-2} y^{-1}$ or 0.96 g C $m^{-2} y^{-1}$ assuming a respiratory quotient of 0.85 (Smith, 1978a). When compared to the particulate organic carbon flux to the bottom (4.2 g C $m^{-2} y^{-1}$) (Rowe and Gardner, 1979), sediment community requirements account for only 23% of this food source. If this source of food is not accumulating in the sediment, then it is reasonable to assume it must be utilized by the various components. If the sediment community can only account for 23%, then the other three components must be significant consumers. Again, it must be emphasized that we are only considering maintenance energy requirements as measured by respiration and are not considering other food energy utilized in, for example, growth and locomotion.

To estimate reasonably the energy requirements of the other components of the benthic boundary layer in the western North Atlantic is difficult due to the small amount of data available. We have measurements of the respiration of a few individual organisms from the benthopelagic community and the plankton. However, in order to compare these rates with those of the sediment community, population density and size class information are needed.

Population information on the benthopelagic animals is difficult to obtain due to the problems inherent in deep-water trawling. The best estimates to date for the western North Atlantic are based on photographic transects made with the submersible *Alvin*. The macrourids, with an estimated density of 4.4 fish 1000 $m^{-2}$, represent one of the dominant groups of benthopelagic animals at the deepest station sampled (Sta. DOS-1, 1830 m) (Grassle *et al.*, 1975). This density estimate was made within 2 m of the bottom. For the time being we will ignore the influence of the benthopelagic fishes in the overlying benthic boundary layer water column, since observations and trapping results suggest an

---

[2] Also from Smith, unpublished data.

exponential decrease in numbers of macrourids with increased distance off the bottom (Smith *et al.*, 1979b).[3]

In order to extrapolate this density value to the deeper station (DOS-2) at 3650 m and use it as a population parameter from which to derive population respiration, we have to make the following assumptions: (1) The total number of macrourids is the same at 1830 m as at 3650 m. This may not be unreasonable since Haedrich and Rowe (1977) found that the total number of demersal fishes decreases from 11.6 fish 1000 m$^{-2}$ at 1830 m to 4.2 fish 1000 m$^{-2}$ at 2790 m. (2) Macrourids are the only benthopelagic animal at 3650 m. *C. armatus* can indeed contribute as much as 80% of the demersal fish biomass on the continental rise in the western North Atlantic (Haedrich and Rowe, 1977). (3) All the macrourids are *C. armatus*. (4) Respiration per unit body weight of *Coryphanenoides* does not vary significantly with depth. This assumption seems plausible since the respiration rates of congeners of *Coryphaenoides* were not significantly different between 1230 and 3650 m (Smith, 1978b). (5) The size range of *C. armatus* caught for respiration measurements (Table 12-3) is representative of the macrourid population at 3650 m. This variation is reasonable considering the size distribution observed from the *Alvin* at Stations ADS and DOS-2[4] and the size range for this species (Iwamoto and Stein, 1974; Marshall and Iwamoto, 1973).

Given the above assumptions, we can take the mean total respiration rate for *C. armatus* (Table 12-3), 2.47 ml O$_2$h$^{-1}$, and multiply it by the density of 4.4 fish 1000 m$^{-2}$ to get a total annual population respiration of 0.095 1 O$_2$m$^{-2}$y$^{-1}$ or 0.04 g C m$^{-2}$y$^{-1}$ for benthopelagic animals at 3650 m. This benthopelagic animal respiration rate is then only 5% of that measured for the sediment community at the same station (Table 12-1).

The plankton of the benthic boundary layer is virtually unstudied and population parameters such as density and size class information are not available. The only plankton group for which we have respiration measurements are the amphipods, *Paralicella caperesca*. This species is quite ubiquitous at Station DOS-2 as determined from direct observation and the numbers collected in baited traps over short intervals of time (Smith and Baldwin, in press). The density of these animals (16 amphipods/m$^2$) was estimated from direct observations from *Alvin* as they emerged from the sediments toward the baited traps suspended roughly one meter above the bottom.[5] The size range of *P. caperesca* captured in the baited traps at the time of the respiration measurements ranged from 2 to 15 mm in overall length. The specimens on which the respiration was measured were in the upper size range, which was, however, the dominant size class.[6]

If we use the above population parameters and the mean total resting respiration rate of 1.25 μl O$_2$h$^{-1}$ (Table 12-4), assuming that rate to be the mean for the entire population, we obtain an annual population estimate of 0.16 1 O$_2$m$^{-2}$y$^{-1}$ or 0.08 g C m$^{-2}$y$^{-1}$. A more realistic estimate of amphipod respiration is probably predominated

[3] Also from Smith, unpublished data.

[4] Smith, unpublished data.

[5] *Ibid.*

[6] R. Hessler and C. Ingram, unpublished data.

by a resting rate interspersed with periods of activity (10 days per year), when episodic food sources are available or when disturbed by predators such as benthopelagic fishes. An estimate generated on these assumptions is 0.25 l $O_2 m^{-2} y^{-1}$ or 0.12 g C $m^{-2} y^{-1}$. This amount accounts for 3% of the particulate organic carbon flux. Again we have ignored the other animals of the plankton which appear significantly less abundant than the amphipods.

Another component of the plankton are the bacteria. Jannasch and Wirsen (1973) determined an *in situ* respiration rate of enriched bacterial samples collected immediately above the sediment-water interface at 1830 m in the western North Atlantic (Sta. DOS-1) (Table 12-5). These values can be extrapolated to the 3650-m station; however, caution must be exercised for the following considerations: (1) the possible inhibitory effects of appreciable enrichments (Williams and Carlucci, 1976), and (2) the increased depth and distance from shore. If we use the respiration value determined for casamino acids (Table 12-5) to be representative of the bottom two meters of the water column, then there is a total utilization of 0.06 g C $m^{-2} y^{-1}$. The combined carbon requirement for the plankton including the amphipods and bacteria is then 0.18 g C $m^{-2} y^{-1}$.

The only component we have not considered is the epibenthic megafauna because there are no reasonable data available as yet even to attempt an estimate of their energetic importance.

The total food energy requirements of all components of the benthic boundary layer, as estimated above from respiration, amounts to 1.18 g C $m^{-2} y^{-1}$. Although we do not know what percentage of each standing crop of available food is utilized by each component, the combined utilization of all the components accounts for only 28% of the particulate organic carbon flux (Fig. 12-7). Since there is only 1.3% organic carbon in the sediments at this station (DOS-2) (there is no indication of accumulation with depth in the sediment), the sediments are not serving as a significant sink for this material (Smith, 1978a). It is not reasonable to assume that the epibenthic megafauna have an energy requirement greater than that of the other components combined. Overestimating of the particulate organic carbon flux by sedimentation traps cannot be discounted due to the possible influence of resuspension and advective processes.

Other food energy sources further increase the difference between estimated supply and demand. Dissolved organic carbon is believed to be an important food source to bacteria (Williams and Carlucci, 1976). Rapid consumption of large parcels of bait placed on

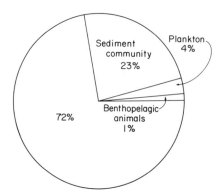

Fig. 12-7. Estimated percent utilization of the total sedimenting particulate organic matter by three components (sediment community, plankton, and benthopelagic animals) of the benthic boundary layer in the western North Atlantic (Sta. DOS-2, 3650-m depth).

ECOLOGICAL ENERGETIC STUDIES

the deep-sea floor by benthopelagic and epibenthic scavengers (Hessler *et al.,* 1972; Isaacs and Schwarzlose, 1975; Jannasch and Wirsen, 1977) suggests the importance of carrion as a food source.

A great number of assumptions were made to obtain the above estimates of food-energy demand by the benthic boundary layer for one station in the western North Atlantic. Hopefully these speculations will serve as a stimulus for further work to elucidate the energetics of this important boundary. There are insufficient data on any one deep-sea station at present to perform a critical analysis of any component of the benthic boundary layer. We are currently conducting extensive *in situ* studies of the benthic boundary layer system of a deep-sea station in the eastern North Pacific in the hopes of filling this void and providing much-needed insights into the functioning of the open-ocean ecosystem.

## ACKNOWLEDGMENTS

Much of the research included in this review was supported by the National Science Foundation (OCE78-08640, OCE76-10728, OCE76-10520). We thank P. Jumars, D. Karl, and P. Williams for comments, and R. Baldwin and S. Hamilton for editorial assistance.

## REFERENCES

Barnes, A. T., Quetin, L. B., Childress, J. J., and Pawson, D. L., 1976, Deep-sea macro-planktonic sea cucumbers—suspended sediment feeders captured from deep submergence vehicle: *Science,* v. 194, p. 1083-1085.

Childress, J. J., 1975, The respiratory rates of midwater crustaceans as a function of depth of occurrence and relation to the oxygen minimum layer off southern California: *Comp. Biochem. Physiol,* v. 50A, p. 787-799.

Grassle, J. F., Sanders, H. L., Hessler, R. R., Rowe, G. T., and McLellan, T., 1975, Pattern and zonation—a study of bathyal megafauna using the research submersible *Alvin: Deep-Sea Res.,* v. 22, p. 457-481.

Haedrich, R. L., and Henderson, N. R., 1974, Pelagic food of *Coryphaenoides armatus,* a deep benthic rattail: *Deep-Sea Res.,* v. 21, p. 739-744.

——, and Rowe, G. T., 1977, Megafaunal biomass in the deep sea: *Nature,* v. 269, p. 141-142.

Hessler, R. R., Isaacs, J. D., and Mills, E. L., 1972, Giant amphipod from the abyssal Pacific Ocean: *Science,* v. 175, p. 636-637.

Hinga, K. R., Sieburth, J. McN., and Heath, G. R., 1979, The supply and use of organic material at the deep-sea floor: *J. Mar. Res.,* v. 37, p. 557-579.

Isaacs, J. D., and Schwarzlose, R. A., 1975, Active animals of the deep-sea floor: *Scient. Amer.* v. 233, p. 84-91.

Iwamoto, T. and Stein, D. L., 1974, A systematic review of the rattail fishes (Macrouridae:Gadiformes) from Oregon and adjacent waters: *Occs. Pap. Calif. Acad. Sci.* #111, 79 p.

Jannasch, H. W., and Wirsen, C. O., 1973, Deep-sea microorganisms—an *in situ* response to nutrient enrichment: *Science,* v. 180, p. 641–643.

——, and Wirsen, C. O., 1977, Microbial life in the deep sea: *Scient. Amer.,* v. 236, p. 42–52.

Kanwisher, J., 1959, Polarographic oxygen electrode: *Limnol. Oceanogr.,* v. 4, p. 210–217.

Marshall, N. B., and Iwamoto, T., 1973, Family Macrouridae: *in Fishes of the Western North Atlantic,* Mem. Sears Found. Mar. Res., v. 1, no. 6, p. 496–537.

Pamatmat, M. M., 1973, Benthic community metabolism on the continental terrace and in the deep-sea in the North Pacific: *Int. Rev. ges. Hydrobiol.,* v. 58, p. 345–368.

Pearcy, W. G., and Ambler, J. W., 1974, Food habits of deep-sea macrourid fishes off the Oregon coast: *Deep-Sea Res.,* v. 21, p. 745–759.

Polloni, P. A., Rowe, G. T., and Teal, J. M., 1975; *Biremis blandi* (Polychaeta: Terebellidae), new genus, new species, caught by DSRV *Alvin* in the Tongue of the Ocean, New Providence, Bahamas: *Mar. Biol.,* v. 20, p. 171–175.

Rowe, G. T., and Gardner, W., 1979, Sedimentation rates in the slope water of the northwest Atlantic Ocean measured directly with sediment traps: *J. Mar. Res.,* v. 37, p. 581–600.

Saunders, R. L., 1963, Respiration of the Atlantic cod: *J. Fish. Res. Bd. Can.,* v, 20, p. 373–386.

Smith, K. L., Jr., 1974, Oxygen demands of San Diego Trough sediments: an *in situ* study: *Limnol. Oceanogr.,* v. 19, p. 939–944.

——, 1978a, Benthic community respiration in the N.W. Atlantic Ocean—*in situ* measurements from 40 to 5200 m: *Mar. Biol.,* v. 47, p. 337–347.

——, 1978b, Metabolism of the abyssopelagic rattail *Coryphaenoides armatus* measured *in situ: Nature,* v. 274, p. 362–364.

——, and Baldwin, R. J., 1982, Scavenging deep-sea amphipods: effects of food odor on $O_2$ consumption and a proposed metabolic strategy: *Mar. Biol.* (in press).

——, Clifford, C. H., Eliason, A. H., Walden, B., Rowe, G. T., and Teal, J. M., 1976, A free vehicle for measuring benthic community metabolism: *Limnol. Oceanogr.,* v. 21, p. 164–170.

——, and Hessler, R. R., 1974, Respiration of benthopelagic fishes—*in situ* measurements at 1230 meters: *Science,* v. 184, p. 72–73.

——, and Teal, J. M., 1973, Deep-sea benthic community respiration—an *in situ* study at 1850 meters: *Science,* v. 179, p. 282–283.

——, White, G. A., and Laver, M. B., 1979a, Oxygen uptake and nutrient exchange of sediments measured *in situ* using a free vehicle grab respirometer: *Deep-Sea Res.,* v. 26A, p. 337–346.

——, White, G. A., Laver, M. B., and Haugsness, J., 1978, Nutrient exchange and $O_2$ consumption by deep-sea benthic communities—preliminary *in situ* measurements: *Limnol. Oceanogr.,* v. 23, p. 997–1005.

——, White, G. A., Laver, M. B., McConnaughey, R. R., and Meador, J. P., 1979b, Free vehicle capture of abyssopelagic animals: *Deep-Sea Res.,* v. 26A, p. 57–64.

Williams, P. M., and Carlucci, A. F., 1976, Bacterial utilization of organic matter in the deep sea: *Nature,* v. 262, p. 810–811.

Stephen H. Wright and Grover C. Stephens
Department of Developmental and Cell Biology
University of California
Irvine, CA 92717

# 13

# TRANSEPIDERMAL TRANSPORT OF AMINO ACIDS IN THE NUTRITION OF MARINE INVERTEBRATES

# ABSTRACT

Dissolved organic material (DOM) in seawater represents an extremely large pool of reduced carbon, and the potential of DOM as a nutritional source for aquatic organisms has been debated since the turn of this century. Several technical advances in the last 20 years led to a series of studies demonstrating that all soft-bodied marine invertebrates are capable of accumulating dissolved free amino acids from dilute solution. Analysis of the nutritional potential of such uptake processes requires detailed information on (1) nutritional requirements of an organism, (2) availability of dissolved substrates in that organism's environment, and (3) the kinetics of transport of the substrate(s) in question. For several intertidal species available data suggest that amino acid uptake can provide a significant nutritional supplement to other modes of feeding. Relevant information for abyssal organisms is more limited but is consistent with the hypothesis that DOM may play a nutritional role in at least some kinds of abyssal habitats.

# INTRODUCTION

Initial interest in the utilization of dissolved organic material (DOM) is classically associated with the name August Pütter (1909). His idea that a major input of organic material other than particulate food was required to account for observed growth and reproduction in aquatic animals was the cause of considerable controversy and research in the first quarter of this century. August Krogh (1931) reviewed the studies testing "Pütter's hypothesis," and came to the conclusion that there was no substantial evidence in support of the contention that aquatic animals utilize DOM. Krogh went on to make significant contributions to this area, developing improved analytical techniques for the determination of dissolved organic compounds in natural waters, and maintaining his interest in the field until his death. Jørgensen (1976) provides an informative discussion of the early work on the utilization of DOM, and a fine account of Krogh's contributions to the field.

Within the last 20 years there has been a resurgence of interest in the idea that DOM may have a nutritional role for some aquatic animals. Of particular importance in the advancement of the field was the development of new analytical techniques, including the introduction of radioactive tracers to the study of biological transport processes, and the development of sensitive chemical assays for various classes of dissolved compounds found in natural waters. In recent years, a substantial literature has accumulated dealing with the transport of amino acids in marine invertebrates. Evidence now indicates that amino acid uptake may provide a significant nutritional supplement for selected organisms. In this chapter we will not attempt an exhaustive review of this literature; the interested reader is directed to several recent reviews on the subject (Jørgensen, 1976; Sepers, 1977; Stephens, 1972). Rather, we will summarize the evidence concerning the nutritional role of amino acid uptake in marine invertebrates, with emphasis on recent studies in the field. Information on the role of DOM in abyssal forms is understandably limited because of the formidable problems encountered in attempting physiological work on these organisms *in situ*. Thus our discussion of the potential importance of DOM in these habitats is necessarily largely speculative.

Most investigations of amino acid transport in marine organisms in the last 20 years have relied on the use of radioactively labeled substrates. Typically, these compounds have been provided in experimental solutions to which test organisms are exposed. The disappearance of radioactivity from the medium and/or the appearance of radioactivity in the animals is monitored (Fig. 13-1). This type of experimental approach has provided a great deal of information on the rates of influx of a variety of amino acids into marine animals. In a few cases, investigators have used radioactively labeled compounds (such as amino acid analogues) to try and estimate rates of loss of amino acid after pre-loading organisms. However, experiments which utilize radiolabeled substrates generally can only provide direct information concerning rates of influx of the substrates, not net exchanges between organisms and environment.

Another category of studies has used the technique of autoradiography to demonstrate uptake and assimilation of radiolabeled amino acids. In these studies, organisms are pulse-labeled with radioactively tagged amino acids and subsequently fixed and sectioned, and autoradiographic techniques are employed to localize radioactive material in the tissues (Fig. 13-2). As in the case of the transport experiments described above, this procedure also leaves unanswered the question of net exchanges of amino acid. It is, however, capable of a qualitative demonstration of assimilation of externally supplied dissolved substrate, and has proven to be extremely informative.

Determination of net exchanges of dissolved substrates requires the use of chemical assays capable of measuring extremely small concentrations of dissolved substrate in seawater. Stephens and Schinske (1961) utilized the ninhydrin technique to demonstrate net uptake of glycine into representatives of 10 phyla of soft-bodied marine invertebrates. However, due to the relative insensitivity of the ninhydrin procedure, they were forced to use artificially high levels of substrate. Ferguson (1971) used the technique of gas chromatography to show net uptake of amino acids into starfish from ambient concentrations as low as 20 $\mu M$. However, he did not examine the kinetics of net accumulation. Within the last few years, the introduction of simple, sensitive fluorometric methods for the determination of amino acids in solution (e.g., the fluorescamine technique; Udenfriend *et al.*, 1972) has made it possible to study conveniently the kinetics of net

Fig. 13-1. Uptake of $^{14}$C-cycloleucine in isolated gill tissue of *Mytilus californianus*. Solid circles are described by the left ordinate, and indicate the gradual decline in radioactivity in a 50 ml test solution containing 40 uniform discs of gill tissue (0.7 cm diam.). Open circles are described by the right ordinate, and indicate the gradual increase to a steady-state level of radioactivity accumulated in the discs of gill tissue. Each open circle represents radioactivity accumulated in a single gill disc, except the point at 120 minutes which is the mean (±1 s.d.) of 5 discs. Data supplied by Wright.

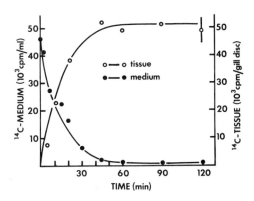

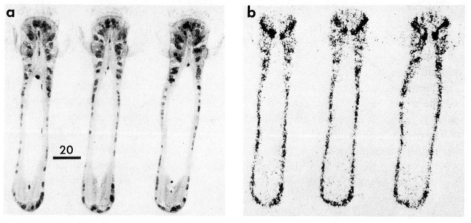

Fig. 13-2. Autoradiographic localization of accumulated and incorporated radioactivity in gill tissue of *Mytilus californianus*. Isolated gills were incubated in artificial seawater containing [3]H-leucine for 5 min and immediately fixed in gluteraldehyde-seawater. Pieces of tissue were embedded in plastic and sectioned at 0.5 $\mu$. Part (a) is a cross-section of three gill filaments, stained with basic fuschin to show the general morphology of the tissue; the frontal surface of the filaments is oriented toward the top of the figure. (b) is a closely adjacent section, unstained, dipped in autoradiographic emulsion and allowed to expose for one week. Heavy deposition of silver grains over the epithelial areas of the filaments is evident.

uptake of amino acids (Fig. 13-3). Several recent reports have combined radiotracer and fluorometric techniques to monitor simultaneously the influx and net flux of amino acids into test organisms (e.g., Crowe *et al.*, 1977; Stephens *et al.*, 1978; Wright and Stephens, 1977, 1978).

There are virtually no recent accounts of experiments attempting a direct demonstration of nutritional effects of dissolved substrates on marine animals. An exception is the work of Shick (1975), who demonstrated that the successful strobilization of the scyphozoan *Aurelia aurita* could be promoted by addition of very low levels of glycine to the ambient medium. Aside from this work, most investigators have avoided attempts to demonstrate a direct nutritional role for DOM. There are two major reasons

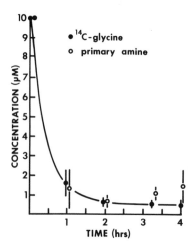

Fig. 13-3. Influx of [14]C-glycine and net flux of total primary amines in the annelid, *Pareurythöe californica*. Solid circles represent the levels of radioactivity remaining in the 50 ml test solution. Open circles indicate the concentration of primary amines, measured using fluorescamine and expressed as glycine-equivalent concentration. Each point is the mean of three separate determinations with individual worms; vertical bars are ±1 s.d.. Data supplied by James Costopulos, San Diego State University.

for this. First, normal substrate levels in the natural environment are very low, making it technically difficult to mimic the natural habitat in the laboratory. Second, and more important, it is extremely difficult to divorce direct nutritional effects of DOM via uptake from dilute solution from indirect effects via microbial assimilation of DOM and the subsequent ingestion of the microorganisms by the animals under study. The result is that the evidence for a nutritional role for dissolved amino acids in marine animals in nature is indirect, based on comparisons of net uptake at estimated environmental concentrations of substrate with nutritional requirements estimated from metabolic rates or other relevant measures.

## GENERAL FEATURES OF AMINO ACID TRANSPORT IN MARINE INVERTEBRATES

The results of the studies utilizing the techniques discussed above permit several general statements about amino acid uptake in marine animals.

1. All soft-bodied marine invertebrates examined have the ability to accumulate some radioactively labeled amino acids from dilute solution. To date, this statement is supported by studies utilizing representatives of some 18 classes in 13 phyla (see Table 13-1).

TABLE 13-1. List of Those Taxa of Marine Invertebrates That Have Had One or More Representatives Examined for the Transepidermal Transport of Amino Acids[a]

| Phylum Class | Phylum Class |
|---|---|
| Porifera | Annelida |
| Demospongiae* | Polychaeta* |
| Cnidaria | Oligochaeta |
| Anthozoa* | Mollusca |
| Scyphozoa | Bivalvia* |
| Hydrozoa | Gastropoda* |
| Platyhelminthes | Hemichordata |
| Turbellaria | Enteropneusta* |
| Nemertea | Pogonophora |
| Anopla* | Echinodermata |
| Brachiopoda | Echinoidea* |
| Articulata | Asteroidea* |
| Ectoprocta | Ophiuroidea |
| Gymnolaemata* | Holothuroidea* |
| Sipuncula* | Chordata |
| | Ascidiacea* |

[a]In every case examined the animals have been capable of showing an influx of radioactively labeled amino acid. Those taxa marked with an asterisk (*) demonstrated the capability of a net influx of amino acid.

Rates of course vary and some forms do not accumulate dicarboxylic amino acids. Those soft-bodied invertebrates that have been tested have also shown the capacity for a net accumulation of amino acids as well. The only group of marine invertebrates tested that is not capable of accumulating amino acids is the Arthropoda; cases of apparent uptake have been demonstrated to be due to the activity of surface microorganisms (Anderson and Stephens, 1969). No vertebrate, with the exception of hagfishes (Stephens, 1972), and no freshwater invertebrate have been shown unambiguously to possess transport capabilities similar to those found in marine invertebrates.

2. The transport process is transepidermal in nature, i.e., it occurs directly across external epithelia. Rates of uptake are unaffected by the ligation of mouth and anus of test animals (e.g., Fisher and Oaks, 1978; Stephens, 1963). Transport appears to occur generally across the body wall, as is the case in annelids (e.g., Chien et al., 1972), or primarily in specialized epithelia, such as mollusk ctenidia (Péquignat, 1973).

3. The transport process is carrier-mediated. Uptake is a saturable process, and typically is adequately described by Michaelis-Menten saturation kinetics (e.g., Wright and Stephens, 1977). Generally, transport systems display specificity for chemically related substrates (e.g., Crowe et al., 1977; DeBurgh, 1978; Schlichter, 1978). Transport rates are related to levels of metabolic energy (e.g., Costopulos et al., 1979).

4. Accumulated substrates are available to general metabolic pathways. Shortly after exposure to $^{14}$C-labeled substrates, $^{14}CO_2$ can be found in the ambient medium (Jørgensen, 1979). Labeled proteins, lipids, and nucleic acids can be isolated from animal tissue after exposure to externally supplied labeled amino acid (Stephens, 1972). Also, the fact that autoradiography routinely demonstrates uptake and assimilation of labeled amino acids is evidence for incorporation of low molecular weight substrates into compounds that can survive fixation procedures using standard histological reagents, such as alcohol and xylene.

The above summary is by no means complete. A number of studies have examined aspects of transepidermal transport processes too specific to be included in such a list of general features. However, this list does serve to point out the remarkable ubiquity of amino acid transport processes in marine invertebrates. That marine invertebrates can accumulate amino acids is no longer in doubt; the physiological role of such accumulation is now the center of interest in this field.


## NUTRITIONAL SIGNIFICANCE OF AMINO ACID TRANSPORT IN MARINE INVERTEBRATES

As pointed out earlier, there have been virtually no modern attempts made to demonstrate a direct nutritional effect of dissolved amino acids in marine invertebrates. The estimates have been indirect. Such estimates require data concerning:

1. Nutritional requirements of the organism,
2. Availability of the substrate(s) in the organism's environment,
3. The rate of acquisition of the substrate in question (i.e., the kinetics of the transport process).

## Nutritional Requirements

In practice, requirements for reduced carbon for marine invertebrates are usually estimated from measurements of metabolic rate, typically, determined as oxygen consumption. The basis for such procedures is the original recognition by Lavoisier and LaPlace
(1780) that there exists a stoichiometric relationship between the consumption of foodstuffs and the consumption of oxygen, production of $CO_2$ and $H_2O$, and the liberation
of heat. Thus values for oxygen consumption by an animal can be converted to the
weight of substrate (protein, fat, carbohydrate, or some suitable mixture thereof) oxidized. This value represents a minimum estimate of food an animal must acquire to
remain in energy balance. Actual food intake must exceed this figure for a variety of
reasons: for example, inefficiency of energy metabolism, loss of organic material by
routes other than metabolism, and anaerobic or partially anaerobic pathways in which
substrates are not fully oxidized. Anaerobiosis is significant in marine invertebrates (e.g.,
de Zwaan, 1977) including forms living in well-aerated water (Hammen, 1979). Therefore, measurement of heat production would be a more satisfactory estimate of metabolism (Hammen, 1979; Pamatmat, 1978). However, relatively few such measurements
have been made.

Estimates of metabolism from oxygen consumption data thus should be interpreted
as merely a benchmark against which rates of acquisition of a potential food source can
be placed. If food intake by a particular pathway provides an input roughly comparable
to the reduced organic material required to support an organism's oxygen consumption,
then the tentative conclusion can be drawn that this pathway may serve a nutritional role
(see Jørgensen, 1966, for a similar discussion of filter feeding). It should be stressed that
the pathway so studied need not be the sole source of food. Many and perhaps most
animals have more than one food resource. In general, DOM appears to be a candidate for
a supplementary source in all but a few cases where it may be the sole food supply (e.g.,
pogonophorans and some animals with a reduced or vestigial digestive tract). Note also
that the argument does not imply that all foodstuffs acquired by the pathway need be
immediately oxidized. Food intake provides raw materials for the anabolic process of
growth and repair as well as substrates for energy metabolism. We have noted that free
amino acids do indeed participate in anabolic pathways after being acquired from the
ambient medium.

## Availability of Substrate: Levels of Amino Acid
in Marine Waters

DOM in the oceans represents an immense pool of reduced carbon outweighing all living
things. Roughly speaking, it is comparable in mass to all the rest of the organic matter
on earth excluding fossil fuels (Woodwell et al., 1978). However, the actual concentration
of DOM is quite low, typically on the order of one to a few milligrams per liter. Of this,
less than 10% are compounds whose chemical identities are known, with free amino
acids averaging approximately 1–2% of the total. The turnover time of the total DOM
pool must of necessity be quite long, simply because it is so very large. Also, portions of
the DOM pool seem very refractory to biological utilization. Williams et al. (1969) report
that the $^{14}$C-age of dissolved organic carbon collected at a depth of 3500 m in the eastern

Pacific is 3400 years. However, the turnover time for selected fractions of the general DOM pool, such as dissolved free amino acids (DFAA), is much shorter. Turnover time for surface DFAA, based on rates of leakage of amino acids from zooplankton, has been estimated at from 1–2 days (Lee and Bada, 1977) to 1 month (Webb and Johannes, 1967). There are other constituents of the DOM (e.g., urea, glucose) which may also have a short half-life in some habitats.

Levels of DFAA in marine waters vary considerably depending upon the specific environment under study. For purposes of this discussion, it is convenient to consider the concentrations of DFAA in two broad categories of habitat: the water column and the sediment. Reports of concentrations of DFAA in the near shore water column generally range from 50 to 300 $\mu g/l$ (or approximately 0.5 to 3.0 $\mu M$). For example, Siegel and Degens (1966) reported an average of approximately 0.6 $\mu M/l$ of DFAA (61 $\mu g/l$) in surface samples from Buzzards Bay, Massachusetts. North (1975) found total primary amines (TPA; measured with fluorescamine and expressed as glycine equivalent concentration) in samples from Newport Bay, California, to range from 0.2 to 2.0 $\mu M/l$. She reported evidence suggesting that FAA accounted for approximately 50% of the TPA in her samples, giving an approximate range for DFAA of 10–100 $\mu g/l$. Crowe *et al.* (1977) reported a range of 2 to 7 $\mu M$ TPA (glycine equivalent concentration) in samples of water above the sediments of Little Sippewissett Marsh, Cape Cod, Massachusetts. While these concentrations of DFAA are low, they are in the same range as the amount of organic material available as phytoplankton in near-shore waters (Jørgensen, 1966).

The concentrations of DFAA in the interstitial water of near shore sediments are one to two orders of magnitude higher than the levels found in the water column, ranging from 2 to 20 mg/l (20 to 200 $\mu M$). Stephens *et al.* (1978) reported a mean value for TPA of 115 $\mu M$ in cores of sandy intertidal sediment collected near Point Mugu, California, and identified the major constituents of the primary amines as amino acids. Crowe *et al.* (1977) found levels of TPA ranging from 50 to 110 $\mu M$ in water samples collected from the top three centimeters of sediments in Little Sippiwisset Marsh. Henrichs and Farrington (1979) used gas chromatography to identify the amino acids in sediment samples from Buzzards Bay, and reported the total concentration of DFAA to be 15 $\mu M$.

Most reports identifying individual amino acids in samples from either the water column or sediment water list glycine, alanine, serine, valine, glutamic acid, and lysine as the major constituents of the DFAA pool, with other amino acids making modest contributions to the total.

The general conclusion to this discussion is that the near-shore environment consists of two very different habitats with respect to the concentration of DFAA. Roughly speaking, animals exposed to the water column will encounter levels of amino acids of a micromole per liter or less. On the other hand, the sediment infauna may be exposed to levels of amino acid of 20 $\mu M$ or more in many cases.

## Kinetics of Amino Acid Transport

Many studies dealing with amino acid transport in marine invertebrates have described the kinetics of the process. Typically, these studies reveal that the rate of uptake at low levels of substrate is a nearly linear function of the concentration of substrate; however, the rate of uptake at high levels is independent of concentration. Fig. 13-4 presents typical data. Such data are adequately described by the relation

$$J^i = \frac{J^i_{max}\,[S]}{K_t + [S]}$$

where $J^i$ is the rate of influx of substrate, $J^i_{max}$ is the maximal rate of influx, $[S]$ is the concentration of substrate, and $K_t$ is the concentration of substrate at which influx is one-half maximal. This will be recognized as being equivalent to the familiar Michaelis-Menten equation describing the concentration dependence of the rate of reaction of substrate in the presence of a fixed amount of enzyme. While several assumptions are required in the theoretical justification of its use to describe transport processes (Christensen, 1976a, b; Schultz and Curran, 1970), experimental data often fit this relation quite well.

The kinetic constants, $J^i_{max}$ and $K_t$, are important parameters in the characterization of amino acid uptake, and are critical in the evaluation of the physiological role of transepidermal transport for a given organism. The $J^i_{max}$ provides a measure of the potential of a transport process to acquire substrate. The $K_t$ is a measure of how well that potential is realized at the concentration of substrate found in the organism's habitat. Values for $J^i_{max}$ and $K_t$ for transepidermal transport processes are quite low compared to values usually reported for other epithelial transport processes in marine invertebrates. For example, $J^i_{max}$ for amino acid uptake in the gut is on the order of 20 to 50 $\mu M$/g wet wt-h, with $K_t$ ranging from approximately 0.5 to 5 m$M$ (e.g., crustaceans, Ahearn, 1976; echinoderms, Bamford and James, 1972). By contrast, $J^i_{max}$'s for transepidermal transport processes range from 0.1 to 10 $\mu M$/g-h, with $K_t$'s typically from 10 to 100 $\mu M$ (Jørgensen, 1976). $K_t$'s for transport processes in certain marine lamellibranchs are as low as 1 $\mu M$ (Wright and Stephens, 1978). Thus, while the maximum potential for transport in the epidermal systems of marine invertebrates is comparatively small, the low values for $K_t$ result in transport processes capable of the efficient accumulation of substrate at the low concentrations characteristic of the marine environment.

Occasionally, reports have been made suggesting that transepidermal uptake is a two component process, consisting of a carrier-mediated process and passive diffusion, the

Fig. 13-4. Effect of increasing concentration on the influx of $^{14}$C-glycine in the worm, *Nereis diversicolor*. Each point is the mean influx at a test concentration determined from separate measurements of influx with 10 worms; bars are ±1 s.d.. The curve describes the predicted concentration dependency of influx using the Michaelis-Menten equation (see text); kinetic constants, $J^i_{max}$ and $K_t$, were calculated from the data using a linear transformation of the Michaelis-Menten equation. Data supplied by Stephens.

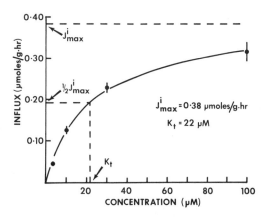

latter becoming the major route of entry at high levels of substrate (e.g., Stewart and Bamford, 1975). Wright and Stephens (1977) made a study of transport of glycine into isolated gills of *Mytilus californianus* in which they noted that apparent influx was described by the relation depicted by the solid line of Fig. 13-5. However, they found that when their data were corrected for the presence of a rapidly labeled external compartment, actual influx into gill tissue became adequately described by a simple single carrier model, as depicted by the dashed line in Fig. 13-5.

It would be most surprising if invertebrate epithelia were permeable to passive transfer of amino acids to any significant extent. Intracellular pools of free amino acids in marine invertebrates are typically greater than $10^{-1}$ $M$. Since environmental levels are at most $10^{-4}$ $M$ and may be as low as $10^{-7}$ $M$, there is always an extremely large gradient promoting loss of amino acid from epithelial cells to the environment. Diffusion is proportional to the concentration gradient. Therefore, if there were any perceptible passive influx at the relatively low levels supplied in the medium, this would imply a continuing

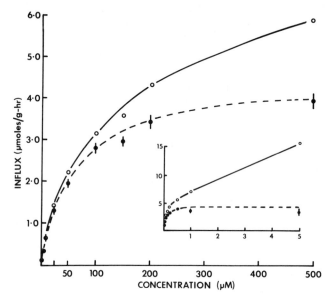

Fig. 13-5. Effect of increasing concentration on the apparent influx of [14]C-cycloleucine in isolated gills of *Mytilus californianus*. Units in the inset are: ordinate, influx ($\mu$m/g-h); abscissa, concentration (m$M$). Open circles and solid lines represent apparent rates of accumulation of radioactivity in discs of gill tissues; data were uncorrected for the presence of a large, rapidly labeled extracellular component. Radioactivity in the extracellular component was estimated from separate measurements of [14]C-inulin space in gill discs, and was subtracted from the total radioactivity per disc. These corrected data are represented by the solid circles. Dashed lines describe influx predicted by the Michaelis-Menten equation, and were calculated using the kinetic constants, $K_t$ (63$\mu$M) and $J'_{max}$ (4.6 $\mu$m/g-h), derived from the corrected data. Vertical bars are ±1 s.e.; $n$ at each point ranges from 12 to 54. Data from Wright and Stephens, 1977.

efflux down these very large concentration gradients, which would be energetically unacceptable.

We may note in passing that the energy requirements for solute transport against such large gradients are quite modest. The thermodynamic work for such transport is described by the relation

$$W = RT \ln(C_i/C_o)$$

where $R$ is the gas constant, $T$ is absolute temperature, $C_i$ and $C_o$ are internal and external concentrations of solute and $W$ is given in energy units per mole. Thus the transport of a mole of solute against a 100,000:1 gradient requires roughly 7 kcal/$M$, a negligible cost compared to the free energy acquired from a reduced carbon substrate. For charged solutes, such as amino acids, this expression is overly simple and an electrical term must be added; however, this does not change the fact that the transport is thermodynamically cheap.

As mentioned earlier, measurements of influx of radiolabeled substrates provide no evidence for net exchanges of amino acid. With this in mind, it has been suggested that uptake of radioactively labeled amino acids in marine invertebrates may actually be accompanied by an equal or even larger carrier-mediated efflux of unlabeled amino acid (Johannes et al., 1969; Johannes and Webb, 1970). Recently it has been possible to couple measurements of influx of radiolabeled substrates with the net uptake of amino acid in marine invertebrates (Fig. 13-3) (Crowe et al., 1977; Costopulos et al., 1979; Stephens et al., 1978; Wright and Stephens, 1977, 1978). These studies have revealed that "exchange diffusion" plays no significant role in the transepidermal transport process. Net influx can be demonstrated in some organisms to occur from ambient levels of amino acid of less than 0.3 $\mu M$ (Fig. 13-6; Wright and Stephens, 1978). Efflux of primary amines does accompany influx by carrier-mediated transport; however, its rate appears to be independent of ambient concentration in the normal range and is slow compared to influx (Wright and Stephens, 1977, 1978).

Estimates of the role that amino acid uptake plays in the nutrition of marine invertebrates must be based on quantitative comparisons of the data from the three categories listed above (i.e., requirements, availability, and kinetics of uptake) for each species considered. Unfortunately, few studies have combined adequate information from all these categories, and thus most of our information is descriptive in nature, consisting largely of demonstrations of the capacity to accumulate amino acids. However, several examples do now exist for which we can collect the information discussed above. The following is a discussion of two such cases. The first is the mussel, *Mytilus californianus,* representing what we have termed the water column habitat. The second is the sand dollar, *Dendraster excentricus,* a sediment dweller in one of its behavioral modes.

### 1. *Mytilus californianus*  The California coastal mussel is widely distributed along the Pacific coast of North America on exposed rocky intertidal areas, as well as in deeper near-shore waters. Although occasionally found completely covered by inshore sandy substrates for periods of several months, it is most familiar as dense aggregations of animals on rocks in the intertidal zone. Animals feed by ciliary-mucoid filtering of large volumes of seawater, collecting and ingesting suspended particulate food.

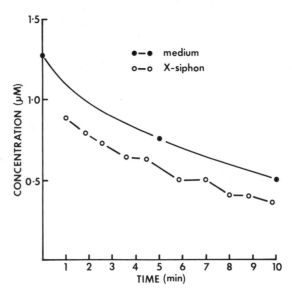

Fig. 13-6. Removal of glycine from solution by an intact specimen of *Mytilus californianus*. Solid circles represent the level of glycine, measured using fluorescamine, remaining in a 400-ml test solution containing an intact, actively pumping 5-g mussel; the solid line describes medium depletion assuming depletion to be a first-order process. The open circles indicate the levels of glycine measured in samples collected from the exhalent water stream (see text for a description of the technique). The average percent clearance of glycine occurring during a single passage of water through the animal was 25%.

*M. californianus* was the subject of a recent study (Bayne *et al.*, 1976) of metabolism and energy balance. These authors reported values for rates of oxygen consumption for actively pumping animals (termed "routine" rates) as a function of body size (Table 13-2). For a mussel of 5 g wet wt, the predicted value for $O_2$ consumption is 0.54 ml/h. The oxidation of mixed protein amino acids (average molecular weight of 100) requires approximately 1 ml $O_2$/l mg amino acid (Stephens, 1972). Thus the oxidative requirements of an active 5-g mussel can be accounted for by the oxidation of approximately 0.54 mg amino acid/h. Although we use these numbers in what follows, note that the recent work of Hammen (1979) indicates there may be a significant anaerobic component in bivalve metabolism.

As discussed earlier, there is general agreement on the concentration of amino acids in near-shore waters, with 0.5–1.0 $\mu M$ total DFAA being the typical range. There are good theoretical reasons to predict that $K_t$'s for amino acid transport systems will be at or near the concentration they normally encounter (Atkinson, 1969; Crowely, 1975). Thus it was disturbing to note that the measured values for $K_t$'s for amino acid uptake into marine bivalves have long been considered to be in the range 10–100 $\mu M$. Wright and Stephens (1977) reported that the $K_t$'s for glycine and cycloleucine transport in gills of *M. californianus* were 34 and 65 $\mu M$, respectively. Bishop (1976) pointed out that, although maximal rates of amino acid transport tend to be large (compared to sediment infauna), the high $K_t$'s precluded these transport processes from acquiring substrate at physiologically meaningful rates from the low concentrations of amino acid characteristic of the water column habitat.

Recently, it has been shown that the large $K_t$'s reported for amino acid uptake in mussels are artifacts resulting from the use of *in vitro* preparations of gill tissue. In most of the literature devoted to transport in bivalves, isolated gill tissue has generally been utilized as an experimental system (recent references include Bamford and Campbell, 1976; Crowe *et al.*, 1977; Stewart, 1977, 1978; Stewart and Bamford, 1976; Wright, 1979; Wright *et al.*, 1980; Wright and Stephens, 1977). This practice has been justified by the demonstration that gill tissue is the primary site of sugar and amino acid uptake in *Mytilus edulis* (Péquignat, 1973). There are obvious advantages to the use of such systems in transport work; exposure time to test solutions can be accurately regulated, and the vagaries of pumping inherent in studies with intact animals can be avoided.

Wright and Stephens (1978) developed a procedure for measuring directly the amount of amino acid accumulated during a single passage of inhalent water across the gill of intact, actively pumping mussels. The technique involves insertion of a small plastic cannula into the exhalent siphon of a pumping mussel, with care taken to avoid touching the sensitive mantle margin. Samples of exhalent water are collected and concentration of radioactive amino acid and total primary amines measured and compared to the levels measured in the surrounding medium. *M. californianus* was found to be capable of a net accumulation of up to 30% of a 1 $\mu M$ solution of glycine during a single passage of water through animals pumping 6 to 10 l/h (Fig. 13-6); slower pumping rates were accompanied by larger efficiencies of clearance. Net uptake was observed to occur from ambient concentrations of less than 0.3 $\mu M$. The $K_t$ for glycine transport determined from such data on intact animals ranged from 1–5 $\mu M$. This is in accord with a brief report (Jørgensen, 1976) of a $K_t$ of 1.7 $\mu M$ for alanine transport for intact *Mytilus edulis*.

The reason for this striking discrepancy in $K_t$'s evaluated from isolated gills versus those found with intact animals is the interference with normal gill perfusion produced at the time of isolation. This creates a large unstirred layer in the isolated gill preparation which in turn is responsible for the high $K_t$'s observed (Wright, 1979; Wright, *et al.* 1980; see also, Winne, 1973).

The data currently in hand permit estimates of the degree to which *M. californianus* can account for its observed oxidative requirements via uptake of amino acids from environmentally realistic concentrations (Table 13-2). For example, a 5-g mussel typically pumps at a rate of approximately 8 l/h. If the concentration of amino acid in the surrounding medium is approximately 1 $\mu M$, a clearance efficiency of 0.25 results in the accumulation of some 2 micromoles per hour. Given an average molecular weight of 100 for mixed amino acids, this rate of accumulation translates to 0.2 mg amino acid per hour. Thus, recalling the estimate of 0.54 mg amino acid/h for the oxidative requirements of a 5-g mussel, amino acid uptake can account for approximately 37% of the oxygen consumption of *M. californianus*.

This calculation is subject to the caveats mentioned earlier and is very likely an overestimate. However, it is clear that *Mytilus* has transepidermal transport systems with the low $K_t$'s appropriate to its habitat. These systems can apparently provide a continuous input of reduced carbon from natural waters at metabolically significant rates.

**2. Dendraster excentricus** The common west coast sand dollar occurs in dense subtidal populations. Individuals are found buried just below the surface of shallow-water inshore sediment or may be found in an inclined position with less than one-third of the test embedded in the sediment and the remaining two-thirds projecting into the overlying water. Timko (1976) has described the mechanism of suspension feeding in animals in the latter orientation. Individuals change from one habit orientation to another (Chia, 1969).

Asteroid and echinoid echinoderms in several genera have been shown to translocate substrates to and from the epidermis very slowly if at all. Labeled amino acids introduced into the epidermis (Ferguson, 1967; Pearse and Pearse, 1973) or into the viscera (Ferguson, 1970) remain in these locations when followed by autoradiography for long periods of time. Translocation from these two major sites is very slow or imper-

TABLE 13-2. Nutritional Role of the Transepidermal Accumulation of Amino Acids in *Mytilus californianus* and *Dendraster excentricus:* Examination of Critical Parameters

|  |  | *Reference* |
|---|---|---|
| *Mytilus californianus* |  |  |
| Oxygen consumption (5 g wet wt) | 0.54 ml $O_2$/h | Bayne *et al.,* 1976 |
| Environmental concentration | $0.2 - 2.0 \mu M$ | North, 1975 |
| Kinetics of transport* (intact animal) |  |  |
| $V_{max}$ | 0.85 $\mu$moles/g-h | Wright and Stephens, 1978 |
| $K_t$ | 1.3 $\mu M$ |  |
| % oxidative requirements accounted for by net uptake of amino acids at environmental level of 1 $\mu M$ | 37 | *Ibid.* |
| *Dendraster excentricus* |  |  |
| Oxygen consumption | 4.3 $\mu$l $O_2$/cm² -h | Stephens *et al.,* 1978 |
| Environmental concentration | $115 \pm 60$ (s.d.) $\mu M$ | *Ibid.* |
| Kinetics of transport* |  |  |
| $V_{max}$ | 215 $\mu$moles/cm² -h | *Ibid.* |
| $K_t$ | 74 $\mu M$ |  |
| Concentration of naturally occurring primary amines at which net uptake accounts for observed oxidative requirements | 35 $\mu M$ | *Ibid.* |

*These values are for influx, measured using radiolabeled substrates; estimates of rates of acquisition at environmental levels of substrate were made using measured values of net flux, and hence are lower than influx.

ceptible for periods of up to three months (Ferguson, 1970). Thus the question of what nutritional input is available to sustain the metabolism of the highly complex and active epidermal organs (spines, pedicellariae, etc.) is of considerable interest.

The report of Stephens *et al.* (1978) provides the following information (Table 13-2). Oxygen consumption of *Dendraster* is directly proportional to surface area of the test. Influx of neutral, dicarboxylic, and polybasic amino acids is carrier-mediated and is also proportional to surface area of the test. The $K_t$ for glycine influx is 74 $\mu M$, and net flux is effectively equivalent to influx at substrate levels in excess of 5 $\mu M$. Primary amines in the interstitial water where these animals were collected ranged in concentration from 17 to 244 $\mu M$, of which 70 to 85% were identified as free amino acids using thin-layer chromatography. Moreover, these naturally occurring amino acids were demonstrated to be available for transepidermal accumulation by decreases in the number of spots and individual spot intensity on chromatograms of natural water samples to which sand dollars were exposed. Comparison of uptake rates of naturally occurring primary amines to measured rates of oxygen consumption indicates that the oxidative requirements of *Dendraster* can be met by rates of accumulation of primary amines occurring at environmental concentrations in excess of 35 $\mu M$; this level was exceeded in 14 of 15 samples of interstitial water analyzed.

The conclusion from this work is that transepidermal transport of free amino acids is quantitatively adequate to sustain the metabolism of sand dollars when they are in the submerged orientation and have access to interstitial water in the sediments where they live.

## THE ROLE OF DOM IN ABYSSAL AND DEEP-WATER COMMUNITIES

We have laid out criteria for judging the potential significance of DOM inputs in the preceding discussion. Information is needed concerning metabolic requirements, availability of dissolved resources, and kinetics of transport systems in the organisms adapted to utilize DOM. Information is available which permits comment on each of these matters for abyssal habitats. However, the quantity, quality, and relevance of such information is less good than that for the shallow-water environment simply because of the formidable difficulties involved in physiological work on abyssal organisms. Indeed, as other chapters of this volume make clear, there are formidable difficulties that impede obtaining reliable information of any sort about the abyssal environment.

One of the things that can be said with some confidence is that conditions in the abyss are diverse. In large areas, there appears to be very little biological activity; what energy flow that is occurring, is proceeding very slowly (Wirsen and Jannasch, 1976). In other areas where there is an import of plant detritus, biological activity is much more striking and community organization appears to be broadly comparable to that in shallow waters (Wolff, 1979). In other quite remarkable and no doubt very unusual areas, biological activity is quite astonishingly high (Corliss *et al.*, 1979). Given this diversity, no simple hypothesis about the role of DOM is possible.

## Metabolic Requirements and Energy Flow in Abyssal Environments

Recent evidence indicates that organisms living in deep-water habitats have metabolic rates an order of magnitude lower than those of closely related shallow-water species. Smith (1978) made *in situ* measurements of benthic community respiration in bottom samples from 40 to 5200 m in depth. He reported a two order of magnitude decrease in abundance and biomass between the shallow and abyssal sediments, and a three order of magnitude decrease in total oxygen uptake in the benthic communities of these sediments compared to shallow communities. The dominant taxa of these communities were polychaetes, bivalves, and crustaceans.

The cause of reduced metabolism in abyssal organisms appears to be related to the increased hydrostatic pressure characteristic of the abyssal environment. In his study, Smith (1978) reported that predictive equations for benthic community respiration showed a stronger correlation with depth than for the parameters of temperature, dissolved oxygen, benthic animal biomass, surface primary productivity, or sediment organic matter. Jannasch and his co-workers have reported results with abyssal bacteria consistent with the idea that hydrostatic pressure has a direct effect on metabolic processes. Jannasch *et al.* (1971) found microbial degradation to be 10 to 100 times slower in the deep sea than in control samples at the same temperature but at surface pressure. Also, Jannasch and Wirsen (1977) found that uptake and incorporation of labeled amino acids in a microbial population in a water sample from a depth of 2600 was lower than in decompressed samples run at the same temperature (3°C).

This view of abyssal community metabolism occurring at very low levels is not supported by all of the information available. The rate at which samples of organic substrates are attacked appears to depend in part on whether they are exposed to mechanical attack by macrofauna (Sieburth and Dietz, 1974; Wirsen and Jannasch, 1976). Seki *et al.* (1974) report high organotrophic rates for an autochthonous abyssal bacterium. Fungal and bacterial attack on wood and cordage proceeds promptly at abyssal depths (Kohlmeyer, 1969; Muraoka, 1971) though fungal reproductive structures appear to develop slowly.

This variability in levels of biological activity probably represents real environmental diversity. Where plant detritus is prominent (eel grass, coconut fronds, etc.) an active assembly of invertebrates is also found (Wolff, 1979); nitrogen content of sediments associated with such material is 10 times higher than in sediments not associated with detritus. Thus it may well be that energy flow in abyssal communities depends on the presence of levels of organic material sufficient to sustain it. Where plant detritus is present, macrofauna promote its decomposition by breaking tissue and increasing available surface, thereby facilitating the activity of the microbial community. Wolff (1979) argues convincingly that community organization and trophic structure in such areas is quite directly comparable to that in shallow waters. That is, plant detritus is available to the macrofauna via microorganisms which appear to be the direct food resource of the detritus feeders rather than the refractory and nutritionally inadequate residues of higher plants (cf., Fenchel, 1970, 1972).

Despite this diversity, it is probably the case that rates of energy flow in deep water are substantially lowered by hydrostatic pressure and the effects of low temperature compared to the rates familiar from shallow waters.

## Concentration of DOM in Abyssal Habitats

The concentration of dissolved organic carbon is about an order of magnitude lower in samples of abyssal water than normally found in shallow- and near-shore water (approximately 0.4 versus 4 mg/l; Degens, 1970). Likewise, the concentration of DFAA in deep water is lower than the 0.5–2.0-$\mu M$ range usually found in surface- and near-shore water. Lee and Bada (1977) reported levels of DFAA ranging from 25 to 50 n$M$ in water samples from 2000 to 4000 m in depth. Dissolved total amino acids (i.e., including short chain polypeptides) were higher, ranging from approximately 75 to 150 n$M$.

Levels of DFAA in some abyssal sediments appear to be comparable to those of shallow sediments. Henrichs and Farrington (1979) reported that total free amino acids in sediment cores from a depth of 4200 m in the Gulf of Maine ranged from 1.5 to 5.2 mg/l (approximately 15 to 52 $\mu M$), with the major constituents of the free pool being alanine, glycine, glutamic acid, and $\beta$-glutamic acid. We have already noted that the organic content of abyssal sediment is greater in areas characterized by the presence of macroscopic plant detritus. Emelyanov (1977) reports great variability in organic carbon content of deep sediments from the western Central Atlantic. Some values are very low, but levels of the order of 1% are quite common and occasional samples are much higher. It is very likely that there is a comparable variation in interstitial DOM.

## Uptake of DOM by Abyssal Animals

Direct evidence of the ability of abyssal organisms to accumulate dissolved organic substrates is very limited. The only studies specifically examining uptake of dissolved nutrients in animals collected from what can be considered "deep-water" habitats (200 to 2000 m) are those of Southward and Southward (pogonophores: 1968, 1970, *et al.*, 1979; polychaetes: 1972) and Little and Gupta (pogonophores: 1968, 1969). All of these studies demonstrated that the organisms tested were capable of the transepidermal accumulation of radioactively labeled sugars and/or amino acids. Though some effort has been made to describe transport kinetics, the problem has been refractory. Such reports would be difficult to interpret, in any event, because all the transport experiments in these studies have been performed on decompressed organisms. It has already been noted that elevated hydrostatic pressures appear to have significant effects on metabolic processes, including transport processes. Also, recent evidence indicates that hydrostatic pressure can have fundamental effects on the kinetics of enzyme-mediated processes (Siebenaller and Somero, 1978). Thus, the existing studies on DOM uptake in deep-water invertebrates should be considered qualitative demonstrations of transepithelial transport in these animals.

DeLaca *et al.* (1981) report uptake of DOM directly across the surface of a large agglutinated foraminiferan, *Notodendrodes antarctikos*. The observations are relevant to the present discussion even though the organisms occur in a relatively shallow-water population (15–40 m). The western side of MacMurdo Sound appears to be more closely similar to the deep sea than to other shallow-water Antarctic areas based on studies of water chemistry and faunal composition. *Notodendrodes* is abundant in the bottom fauna of this region. Since foraminiferans represent the highest test-free biomass of all eucaryotic organisms in a number of deep-sea communities (Smith, 1978) and since this area presents an accessible analog of deep-sea benthic conditions, the fact that uptake of amino acids and glucose as well as rapid assimilation of these substrates occurred is of great interest. Concentration of FAA in interstitial water of the sediment in MacMurdo Sound is 20–30 $\mu M$. It was possible to exclude an indirect bacterial pathway for entry of DOM in these observations. Since these organisms survive well in the laboratory when maintained at low temperature, one can realistically hope for much more extensive information from future studies concerning the role of DOM in their nutrition. Despite its close similarity to deep-sea benthic habitats in most respects, this remains a shallow-water community and thus offers no insight into possible effects of high hydrostatic pressure on the uptake process.

While transport studies utilizing deep-water animals are limited, we can make some predictions about the general characteristics of transepithelial transport in abyssal organisms. First, it is quite probable that all soft-bodied invertebrates dwelling in abyssal habitats possess the capability to accumulate DOM. The ubiquity of transepithelial transport systems for amino acids in shallow-water invertebrates (refer to Table 13-1), along with positive findings of DOM uptake in the studies cited above utilizing deep-water animals, suggest that such processes are a general phenomenon possessed to one degree or another by all soft-bodied marine invertebrates, regardless of habitat.

There is some basis for speculation concerning the kinetics of transport in abyssal animals. Shick (1975) found that $Q_{10}$ for glycine uptake into polyps of *Aurelia* was in excess of 12 over the range 12° to 22°C. Wright *et al.* (1980) found that glycine uptake into gills of *Mytilus* had a $Q_{10}$ of 5 between 7° and 23°C. The very low temperatures characteristic of the abyssal environment (*ca.* 3°C) may well result in concomitant decreases in the activity of rate processes such as membrane transport. Thus maximum rates of transport ($J^i_{max}$) are probably quite low in abyssal animals by virtue of these temperature effects.

$K_t$'s are also probably low. Wright *et al.* (1980) studied the effects of temperature on $K_t$ in isolated gill preparations from the mussel, *Mytilus*. In this preparation, unstirred layers separating the transporting tissue from the bulk phase of the medium play a prominent role in determining the observed kinetics as we have discussed. At lower temperatures, the effect is reduced. The decrease in ($J^i_{max}$) as a result of the lower temperature decreases the discrepancy between concentrations at the transport surface and those present in the bulk medium, and has as a result a dramatic decrease in $K_t$ with temperature (see Thomson and Dietschy, 1977, for a theoretical account of this effect). Stephens[1] has made similar observations on the polychaete, *Nereis diversicolor*. It appears likely that the "true" $K_t$ at the transporting surface is very low in marine invertebrates generally. Thus the low temperatures in abyssal environments will also have the effect of decreasing the observed $K_t$'s for transport occurring under such conditions.

[1] Stephens, unpublished.

Evidence for a substantial role of DOM in shallow-water marine communities is now quite impressive. However, the same cannot be said for DOM in abyssal communities. What little information we have is consistent with the hypothesis that DOM may play a role in at least some kinds of abyssal habitats. We do know that abyssal invertebrates possess transport systems for amino acids. We also know that DFAA are available in selected deep-water locations. We can speculate with some confidence that maximum rates of influx are likely to be low compared with those found in shallow-water forms. On the other hand, there is also good reason to believe that the systems will be adapted to deal with low substrate concentrations (i.e., by way of having low $K_t$'s). This adaptation would compensate to some extent for the likelihood that resource levels may be lower. Finally, the lower energy flow which seems to characterize abyssal communities demands correspondingly lower rates of input for sustenance. This scarcely makes a convincing case. However, nothing we know makes tentative postulation of a nutritional role for DOM in deep-water animals unreasonable.

The utilization of DOM by abyssal animals is in some ways a particularly attractive possibility. Food resources in abyssal habitats appear to be generally very limited. With metabolic requirements being low, a continuous, if modest, input of reduced carbon which DOM can provide may be particularly important. We have pointed out that the thermodynamic costs of transport are modest compared to the profits. Activity costs for food aquisition via transepidermal pathways are zero.

The role, if any, of DOM in the abyss is a question to be settled in the future. We suggest that the analysis carried out on shallow-water forms provides a model for the required investigation of the issue. It is now technically possible to undertake such investigations and the matter seems sufficiently promising and important to justify doing so.

# REFERENCES

Ahearn, G. A., 1976, Co-transport of glycine and sodium across the mucosal border of the midgut epithelium in the marine shrimp, *Penaeus marginatus: J. Physiol.,* v. 258, p. 499–520.

Anderson, J. W., and Stephens, G. C., 1969, Uptake of organic material by aquatic invertebrates, VI. Role of epiflora in apparent uptake of glycine by marine crustaceans: *Mar. Biol.,* v. 4 p. 243–249.

Atkinson, D. E., 1969, Limitation of metabolite concentrations and the conservation of solvent capacity in the living cell, *in* Horecker, B. L., and Stadtman, E. R., eds., *Current Topics in Cellular Regulation, Vol. 1:* New York, Academic Press, p. 29–43.

Bamford, D. R., and James, D., 1972, An *in vitro* study of amino acid and sugar absorption in the gut of *Echinus esculentus: Comp. Biochem. Physiol.,* v. 42A, p. 579–590.

——, and Campbell, E., 1976, The effect of environmental factors on the absorption of L-phenylalanine by the gill of *Mytilus edulis: Comp. Biochem. Physiol.,* v. 53A, p. 295–299.

Bayne, B. L., Bayne, C. J., Carefoot, T. C., and Thompson, R. J., 1976, The physiological ecology of *Mytilus californianus* Conrad, 1. Metabolism and energy balance: *Oecoligia,* v. 22, p. 211–228.

Bishop, S. H., 1976, Nitrogen metabolism and excretion—regulation of intracellular amino acid concentrations, *in* Wiley, M., ed., *Estuarine Processes, Vol. 1:* New York, Academic, p. 414–431.

Chia, F. S., 1969, Some observations on the locomotion and feeding of the sand dollar, *Dendraster excentricus: J. Exp. Mar. Biol. Ecol.,* v. 3, p. 162–170.

Chien, P. K., Stephens, G. C., and Healey, P. L., 1972, The role of ultra-structure and physiological differentiation of epithelia in amino acid uptake by the bloodworm, *Glycera: Biol. Bull.,* v. 142, p. 219–235.

Christensen, H. N., 1976a, *Biological Transport, 2nd Ed.:* New York, W. A. Benjamen Co., 514 p.

——, 1976b, Towards a sharper definition of energetic coupling through integration of membrane transport into bioenergetics: *J. Theor. Biol.,* v. 57, p. 419–431.

Corliss, J. B., Dymond, J., Gordon, L. I., Edmond, J. M., von Herzen, R. P., Ballard, R. D., Green, K., Williams, D., Bainbridge, A., Crane, K., and van Andel, T. H., 1979, Submarine thermal springs on the Galapogos Rift: *Science,* v. 203, p. 1073–1083.

Costopulos, J. J., Stephens G. C., and Wright, S. H., 1979, Uptake of amino acids by marine polychaetes under anoxic conditions: *Biol. Bull.,* v. 157, p. 434–444.

Crowe, J. H., Dickson, K. A., Otto, J. L., Colon, R. D., and Farley, K. K., 1977, Uptake of amino acids by the mussel, *Modiolus demissus: J. Exp. Zool.,* v. 202, p. 323–332.

Crowely, P. H., 1975, Natural selection and the Michaelis constant: *J. Theor. Biol.,* v. 50, p. 461–475.

DeBurgh, M. E., 1978, Specificity of L-analine transport in the spine epithelium of *Paracentrotus lividus* (Echinoidea): *J. Mar. Biol. Assoc. U. K.,* v. 58, p. 425–440.

Degens, E. T., 1970, Molecular nature of nitrogenous compounds in sea water and recent marine sediments, *in* Hood, D. W., ed., *Organic Matter in Natural Waters:* Institute of Mar. Sci., Occasional Publ., no. 1, Univ. Alaska, p. 77–106.

DeLaca, T. E., Karl, D. M., and Lipps, J. H., 1981, The direct utilization of dissolved organic carbon by agglutinated benthic foraminifera: *Nature,* v. 289, p. 287–289.

Emelyanov, E. M., 1977, 14. Geochemistry of sediments in the Western Central Atlantic, DSDP Leg 39, *in, Initial Reports of the Deep Sea Drilling Project,* U.S.G.P.O., Washington, D.C., v. 39, p. 477–492.

Fenchel, T., 1970, Studies on the decomposition of organic detritus derived from the turtle grass *Thalassia testudinum: Limnol. Oceanogr.,* v. 15, p. 14–20.

——, 1972, Aspects of decomposer food chains in marine benthos: *Verh. dt. zool. Ges.,* v. 65, p. 14–22.

Ferguson, J. C., 1967, An autoradiographic study of the utilization of free exogenous amino acids by starfishes: *Biol. Bull.,* v. 133, p. 317–329.

——, 1970, An autoradiographic study of the translocation and utilization of amino acids by starfishes: *Biol. Bull.,* v. 138, p. 14–25.

——, 1971, Uptake and release of free amino acids by starfishes: *Biol. Bull.,* v. 141, p. 122–129.

Fisher, F. M., and Oaks, J. A., 1978, Evidence for a nonintestinal nutritional mechanism in the Rhynchocoelan, *Lineus ruber: Biol. Bull.*, v. 154, p. 213–225.

Hammen, C. S., 1979, Metabolic rates of marine bivalve molluscs determined by calorimetry: *Comp. Biochem. Physiol.*, v. 62A, p. 955–959.

Henrichs, S. M., and Farrington, J. W., 1979, Amino acids in interstitial waters of marine sediments: *Nature*, v. 279, p. 319–322.

Jannasch, H. W., Eimhjellen, K., Wirsen, C. O., and Farmanfarmaian, A., 1971, Microbial degradation of organic matter in the deep sea: *Science*, v. 171, p. 673–675.

——, and Wirsen, C. O., 1977, Retrieval of concentrated and undecompressed microbial populations from the deep sea: *Appl. and Environ. Microbiol.*, v. 33, p. 642–646.

Johannes, R. E., Coward, S. J., and Webb, K. L., 1969, Are dissolved amino acids an energy source for marine invertebrates? *Comp. Biochem. Physiol.*, v. 29, p. 283–288.

——, and Webb, K. L., 1970, Release of dissolved organic compounds by marine and freshwater invertebrates, *in* Hood, D. W., ed., *Organic Matter in Natural Waters:* Institute of Mar. Sci., Occasional Publ., no. 1, Univ. Alaska, p. 257–273.

Jørgenson, C. B., 1966, *Biology of Suspension Feeding:* Oxford, Pergamon, 357 p.

——, 1976, August Pütter, August Krogh, and modern ideas on the use of dissolved organic matter in aquatic environments: *Biol. Rev.*, v. 51, p. 291–328.

Jørgenson, N. O. G., 1979, Uptake of L-valine and other amino acids by the Polychaete *Nereis virens: Mar. Biol.*, v. 52, p. 42–52.

Kohlmeyer, J., 1969, Deterioration of wood by marine fungi in the deep sea, *in, Material Performance and the Deep Sea,* ASTM STP 445: Amer. Soc. for Testing Materials, p. 20–30.

Krogh, A., 1931, Dissolved substances as food of aquatic organisms: *Biol. Rev.*, v. 6 p. 412–442.

Lavoisier, A., and LaPlace, P., 1780, Memoire sur la chaleur. Mem. Acad. Sci., Paris, *in* Gabriel, M. L., and Fogel, S., eds., *Great Experiments in Biology:* Prentice-Hall, Englewood Cliffs, N.J., (1955), p. 85–93.

Lee, C., and Bada, J. L., 1977, Dissolved amino acids in the equatorial Pacific, the Sargasso Sea, and Biscayne Bay: *Limnol. Oceanogr.*, v. 22, p. 502–510.

Little, C., and Gupta, B. L., 1968, Pogonophora—uptake of dissolved nutrients: *Nature,* v. 218, p. 873–874.

——, and Gupta, B. L., 1969, Studies on Pogonophora, III. Uptake of nutrients: *J. Exp. Biol.*, v. 51, p. 559–573.

Muraokoa, J. S., 1971, Deep-ocean biodeterioration of materials, *Materials for the Sea, Part 8: Ocean Ind.*, March 1971, p. 44–46.

North, B. B., 1975, Primary amines in California coastal waters—utilization by phytoplankton: *Limnol. Oceanogr.*, v. 20, p. 20–26.

Pamatmat, M. M., 1978, Oxygen uptake and heat production in a metabolic conformer (*Littorina irrorata*) and a metabolic regulator (*Uca pugnax*): *Mar. Biol.*, v. 48, p. 317–325.

Pearse, J. S., and Pearse, V. B., 1973, Removal of glycine from solution by the sea urchin *Strongylocentrotus purpuratus: Mar. Biol.*, v. 19, p. 282–284.

Péquignat, E., 1973, A kinetic and autoradiographic study of the direct assimilation of amino acids and glucose by organs of the mussel *Mytilus edulis: Mar. Biol.*, v. 19, p. 227–244.

Pütter, A., 1909, *Die Ernährung der Wassertiere und der Stoffhaushalt der Gewässer:* Fisher, Jena, 168 p.

Schlichter, D., 1978, On the ability of *Anemonia sulcata* (Coelenterata:Anthozoa) to absorb charged and neutral amino acids simultaneously: *Mar. Biol.,* v. 45, p. 97–104.

Schultz, S. G., and Curran, P. F., 1970, Coupled transport of sodium and organic solutes: *Physiol. Rev.,* v. 50, p. 637–718.

Seki, H., Wada, E., Koike, I., and Hattori, A., 1974, Evidence of high organotrophic potentiality of bacteria in the deep ocean: *Mar. Biol.,* v. 26, p. 1–4.

Sepers, A. B. J., 1977, The utilization of dissolved organic compounds in aquatic environments: *Hydrobologia,* v. 52, p. 39–54.

Shick, J. M., 1975, Uptake and utilization of dissolved glycine by *Aurelia aurita* scyphistomae—temperature effects on the uptake process; nutritional role of dissolved amino acids: *Biol. Bull.,* v. 148, p. 117–140.

Siebenaller, J., and Somero, G. N., 1978, Pressure-adaptive differences in lactate dehydrogenases of congeneric fishes living at different depths: *Science,* v. 201, p. 255–257.

Sieburth, J. Mcn., and Dietz, A. S., 1974, Biodeterioration in the sea and its inhibition, *in,* Colwell, R. R., and Morita, R. Y., eds., *Effect of the Ocean Environment on Microbial Activities:* Baltimore, University Park Press, p. 318–326.

Siegel, A., and Degens, E. T., 1966, Concentration of dissolved amino acids from saline waters by ligand-exchange chromatography: *Science,* v. 151; p. 1098–1101.

Smith, K. L., Jr., 1978, Benthic community respiration in the N.W. Atlantic Ocean—*in situ* measurements from 40 to 5200 m: *Mar. Biol.,* v. 47, p. 337–347.

Southward, A. J., and Southward, E. C., 1968, Uptake and incorporation of labelled glycine by Pogonophores: *Nature,* v. 218, p. 875–876.

——, and Southward, E. C., 1970, Observations on the role of dissolved organic compounds in the nutrition of benthic invertebrates—Experiments on three species of Pogonophora: *Sarsia,* v. 45, p. 69–95.

——, and Southward, E. C., 1972, Observations on the role of dissolved organic compounds in the nutrition of benthic invertebrates, II. Uptake by other animals living in the same habitat as pogonophores, and by some littoral polychaeta: *Sarsia,* v. 48, p. 61–70.

——, Southward, E. C., and Brattegard, T., and Bakke, T., 1979, Further experiments on the value of dissolved organic matter as food for *Siboglinum fiordicum* (Pogonophora): *J. Mar. Biol. Assoc. U. K.,* v. 59, p. 133–148.

Stephens, G. C., 1963, Uptake of organic material by aquatic invertebrates, II. Accumulation of amino acids by the bamboo worm, *Clymenella torquata: Comp. Biochem. Physiol.,* v. 10, p. 191–202.

——, 1972, Amino acid accumulation and assimilation in marine organisms, *in* Campbell, J. W., and Goldstein, L., eds., *Nitrogen Metabolism and the Environment:* New York, Academic Press, p. 155–184.

——, 1975, Uptake of naturally occurring primary amines by marine annelids: *Biol. Bull.,* v. 149, p. 397–407.

——, and Schinske, R. A., 1961, Uptake of amino acids by marine invertebrates: *Limnol. Oceanogr.,* v. 6, p. 175–181.

——, Volk, M J., and Wright, S. H., and Backlund, P. S., 1978, Transepidermal transport of naturally occurring amino acids in the sand dollar, *Dendraster excentricus: Biol. Bull.,* v. 154, p. 335–347.

Stewart, M. G., 1977, The uptake and utilization of dissolved amino acids by the bivalve *Mya arenaria* (L.), *in,* McLusky, D. S., and Berry, A. J., eds., *Physiology and Behavior of Marine Organisms:* New York, Pergamon, p. 165–176.

——, 1978, Kinetics of neutral amino-acid transport by isolated gill tissue of the bivalve *Mya arenaria* (L.): *J. Exp. Mar. Biol. Ecol.,* v. 32, p. 39–52.

——, and Bamford, D. R., 1975, Kinetics of alanine uptake by the gills of the soft shelled clam *Mya arenaria: Comp. Biochem. Physiol.,* v. 52A, p. 67–74.

——, and Bamford, D. R., 1976, The effect of environmental factors on the absorption of amino acids by isolated gill tissue of the bivalve, *Mya arenaria* (L.): *J. Exp. Mar. Biol. Ecol.,* v. 24, p. 205–212.

Thomson, A. B. R., and Dietschy, J. M., 1977, Derivation of the equations that describe the effects of unstirred layers on the kinetic parameters of active transport processes in the intestine: *J. Theor. Biol.,* v. 64, p. 277–294.

Timko, P. L., 1976, Sand dollars as suspension feeders—a new description of feeding in *Dendraster excentricus: Biol. Bull.,* v. 151, p. 247–259.

Udenfriend, S., Stein, S., Bohlen, P., Dairman, W., Leimgruber, W., and Weigele, M., 1972, Fluorescamine—a reagent for assay of amino acids, peptides, proteins, and primary amines in the picomole range: *Science,* v. 178, p. 871–872.

Webb, K. L., and Johannes, R. E., 1967, Studies of the release of dissolved free amino acids by marine zooplankton: *Limnol. Oceanogr.,* v. 12, p. 376–382.

Williams, P. M., Oeschager, H., and Kinney, P., 1969, Natural radiocarbon activity of the dissolved organic carbon in the northeast Pacific Ocean: *Nature,* v. 224, p. 256–258.

Winne, D., 1973, Unstirred layer, source of biased Michaelis constant in membrane transport: *Biochim. Biophys. Acta,* v. 298, p. 27–31.

Wirsen, C. O., and Jannasch, H. W., 1976, Decomposition of solid organic materials in the deep sea: *Environ. Sci. Tech.,* v. 10, p. 880–886.

Wolff, T., 1979, Macrofaunal utilization of plant remains in the deep sea: *Sarsia,* v. 64, p. 117–136.

Woodwell, G. M., Whittaker, R. H., Reiners, W. A., Likens, G. E., Delwiche, C. C., and Botkin, D. B., 1978, The biota and the world carbon budget: *Science,* v. 199, p. 141–146.

Wright, S. H., 1979, Effect of activity of lateral cilia on transport of amino acids in gills of *Mytilus californianus: J. Exp. Zool.,* v. 209, p. 209–220.

——, Becker, S. A. and Stephens, G. C., 1980, Influence of temperature and unstirred layers on the kinetics of glycine transport in isolated gills of *Mytilus californianus: J. Exp. Zool.,* v. 214, p. 27–35.

——, and Stephens, G. C., 1977, Characteristics of influx and net influx of amino acids in *Mytilus californianus: Biol. Bull.,* v. 152, p. 295–310.

——, and Stephens, G. C., 1978, Removal of amino acid during a single passage of water across the gill of marine mussels: *J. Exp. Zool.,* v. 205, p. 337–352.

Zwaan, A. de, 1977, Anaerobic energy metabolism in bivalve molluscs: *Oceanogr. Mar. Biol. Ann. Rev.,* v. 15, p. 103–187.

Jere H. Lipps
Department of Geology
University of California
Davis, California 95616

Carole S. Hickman
Department of Paleontology
University of California
Berkeley, California 94720

# 14

# ORIGIN, AGE, AND EVOLUTION OF ANTARCTIC AND DEEP-SEA FAUNAS

# ABSTRACT

Shallow-water Antarctic and deep-sea environments have many important physical and biotic similarities. These similarities have stimulated the formulation of conflicting hypotheses to account for faunal origins, ages, and evolutionary processes. From faunistic and taxonomic analyses, systematists have suggested alternatively that the modern deep-sea fauna originated in the shallow Antarctic, that the shallow Antarctic fauna originated in the deep sea, that both faunas migrated into their present sites from other regions, or that each fauna evolved independently in place. The suggested timing of events, and hence inferred ages of faunas, range over the entire geologic time scale. Faunal evolution has not been considered in detail heretofore, although ecologists have developed a number of hypotheses to account for high species diversities in the deep sea. Alternative mechanisms for generating and maintaining high diversity include environmental stability or predictability over long periods of time, biological disturbance, spatial heterogeneity, and the nature of trophic regimes.

Identification and analysis of a series of morphological peculiarities of organisms in these environments (in particular those related to gigantism and dwarfing) provide a new approach to testing and eliminating hypotheses. Evidence from the fossil record and geologic and paleoceanographic data also eliminate some hypotheses. The evidence supports hypotheses that faunas have evolved primarily in place; that faunas have occupied both regions as far back as the fossil record goes (Mesozoic) and probably much earlier; and that evolutionary processes have emphasized constructional, ecological, life history, and physiological adaptations to predictable environments with low levels of trophic resources. Efficient use of resources and skeletal construction with little or no calcium carbonate are recurring themes. Further studies of microhabitat diversity and specialization may shed considerable light on evolution of biotic diversity in both regions.

# INTRODUCTION

The deep-sea and Antarctic marine realms share several environmental and biotic attributes that have suggested an evolutionary relationship between the faunas in these regions. This inferred relationship has been the subject of much speculation, and a number of hypotheses have been proposed to explain the perceived similarities. The controversy concerns the site of origin of these faunas, their age, and the evolutionary processes involved. Herein we review for the nonspecialist the major categories of hypotheses for each of these topics, we examine the available evidence, we propose tests of the hypotheses, and we discuss those hypotheses not eliminated.

Much has been written regarding the age, origin, and evolution of deep-sea and Antarctic marine biotas. We will not consider all of this literature in this general review, but instead cite syntheses that will provide entry into the more detailed discussions for those who need to pursue the subjects.

Before considering evolution in the deep sea and Antarctica, we must define the limits of these regions. Various boundaries have been selected by previous workers on a variety of criteria, ranging from the physical characteristics to the systematic affinities of each region. We prefer to distinguish these regions on the basis of faunal change.

The term "deep sea" commonly suggests rather deep depths; for example, Kussakin (1973) considered the deep sea to be deeper than 2000 m and Menzies *et al.* (1973) indicate a distinctive break in faunal composition between the bathyal or archibenthal and the abyssal zones. Hessler and Thistle (1975) point out, however, that such breaks are artificial: Kussakin gave no reasons for using 2000 m as his boundary, and Menzies *et al.*'s faunal boundary seems to be an artifact of differential sampling intervals. The greatest faunal and physical breaks lie at the edge of the continental shelf between 200 and 400 m. Because of these changes, Hessler and Thistle (1975) follow some previous workers and accept this general depth range. We accept 200 to 400 m also as marking the start of the "deep sea."

The Antarctic marine environment has been separated into several biogeographic schemes (see Dell, 1972, and Hedgpeth, 1969, for thorough reviews). These schemes, based on a variety of organisms, distinguish a high Antarctic fauna or province with one or more subregions or provinces. Hedgpeth (1969) proposed that the entire area surrounding Antarctica south of the Antarctic Convergence should be united into a single biogeographic division, the Antarctic Region. This Region was divided into a Continental or High Antarctic Province that included the area south of the Antarctic Divergence, and within this province, the northern Scotia Subregion along the Antarctic Peninsula and the South Georgia District at South Georgia Island. Hedgpeth's conclusion is supported by more recent distributional data for fishes (Daniels and Lipps, in press), which also suggest a possible distinction between the Scotia Subregion including South Georgia Island, and the remainder of Antarctica. Although animal groups are common to the Scotia Subregion and areas farther south, there is a significant change in shallow-water communities from plant-dominated to animal-dominated ones (Hedgpeth, 1971; Lipps and Delaca, 1976). In this paper, we consider the Antarctic marine faunas to correspond to the Antarctic Region of Hedgpeth (1969) without distinction of the more northerly subdivisions.

The deep-sea and shallow Antarctic marine environments share many of the same physical, chemical, and biotic characteristics. Furthermore, the Antarctic region is the source of most of the world's deep-sea water. Both environments have constant physical conditions (except light varies in shallow Antarctic seas), cold temperatures, similar salinities, and low terrestrial sedimentation rates. Temperatures are $2°C$ or less, salinities are between 34.6 and 34.9‰, and there is little input of terrestrial sediment in most places around Antarctica because of its ice-cover and in the deep sea because of its distance from continents and the presence of sediment traps adjacent to the continents. The single greatest difference between Antarctica and most of the deep sea is the great seasonal productivity around Antarctica during the austral spring and summer. In some places adjacent to the Antarctic continent where ice covers the sea throughout the year, productivity may also be exceedingly low (Dayton and Oliver, 1977; Lipps *et al.,* 1979). During the austral winter productivity around Antarctica is very low, producing oligotrophic conditions in the water column. Much of the deep sea is permanently oligotrophic.

A variety of hypotheses have been developed to account for the origin, age, and evolution of deep-sea and Antarctic marine biotas. Commonly, the two realms have been linked in their biotic development because of the similarities in environment and biota. Also, the questions of origin and age often have been considered together. Most of these hypotheses have gone untested because evidence has not been available. We separate the hypotheses to account for origin, age, and evolution in this discussion.

## Origin

Hypotheses to account for the origin of the faunas can be placed in four categories. Details and variations of each hypothesis are reviewed by Menzies *et al.* (1973), Hessler and Thistle (1975), Dell (1972) and Kussakin (1973), among others.

1. The modern deep-sea fauna originated in the Antarctic shallow water regions. Menzies *et al.* (1973) explore this possibility in detail. Their model suggests that a deep-sea fauna existed for a long geologic time, since perhaps the Precambrian, and that the fauna underwent periodic extinctions in response to changing bottom temperatures. The present biota, for the most part, invaded the deep sea from shallow Antarctic areas since the last period of polar cooling began in the Miocene, but, especially, during the Pleistocene glacial ages.

2. The Antarctic fauna is simply an extension of the deep-sea fauna that has followed cool temperatures into shallow waters. Certainly, evidence from isopods suggest that at least a few families have penetrated the shallow Antarctic waters from the deep sea (Hessler and Thistle, 1975), but this cannot be a general conclusion for the majority of taxa.

3. The deep-sea (Douglas and Woodruff, 1981; Hessler and Thistle, 1975) and the Antarctic (Dell, 1972; Zinsmeister, 1979) faunas have undergone extensive evolution in place. Hessler and Thistle supported this hypothesis on the basis of an analysis of deep-sea isopod taxa, and Douglas and Woodruff infer it from fossil evidence.

For Antarctica, Zinsmeister (1979) proposed that the modern fauna was derived from the Late Cretaceous and Early Tertiary Weddellian Province that extended around Antarctica from at least the tip of South America and the Antarctic Peninsula to Australia. The modern fauna developed along Antarctica as that province was dismembered through continental fragmentation and climatic deterioration.

4. The two faunas, although similar in many respects, originated in other regions and subsequently invaded Antarctica (Dell, 1972) and the deep sea (Menzies *et al.*, 1973). Dell (1972) summarized zoogeographic evidence that some of the Antarctic fauna invaded the continent's shelves from South America via the Scotia Arc. Likewise, Menzies *et al.* (1973) suggested that a large number of deep-sea species moved down slope from the ancient Tethys seaway during the Cretaceous and Early Tertiary, and that these species' ancestors remain in the deep sea as relics.

## Age

The ages of the deep-sea and Antarctic faunas have been highly speculative because, until recently, there has been no good fossil evidence. The speculations have rested on systematic affinities and age of genera found today in the two areas.

For the Antarctic, two speculations have predominated (see Dell, 1972). One, that many of the organisms invaded the region in pre-Pleistocene times, probably during the Middle Tertiary when Antarctica began to cool; and, two, that the shallow-water element in the biota was eliminated during Pleistocene glaciations that presumably obliterated the shelf, and that this element of the modern biota reinvaded the shelves from deeper water. Based on fossil occurrences, Zinsmeister (1979) suggested that his Weddellian Province biota was in place on Antarctica in the Early Tertiary. The fossils, however, have little in common with the living fauna; for example, several species of sharks and rays are known from Eocene deposits on the Antarctic Peninsula (Welton and Zinsmeister, 1980), yet the modern fishes of the region include no sharks and only one rare ray (Daniels and Lipps, in press; DeWitt, 1971). Thus, Zinsmeister (1979) concluded that the modern Antarctic biota originated sometime in the later Tertiary.

For the deep-sea fauna, a variety of ages has been proposed, which Menzies *et al.* (1973) reviewed in detail. Some of these age estimates were made over a century ago (see Madsen, 1961). The faunal ages include: "very early" (Precambrian), suggesting that a fauna has always resided in the deep sea even though it may have been periodically wiped out by environmental changes on the bottom; the Paleozoic, based on similarities of generic morphology; the Cretaceous; various times during the Tertiary; and finally, the Pleistocene, as cool water taxa invaded during the ice ages from Antarctica. Among foraminifera, genera with at least three different age aspects about them exist in the deep sea: early Paleozoic agglutinated genera, pre-Cretaceous agglutinated and calcareous genera, and Eocene genera (Douglas and Woodruff, 1981). Douglas and Woodruff warn, however, that convergence in morphology may be an important factor in the similarities observed. For foraminiferal species, most have evolved in the deep sea since the beginning of the Miocene (Douglas and Woodruff, 1981). Thus, there are guesses and estimates made on various criteria that cover the entire range of geologic time.

## Evolution

Since the discovery that the fauna of the deep sea is highly diverse in terms of species (Hessler and Sanders, 1967), several hypotheses have been proposed to account for it. These hypotheses do not necessarily suggest that the species evolved in the deep sea, but suggest means by which numerous species may coexist. All of these hypotheses, nevertheless, imply or state that speciation does take place that gives rise to the high species diversity, and so we consider them here. For Antarctica, the problem of evolution or speciation processes has not been singled out for discussion, presumably because species diversity around Antarctica is considered relatively low. Four hypotheses have been proposed and discussed in the last decade or so.

1. Time-stability hypothesis (Sanders, 1968): Sanders (1968) proposed that high deep-sea species diversity was made possible by an environment physically stable over

periods of time long enough that species originating or immigrating into a community could become biologically accommodated to one another. Each species within such a community necessarily would have to occupy an increasingly narrow and specialized niche.

2. Biological disturbance (Dayton and Hessler, 1972): Under this hypothesis, high species diversity is the result of cropping pressure on smaller organisms that reduce competition between coexisting species even though they may have considerable niche overlap especially in food preference. Thus in the oligotrophic deep sea where diversity is very high, food is not limiting on smaller species because their population sizes are limited by cropping pressure of larger animals, which themselves may be restricted by food availability.

3. Trophic resource stability (Valentine, 1973): Using the high diversity of the deep sea for evidence, Valentine (1973) suggested that where trophic resources are stable through time, species compete strongly for food and species diversification through trophic specialization occurs. Where trophic resources fluctuate, species are opportunistic and thus generalists, few of which can be accommodated within a single community. This hypothesis singles out trophic resource stability from Sanders' more general hypothesis because energetic requirements seem a fundamental aspect of community structure (Valentine, 1973). Valentine (1973) also suggested that oligotrophic regimes should have higher species diversity as species are forced to specialize in order to acquire food, and that eutrophic regimes should have lower species diversity. For much of the deep sea, oligotrophic conditions predominate, and indeed these parts contain the most nutrient-poor, large community that exists on earth (Hessler and Jumars, 1974). Hessler and Thistle (1975) suggest that morphologic, behavioral, and physiologic adaptations to these very low food supplies must be an important means of speciation. A similar case can be made for Antarctic communities that depend on water column or benthic primary productivity during the winter. Suspension-feeding animals and grazers that must consume growing plant material are subjected to long periods of very low productivity when the sun is low or absent.

4. "Environmental grain" or microhabitat specialization (Jumars, 1975): Jumars (1975, 1976) found that deep-sea species diversity is controlled by processes operating on a scale of less than .01 $m^2$, an area approximating the activity sphere of a single macrofaunal individual. Such differences also seem to be persistent through some time, at least for foraminifera (Bernstein and Meador, 1979). Spatial heterogeneity, then, developed largely by biological interactions, allows speciation or immigration through microhabitat specialization.

# EVIDENCE

The hypotheses enumerated above can be tested in several ways. These include an examination of the fossil record of the environments, the geologic and oceanographic history of the environments, the systematic affinities of the biotas, and morphologic similarities in the environments regardless of phyletic relationships. The first two lines of evidence have a direct historical component while the latter two have an inferred historical component. These are important to keep in mind while trying to decipher the evolution of the faunas.

## The Fossil Record

**Antarctica** Fossil deposits are generally scarce in Antarctica. On the Antarctic Peninsula, Cretaceous and Early Tertiary deposits are known that contain marine fossils (Zinsmeister, 1979; Zinsmeister and Camacho, 1980). A single, glacial erratic boulder containing Early Tertiary gastropods is also known from Cape Crozier, Ross Island, in the Ross Sea sector of Antarctica (Hertlein, 1969). These assemblages, however, are unlike the modern Antarctic fauna. Middle and later Tertiary fossils are unknown from Antarctica, other than in drill holes in the Ross Sea (Hayes *et al.,* 1975) and under the Ross Ice Shelf (Webb *et al.,* 1979) and from a locality in the Wright Valley on the west side of McMurdo Sound. From the Deep Sea Drilling Project (DSDP) sites in the Ross Sea, scanty late-Oligocene through mid- to late-Miocene mollusks were found. As a whole, these mollusks indicate somewhat warmer seas, as a more northerly element is present, but also that the Antarctic fauna was established, at least in part, by the Middle Tertiary (Dell and Fleming, 1975). Microfossil assemblages from possibly early- and middle-Miocene sediments cored under the Ross Ice Shelf, several hundred kilometers south of the DSDP sites, contain a large number of diatoms (Brady and Martin, 1979), yet the sediments are of glacial origin (Webb *et al.,* 1979). These facts suggest that Antarctica was indeed partially glaciated although the surrounding seas were at least seasonally free of solid ice-cover. Pliocene fossils from the Wright Valley, adjacent to the Ross Sea, contain species that are, for the most part, present in the Recent fauna of the area (Webb, 1974).

Quaternary fossils have been cored from various places around Antarctica, but, other than distributional changes, are closely similar to the Recent assemblages. Thus the critical change to a modern type Antarctic marine fauna occurred during the Oligocene or Miocene, times that are not, or are poorly, documented by fossils.

**Deep sea** The fossil record of the deep sea is still obscure because larger fossils have not been commonly recovered or identified from known deep-sea deposits. These deposits may exist in subduction complexes now exposed on continents, but the faunas have not been subjected to analyses that would suggest their deep-sea origin, or they represent marginal marine basins. For example, one of us has described gastropod assemblages from the Oregon Paleogene that are of deep-water origin, but these were most likely deposited in marginal basins (Hickman, 1976, 1980a) that were not characteristic of the deep sea generally. Similarly, foraminifera from marginal deep-water basins (see McDougall, 1980, for a western North American example) are also unlikely to be typical of truly deep-sea situations. The record of the deep sea must therefore be recovered from cores taken in the deep-sea regions, and larger fossils are not commonly retrieved in cores. One good Lower Cretaceous deep-sea mollusk assemblage, however, is reported from the central Pacific by Kauffman (1976). Genera contained in this assemblage are identical or related to modern deep-sea genera. Foraminifera provide the only good record of the deep-sea fauna through the past 150 million years, although even knowledge of them must be considered rudimentary. A good summary of the existing data is given by Douglas and Woodruff (1981).

The evolution of foraminifera shows two patterns, one at the generic level and another at the specific level (Douglas and Woodruff, 1981). Genera of foraminifera are divided into three age groups by marked changes in generic composition in the mid-Cretaceous and the Eocene. Before the mid-Cretaceous (approximately 95 mya), an assemblage is known from the Atlantic, Pacific, and Indian Oceans that represents genera of Paleozoic or early Mesozoic age. During the mid-Cretaceous the number of genera increased threefold among calcareous genera (excluding nodosariids). These genera were largely additions to the deep-sea assemblages, with the pre-existing older genera persisting as well. This mid-Cretaceous complex of genera continued into the Early Tertiary essentially unaffected by the terminal Cretaceous extinctions that decimated planktonic and some shallow benthic foraminiferal and invertebrate faunas. From the end of the Paleocene through the Eocene, many new genera originated but they seem to have replaced genera already present. Thus the generic diversity of deep-sea foraminifera has increased through time, and genera representing each of the three time intervals (pre-Middle Cretaceous, Late Cretaceous–Early Tertiary, and post-Eocene) are still present today (Fig. 14-1). Of the modern calcareous genera, about 20% remain from the pre-Middle Cretaceous, 25% from the later Cretaceous, and nearly 50% from the Tertiary, especially the Eocene. Agglutinated genera, on the other hand, appear to have resided in the deep sea since the early part of the Mesozoic or even the Paleozoic.

Species evolution shows a different pattern (Douglas and Woodruff, 1981). In the Cenozoic of the Pacific Ocean, three time-stratigraphic groups of species are present in the Paleocene–mid-Eocene, mid-Eocene to late Oligocene, and the early Miocene and later. Within each of these time-stratigraphic groups, considerable evolutionary change took place, involving as many as 60% of the species at one time during the Eocene. Species living in the deep sea today mostly originated during the post-early Miocene period, and thus almost all species are less than 20 million years old. Ostracodes from the deep sea show a similar trend, with a Late Cretaceous–Early Tertiary group of species which became extinct at the end of the Eocene, and an Oligocene and later group, many of which still live in the deep sea (Benson, 1975).

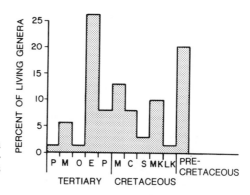

Fig. 14-1.    Percentage of living deep-sea foraminiferal genera originating during various times in the past. Data from Douglas and Woodruff (1981).

## Geologic and Oceanographic History

The histories of the modern Antarctic and deep-sea biotas are closely linked because of the oceanographic similarities and potential faunal pathways between the two regions. The deep-sea waters are mostly derived from around Antarctica, and the water characteristics are therefore related. Of importance to our consideration of the evolution and relationships between the two faunas is the timing of the initial formation of Antarctic bottom water and its subsequent history. Other sources of deep bottom water may have dominated prior to the onset of the modern-style, deep-water circulation.

Oxygen isotopic temperature estimates have been determined for Antarctic deep water and Pacific deep bottom water back to at least the Cretaceous. These paleo-temperatures provide a basis for a reasonable history (Kennett, 1977; Savin, 1977). Figure 14-2 shows bottom-water paleotemperatures (in terms of modern temperatures; Tertiary temperatures may have been several degrees cooler) for the subantarctic region. As these temperatures occurred in waters that fed the deep sea elsewhere in the world, bottomwater temperatures parallel major features (see Douglas and Woodruff, 1981, for a recent compilation), but may differ in detail.

During the Cretaceous, Antarctica, Australia, and probably the southern part of South America were joined closely enough to prevent deep-water and probably even shallow-water circulation between them. Southern hemisphere circulation was characterized by a component of north-south currents that insured mixing of low and high latitude water, providing a more homogeneous horizontal and vertical thermal gradient. In the mid-Cretaceous, the deep sea apparently became anoxic (Schlanger and Jenkyns, 1976). The bottom waters began to cool from temperatures greater than 15°C, to perhaps as low as 10°C at the end of the Cretaceous and the early Paleocene. In the late Paleocene and early Eocene temperatures again rose to about 15°C (Fig. 14-2), but soon began to cool as Australia began to separate from Antarctica and move northward. At the close of the Eocene, some 38 mya, temperatures dropped about five degrees in Antarctic and deep-sea waters. This cooling produced glacial conditions throughout Antarctica and allowed sea ice to form in adjacent seas (Fig. 14-2; Kennett, 1977, 1978). Antarctic seas were cool enough at this time to permit the initiation of thermohaline deep circulation. No circum-Antarctic currents existed between Australia and Antarctica, although they were well separated, and there were no ice caps on Antarctica. By the later Oligocene, the circum-Antarctic current was well developed, and sea ice and ice-rafting were common (Kennett, 1978). A slight warming occurred during the early Miocene, followed by another decrease in Antarctic and deep-sea temperatures. Antarctica was surrounded by pack ice but not by permanent ice shelves (Webb *et al.*, 1979). Middle Miocene temperatures began to decrease markedly about 16 mya followed by a 3-my period of cyclic temperature fluctuations, similar to the Pleistocene record (Woodruff *et al.*, 1981). During this period, the modern glacial mode developed on the earth, with the emplacement of an ice cap on Antarctica, perhaps of ice shelves surrounding the continent, and with the intensification of circulation and cooling of deep bottom water. The cooling and associated effects continued to the Recent with another period of cyclicity between warming and cooling in the Pleistocene. The isotopic paleotemperatures are somewhat controversial for ages later than mid-Miocene because variations in ice volume also affect the ratio of oxygen isotopes (Kennett, 1977; Woodruff *et al.*, 1981).

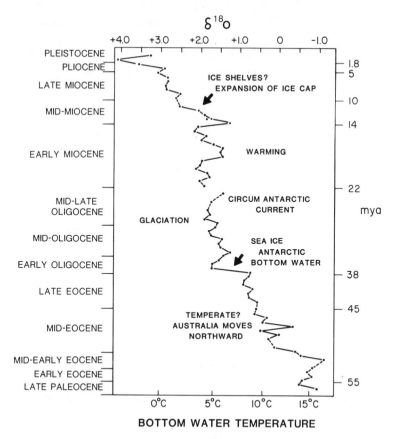

$\delta^{18}o$

BOTTOM WATER TEMPERATURE

Fig. 14-2.    Oxygen isotopic values of benthonic foraminifera and early and middle Cenozoic paleotemperatures for the Subantarctic south of Australia. Major paleoclimatic and geologic events in the Antarctic region are shown at the approximate times they occurred; the most significant cooling events are marked by the dark arrows. Prior to the middle Miocene the oxygen isotope values are reasonably equated with paleotemperatures, but after that time, the effects of ice volume become significant and the values are not related solely to temperature. Various workers interpret this part of the isotopic curve differently (Kennett, 1977; Woodruff, *et al.*, 1981), but all agree that cooling of Antarctic and deep-bottom waters continued with fluctuations to modern values. Modified from Shackleton and Kennett, 1975.

## Systematic Affinities

The inferred phylogenetic relations and geologic age of related taxa have provided indirect evidence bearing on the questions of age, origin, and evolution of deep-sea faunas, in particular. Antarctic assemblages have not always been treated this way, but usually an "old" element is noted together with species that are derivable from nearby regions or that evolved in place (see Dell, 1972).

Menzies *et al.,* (1973) reviewed previous phylogenetic inferences for deep-sea organisms. In general, species living in the deep sea today belong to taxa that have very old (Paleozoic or early Mesozoic), late Mesozoic, or Tertiary origins (see also Madsen,

1961). Clarke (1962), however, concluded that abyssal mollusks, at least, were of relatively recent origin.

A very good example of inference of origin from phylogeny is Hessler and Thistle's (1975) analysis of paraselloidean iospods. Similar reasoning has been attempted for other groups as well; these are reviewed by Menzies *et al.* (1973) and Madsen (1961). Among the isopods, all but one of the 13 primarily deep-sea families lack eyes, blindness being an adaptation to the deep sea (Hessler and Thistle, 1975). The shallow-water representatives are also blind, even though eyes would be of adaptive value, thus suggesting a migration of some taxa into shallow water from a diversification center in the deep sea. Hessler and Thistle (1975) note that among the Ilyarachnidae, the most primitive morphologies in the family are in species that are distributed in the deep sea, and that have given rise to other species there, indicating a deep-sea origin and radiation of the family (except for a few species that invaded shallow polar regions). The ultimate origin of these isopods is inferred to be shallow waters because "the presence of eyes in the most primitive, shallow-water families demands an initial shallow-water origin" (Hessler and Thistle, 1975). No age estimate for these events was given by Hessler and Thistle, other than a suggestion that the morphology of deep-sea families indicated a long evolutionary history.

## Common Characteristics and Morphologies

Antarctic and deep-sea organisms display a number of life-history, physiological, and ecological characteristics in common. Aberrant morphologies also are shared by numerous species from both environments—in particular, gigantism and dwarfing are common. Recognition and understanding of these phenomena has been retarded in part by searching for *causal* explanations.

**Common species characteristics** Life-history characteristics apparently common in both the deep sea and in shallow Antarctic benthos include:

1. Low reproductive potential, with eggs few in number, relatively large, and yolkrich;
2. High incidence of direct development, brooding, and/or vivipary;
3. Slow gametogenesis;
4. Semelparous reproduction;
5. Late reproductive maturity;
6. Low growth rates;
7. Indeterminate growth and high longevity;
8. Slow embryonic development; and
9. Reduced gonadal volume.

Physiological characteristics include:

1. Low basal metabolic rates as measured by $O_2$ uptake; and
2. Low activity levels, lethargy, and vertical orientation.

Ecological characteristics, resulting in part from life-history traits and physiological characteristics, include:

1. Low rates of dispersal,
2. Slow rates of colonization,
3. Low population densities, and
4. Lowered mortality as a result of low predation pressure.

Low growth rates are well documented as a common life-history feature. Slow growth in the deep sea has been reported by Tipper (1968), Sanders and Allen (1973), and Turekian *et al.* (1975). For the Antarctic, Arnaud (1977) and Clarke (1979, 1980) report slow growth in a variety of invertebrates. Most of the data are for mollusks (Everson, 1977; Picken, 1979b; Seager, 1979) or for the more abundant species, such as the starfish *Odontaster* (Pearse, 1965). Low growth rates in the Antarctic may have a strong seasonal component in some species (Clarke, 1980) which is related to food availability.

Other life-history characteristics may be less universal than slow growth. Scheltema (1972) has, for example, indicated a wide range of reproductive potentials in deep-sea invertebrates even though they are low for the majority of species (Grassle, 1978). One exception to the predominant pattern of low reproductive potential is an opportunistic deep-sea bivalve (Turner, 1973) associated with wood, a food source of terrestrial origin that is most abundant in the more eutrophic deep-sea settings.

Generalizations about the frequency and timing of reproduction are difficult to make. Rokop (1974) found continuous reproduction in 9 out of 11 species that he examined from bathyal depths and concluded that this is a more common pattern in the deep sea than the synchronous annual cycles typical of many shallow-water invertebrates. This conclusion stands in marked contrast to the suggestion by Allen (1979) that late-maturing deep-sea bivalves may be semelparous, reproducing only once. In the shallow-water Antarctic the pattern is clearer, with ample evidence that reproduction is seasonal and corresponds with maximum primary production (Arnaud, 1977; Pearse, 1965).

Direct development and brooding as correlates of reduced reproductive effort appear to be more universal in Antarctica than in the deep sea and are coupled with slow gametogenesis and slow embryonic development (Clarke, 1980). Documentation of brooding and nonpelagic development for a variety of Antarctic marine invertebrates is provided by Thorson (1936, 1950) and Dell (1972), for bivalves by Soot-Ryen (1951), for gastropods by Picken (1979a), and for amphipods by Bone (1972), Bregazzi (1972), and Thurston (1972). In the deep sea, direct development and brood protection are common (Knudsen, 1970, 1979), but there is increasing evidence for at least a brief pelagic larval stage in many species (Bouchet and Waren, 1979).

Data on physiological adaptations are less numerous. There are a number of lines of evidence suggesting lowered basal metabolic rates in both shallow-water Antarctic and deep-sea organisms. The most convincing evidence comes from *in situ* measurements of oxygen uptake. Smith and Teal (1973) demonstrated reduced rates of respiration by meiofauna in abyssal sediments, and Smith and Hessler (1974) documented lowered rates of respiration in deep-sea fish. The rates of metabolic activities related to growth (particularly where calcium carbonate is concerned), feeding, and reproduction appear to be low, but they are difficult to assess in terms of their relative importance. Some organisms may be consistently lethargic, while others may alternate between periods of very low and normal metabolic activity.

In shallow-water Antarctic settings, lethargy and presumed drastic reduction in metabolic rates may be even more widespread, but as a seasonal phenomenon only (Barham, 1971). This Antarctic lethargy has been characterized by Arnaud (1977) as a vegetative state resembling terrestrial "hibernation" in response to winter oligotrophy.

Ecological characteristics are also difficult to document, although there are experimental data to confirm low rates of colonization in the deep sea (Grassle, 1977). Low dispersal and colonization rates are to be expected from the generally low rates of population process and the tendency toward direct development discussed above.

**Gigantism and bizarre morphologies** Gigantism, together with bizarreness, and dwarfing are two striking features of the Antarctic and deep-sea biotas. We focus here on the details of form and pattern, and not on evolutionary and ecologic theory, in order to provide better understanding and appreciation of the morphologies (Hickman, 1980b).

Gigantism implies larger than normal size. It is generally thought of in terms of achieving a large body volume relative to surface area and is interpreted as an adaptation for retaining heat (Bergman's Rule; Bergman, 1847). However, size measurements of both vertebrates and invertebrates are recorded as linear measurements. In these terms, size increase is easily achieved in other ways, although some of the logical possibilities are biologically impossible and others are bizarre. Bizarreness implies some standard of normalcy. Organisms that are termed bizarre usually deviate from the standard body plans or morphologies of familiar or typical living relatives. They traditionally have been considered holdovers from the geologic past (living fossils) or, alternatively, as evolutionarily highly specialized organisms.

Six constructional mechanisms for attaining and maintaining large size recur in both the deep-sea and shallow Antarctic biotas: (1) Elongation of appendages and sense organs, (2) extreme flattening, (3) mineralized lattice construction, (4) stalk elongation, (5) agglutinization, and (6) big bag construction.

1. Elongation of appendages and sense organs: An organism can increase its effective size with relatively little investment in building materials through development of long, thin appendages or processes, such as walking legs, sense organs, or spines. These can be arrayed in such a manner as to occupy three-dimensional space with very little biomass. Pycnogonids provide an excellent example of this form of construction (Fig. 14–3). They are primarily small and inconspicuous animals, the exceptions occurring in the deep sea and Antarctica (Dell, 1972; Hedgpeth, 1969). Gigantism is, however, measured in terms of occupation of space rather than extraordinary biomass. The abyssal pycnogonid *Colossendeis colossea* may have a leg span of 60 cm, but the body is little more than 5 cm in length, and the proboscis and trunk are of the same small-diameter construction as the long spindly legs. Giant pycnogonids are bizarre because they occupy a great deal of three-dimensional space with virtually nothing. Pycnogonids are also notably slow-moving organisms.

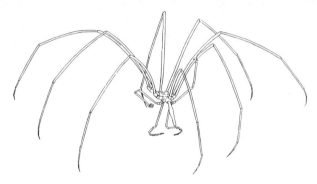

Fig. 14-3.    Appendage elongation. Gigantism resulting from leg elongation in the deep-sea pycnogonid *Colossendeis colossea*. Drawing by J. W. Hedgpeth from underwater photographs and living specimens.

Elongation of legs and sense organs also occurs in a number of deep-sea crustaceans (Wolff, 1961, 1962). Perhaps the most remarkable examples are the elongation of antennae and third and fourth pairs of walking legs in benthic aselloted isopods (Fig. 14-4). In *Munopsis latifrons* the walking legs are 116 mm long, more than seven times longer than the body. Other examples of elongation of legs and sensory appendages occur in deep-sea amphipods (Wolff, 1962).

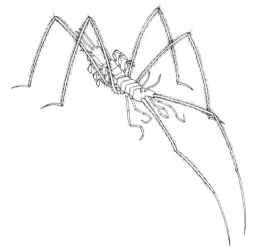

Fig. 14-4.    Appendage elongation. An aselloted isopod of the genus *Munopsis* in which large size is attained through elongation of walking legs and antennae. Redrawn from Wolff, 1961.

2. Extreme flattening: Organisms may also increase effective size, at the expense of volume, by flattening or expanding to occupy a greater amount of two-dimensional space. For organisms living on fine-grained soft substrates, flattening may serve the additional function of increasing surface area and minimizing the potential for sinking into the substrate. Isopods have this morphology, producing some of the most pro-

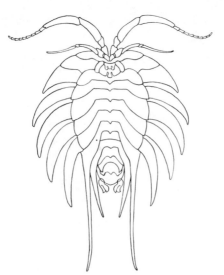

Fig. 14-5. Extreme flattening. Gigantism in the isopod *Serolis*, achieved through dorso-ventral flattening.

nounced examples of dorso-ventral flattening (Fig. 14–5). Gigantism is frequently accompanied by the development of spinosity, which further serves to increase effective size economically in terms of linear measurements.

3. Mineralized lattice construction: By constructing a mineralized skeleton, an organism can achieve large size with a one-time energetic investment that requires little or no maintenance. This cannot be accomplished in very cold water with calcium carbonate, but large size can be attained using silica lattices (Fig. 14-6), both in the deep sea and shallow Antarctic. Hexactinellid sponges are the dominant organisms in many regions of the abyssal ocean floor, and the largest individuals may exceed 2m in height. The

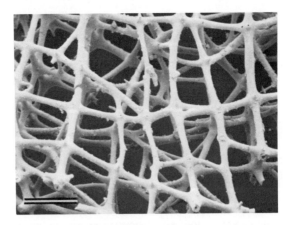

Fig. 14-6. Mineralized lattice construction. Scanning electron micrograph of a portion of the spicular lattice of a late Eocene deep-water hexactinellid sponge. Bar is 400 μm.

lattice-work of spicules is not a solid lattice, but rather a delicate mesh that defines the walls of a cylinder or bowl. The cylinder or bowl describes a volume that is occupied by water (Fig. 14-7). Large glass sponges are particularly conspicuous in the Antarctic, both in terms of their number and large size (Burton, 1929; Dell, 1972; Koltun, 1970), with gigantism most pronounced in members of the family Rossellidae.

Fig. 14-7. Mineralized lattice construction. A hexactinellid sponge in which a delicate spicular lattice describes a bowl-shaped space with a light-weight, low-energy skeletal construction.

Glass sponges in these cold environments are not only using a building material that is not in limited supply, but they are producing a relatively open, light superstructure that, once produced, does not require maintenance in the same fashion as calcium carbonate. Even long after death and mechanical destruction of the superstructure, there is a persistence of the spicules. Thick mats of sponge spicules form a unique benthic substrate in some environments and are noted for their rich infauna (Dayton *et al.*, 1970).

Indeterminate growth and great longevity may contribute to the large size that can be attained through this mode of construction and architecture.

4. Stalk elongation: Many passive suspension-feeding organisms have created a large effective size or length by developing an elongate slender stalk that generally serves to elevate the feeding mechanism above the sediment-water interface where the potential for clogging may be high. Such elevation may serve the additional function of placing the organism in the ambient current regime for passing the maximum amount of water over the feeding mechanism (Vogel, 1974; Vogel and Bretz, 1972). Long slender stalks are particularly common in deep-water siliceous sponges (Fig. 14-8). They are light-weight constructions, up to a meter in length, consisting of long, fine siliceous threads that run parallel or are slightly coiled.

5. Agglutinization: Several groups of marine organisms utilize prefabricated building blocks by removing particles from the sedimentary environment or from the water column and incorporating them into their skeletons. In both the deep sea and Antarctica there is a paucity of calcareous foraminifera and an increase in the prominence of ag-

Fig. 14-8. Stalk elongation. A generalized stalked hexactinellid sponge in which size increase of the basic body plan is achieved through elongation on a siliceous stalk.

glutinated forms. Furthermore, many of the agglutinated foraminifera are able to achieve large size with this constructional mode.

Two large agglutinated foraminifera (Fig. 14-9) from Antarctica exhibit a particularly bizarre arborescent morphology (DeLaca *et al.,* 1980; Lipps and DeLaca, 1980). It is striking not only in that the portion of the branching complex extending above the sediment-water interface may contain very little protoplasm, but also in the life-history characteristics that appear to be associated with it. Growth is apparently very slow and longevity extraordinarily great for a unicellular organism; metabolic rates are low; and

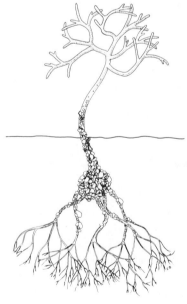

Fig. 14-9. Agglutinization and gigantism in foraminifera. *Notodendrodes antarctikos* from oligotrophic shallow-water environments in McMurdo Sound, Antarctica. The test contains little protoplasm in either the branches in the water or the roots in the sediment. Drawn from preserved specimens by Ellen Bailey.

Fig. 14-10. Agglutinization in xenophyophorians. The deep-sea *Ammolynthus haliphysema* utilizes foraminiferal tests to construct its wall. The species is 15–20 mm in length and 5–8 mm in width. Redrawn from Tendal, 1972.

dissolved organic matter is incorporated directly and may represent a substantial nutritional contribution (DeLaca *et al.*, 1981).

Use of prefabricated building blocks and gigantism are also combined by agglutinating rhizopods in the deep sea, where protozoans are not only dominant members of the meiofauna (Hessler, 1974; Thiel, 1975; Wolff, 1977), but two groups, the xenophyophores (Fig. 14-10) and komokiaceans (Fig. 14-11), are conspicuous members of the macrofauna.

Xenophyophores are large not only in the sense that the agglutinated test may be several centimeters in diameter, but also in pseudopodial arrays extending 10–15 cm (Lemche *et al.*, 1976; Tendal, 1972). The Komokiacea, dendritic or anastamosing clusters of agglutinated tubules (Fig. 14-11), are abundant in hadal trenches and oligotrophic settings in the abyssal North Pacific, although they were overlooked for many years (Tendal and Hessler, 1977). Agglutinated foraminifera are also dominant members of the smaller meiofauna in many portions of the deep sea, and slow growth, increased longevity, and low metabolic rates probably will also be associated consistently with this constructional mode.

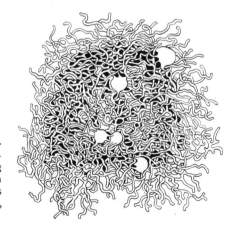

Fig. 14-11. Agglutinization. A giant deep-sea komokiacean foraminifera, *Lena neglecta*, in which dendritic and anastomosing tubules form a large low-cost construction from inorganic prefabricated building blocks (clay). Redrawn from Tendal and Hessler, 1977.

**6.** Big bag construction: A sixth constructional method for attaining large sizes is to assume the form of a gelatinous or mucus bag that may be filled primarily with water. Holothurians, which are a conspicuous element of the deep-sea megafauna, are able to attain large size through use of the gut cavity as a fluid internal skeleton. The ultimate in this mode of construction is described by Barnes *et al.* (1976) for the deep-sea elasipod holothurian, *Peniagone diaphana* (Fig. 14-12). These are translucent, delicate free-swimming forms that orient vertically and live in large populations, oral tentacles hanging downward, in a lethargic state near the bottom. Barnes *et al.* (1976) conclude that metabolic rate is low. Observations of *P. diaphana* from a submersible confirm that it is a slow-moving organism, even when disturbed. Analyses of chemical composition and caloric content further indicate that organic matter is extremely low in this animal.

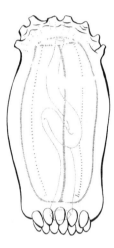

Fig. 14-12. Big bag construction. The translucent deep-sea elapsoid holothurian, *Peniagone diaphana*, in which size is gigantic relative to caloric content. Drawn in swimming-feeding position with the oral tentacles and mouth down (see Barnes *et al.*, 1976).

Like *Peniagone,* a number of lethargic pelagic organisms in Antarctica and the deep sea assume a vertical orientation (Barham, 1971—fish; Roper, 1981[1]—squid). Lethargic, vertically orienting deep-sea squid have been observed in aquaria in both head-up and head-down positions, although head-down orientation is the only one that has been observed in nature. Many of the deep-sea forms resemble *Peniagone* further in that they are gelatinous, transparent, and sac-like in construction. They are presumably also of relatively low caloric value.[2]

Another bizarre form of large structure to which the term "giant" has been applied is the so-called "feeding house" constructed by deep-dwelling planktonic larvaceans (Barham, 1979). In the deep-sea, larvacean feeding structures are extraordinarily large (up to 100 cm across), although they are exceptionally fragile "filmy sacs" of gelatinous or mucus construction. The animal itself is an order of magnitude smaller than its mucus house. But the term "giant" remains appropriate for considering the elaborate apparatus with its intricate intake screens and internal filters, and mucus is a constructional medium that can define a large space at relatively low cost.

[1] C. F. E. Roper, written communication, 1981.
[2] *Ibid.*

These constructional mechanisms fall outside of the factors that are normally considered as determinants of size, because size increase is traditionally conceived of as simply enlarging an organism without drastically altering its body plan. In this traditional sense, the size of any organism results from a complex interaction of internal and external factors. At the level of genetic and physiological internal factors, large size may be achieved either by increasing the rate of growth or by increasing longevity. Maximal size will be obtained when these two factors are combined by an organism with indeterminate growth. External environmental factors that may promote size increase for a given morphological plan are increased temperature, which tends to increase metabolic rates; unlimited and unfluctuating availability of resources required for growth, feeding, and reproduction; and low predation pressure or insurance that predation will not be the cause of death.

**Small body size** Small adult body size is also common in many organisms in both shallow Antarctic and deep-sea environments. In both of these environmental settings there are two distinct types of small organisms. The first type includes an extraordinary diversity of benthic interstitial organisms that are normally very small (less than 1 mm in maximum diameter), the members of the meiofauna. The second group, and the group of primary interest in this discussion, consists of organisms that are dwarfs. As adults, these organisms are much smaller than is characteristic of their close relatives in other environments.

Discussions of the biology of the deep sea repeatedly emphasize the predominance of small organisms (Allen, 1979; Grassle, 1978; Sanders *et al.*, 1965), and Thiel (1975) characterized the deep sea as a "small organism habitat." Intensive sampling of the deep sea with anchor dredge (Sanders *et al.*, 1965), epibenthic sled (Hessler and Sanders, 1967), and box corer (Hessler and Jumars, 1974) followed by careful processing through fine mesh screens and sieves revealed that the high diversity in the deep sea resides primarily in the meiofauna. Meiofaunal organisms predominate in part because macrofaunal taxa drop out with increasing depth. There is no indication that the deep-sea meiofauna is an ecologically stunted or miniaturized macrofauna or that it differs substantially in its size structure from the shallow-water meiofauna.

Of more immediate interest from a morphological point of view are the organisms in the deep sea and Antarctica that are dwarfed. We define dwarfs as organisms which fail to reach an adult size that is at least half the mean size for the group to which it belongs (De Broyer, 1977). It may occur either as a plastic (purely phenotypic) or an evolutionarily fixed (genotypic) response achieved in a variety of ways ranging from lowered metabolism and slow growth at one extreme to precocious maturation through progenesis at the other extreme. Dwarfing in the deep-sea and shallow Antarctic waters occurs primarily in benthic organisms that secrete calcium carbonate. It is these calcareous dwarfs that we will consider in more detail.

Dwarfing in the deep sea and Antarctica requires a great deal of additional study in order to document the ways in which small size is accomplished, particularly in view of the apparent differences between dwarfs in these environments and the kinds of dwarfing that have been studied better in other kinds of organisms and environmental settings. Mammalogists, for example, explore the phenomenon from an allometric point of view, with size, shape, and time as the sources of data. Patterns of dwarfing allometry within many mammalian lineages tend to follow the reverse pattern of developmental allometry within species (Marshall and Corruccini, 1978), implicating paedomorphosis (Gould, 1975), but not excluding the possibility that rapid dwarfing may involve a predominantly plastic component (Bonner, 1968).

For marine invertebrates, there are a number of morphological aspects of dwarfing that are more striking and which demand more immediate attention than simple allometric relationships. Dwarfing as an adaptation (rather than a plastic response to unfavorable conditions) may involve morphological manifestations of simplification or degeneration, such as loss of gills or digestive system. It may also be achieved through "miniaturization." The term "miniature" has been employed in a variety of ways, usually implying that the adult is a scaled-down replica of the normal adult and sometimes implying that it has all the same structures as the normal adult. One way to accomplish miniaturization, below the level of gross form, is through reduction in cell size. Sanders and Allen (1973) have shown that some of the most minute bivalves in the deep sea (species of the protobranch genus *Microgloma,* which attain maximum adult lengths of 1.1 mm) are morphologically complex miniatures in which size reduction is indeed accomplished through reduction in cell size. It is not known, however, how common this phenomenon is in dwarf bivalves.

Even where size reduction does not occur, there may be a reduction in biomass. Oliver (1979) has shown that abyssal species of the bivalve genus *Limopsis* have a small visceral mass relative to the size of the mantle cavity and gills. In these species there is a reduction of the biomass/volume ratio. Again, we do not know how common this phenomenon is.

In both Antarctica and the deep sea, dwarfing in the bivalves has been documented most thoroughly. Although there are some exceptionally large bivalves in the deep sea, most do not exceed 5 mm in maximum adult size (Allen, 1979). Nicol (1966, 1967, 1970) and Dell (1972) have commented on the small size of bivalves in the modern shallow-water Antarctic fauna. In the Antarctic, 61% of the bivalves fail to attain maximum lengths or heights greater than 10 mm. Furthermore, the shells are generally thin and lacking in spines or conspicuous ornamentation. Growth lines are closely spaced, suggesting slow growth.

Arnaud (1977) noted that Antarctic calcareous organisms as a group are always exceedingly fragile and that dwarfing occurs primarily in the calcareous foraminifera, prosobranch gastropods, bivalves, scaphopods, and brachiopods.


## TESTING THE HYPOTHESES


We now use the evidence above to test the hypotheses enumerated earlier. Obviously, the evidence is incomplete, partly because it does not exist and partly because it is difficult to acquire from these remote environments. Thus, we cannot test rigorously these hypotheses, but we can restrict choices and direct further inquiry, a good strategy when pursuing high-cost, logistically difficult research.

# Origins

None of the four hypotheses to account for the origin of the deep-sea or Antarctic marine faunas can be eliminated. Indeed, fossil evidence and phylogentic inference for both environments suggests that the ancestors of some of the modern species were established in the areas before the Cretaceous, some 150 mya, and probably even much earlier. The inferred relationships of modern isopods suggests that the deep-sea species were derived ultimately from shallow-water taxa long ago, but that the majority of them alive today radiated in the deep sea itself. The foraminiferal fossil record indicates also that species diversification took place largely in the deep sea, with little immigration from shallower areas, at least since the mid-Cretaceous. For bivalves, Knudsen (1979) concluded also that the deep sea was "not a refugium for primitive forms which have been expelled from more dynamic shallow-water habitats," but was a site of active bivalve speciation. Even taxa of foraminifera that appear to be very ancient (pre-Cretaceous) may be convergent, and thus not be indicative of origin.

The very fragmentary fossil data for Antarctica indicate that the modern fauna is derived largely from a preexisting provincial assemblage. Some modern elements of the Antarctic fauna migrated into the region through the Scotia Arc from South America, and some emerged from the deep sea.

In summary, the question of the site or sites of origin for the deep-sea and Antarctic faunas has rated too much discussion, probably because each fauna can be neatly circumscribed by its geographical isolation and uniqueness. But like any fauna anywhere, elements of deep-sea and Antarctic faunas have accumulated by several processes, and these require understanding. The majority of species in both environments seem to have originated in place, a far smaller number of species migrated between the two regions in one direction or the other, and probably still fewer immigrated from other nearby sites. These conclusions suggest to us that study of the adaptations of the organisms and their interactions will be a more fruitful line of inquiry in the future.

# Age

Hypotheses for a single age of the Antarctic and deep-sea faunas can be eliminated. In each area, fossil evidence shows that faunas have been present in those regions for a very long time, and that there have been several periods of diversification and/or change. This record in combination with the geologic and oceanographic evidence indicates that both environments have been very dynamic places for at least the last 100 million years. Neither one has been a stable, unchanging environment. The faunas have responded to these changes in subtle ways. Like the world's marine fauna in general, the modern species composition was established relatively recently, since the beginning of the Miocene some 22 mya.

Generic antiquity is variable for both regions, including a rather ancient element, a Late Cretaceous–Early Tertiary element, and a later Tertiary element. The ancient-appearing elements may be convergent themselves, and not identical with modern taxa.

Detailed morphologic examination of critical groups has not been completed, so that this possibility cannot be resolved.


## Evolution

Evolutionary processes in the deep sea and the Antarctic are at least as complex as in any other region. We do not believe that special processes need be called upon in these environmental settings. None of the hypotheses we reviewed above can be eliminated, because each probably is true to a degree. We can eliminate certain older ideas; for example, that a single factor such as cold, pressure, or oligotrophy alone will account for the fossil records and/or modern faunas.

The single factors can be eliminated because the evolutionary similarities observed in the two areas do not correlate uniformly with them. For example, abyssal gigantism in isopods has been attributed to physiological responses to pronounced hydrostatic pressure (see discussion by Menzies *et al.,* 1973, p. 238); however, there are no data to demonstrate that any of the observed patterns are a direct physiological response to pressure. In fact, Menzies and George (1967) indicate that the largest isopods in the genus *Storthyngura* live in environments of moderate hydrostatic pressure. Gigantism in shallow-water Antarctic environments further suggests a different underlying control. Neither does the fossil record of foraminifera show a change in the rate of total or net faunal change with the rather large and abrupt temperature changes documented for the Cenozoic (Douglas and Woodruff, 1981). Pressure can be eliminated as a common factor because the shallow Antarctic and the deep sea vary significantly; cold can be eliminated because the diversity seen in the deep sea is paralleled in shallow tropical regions, foraminifera in the deep sea do not respond to it *per se,* and the diversity is not matched in Antarctic waters; and oligotrophy is not a common factor either because at least part of the Antarctic ecosystem (detrital feeders) is adapted to eutrophy on a year-round basis. These factors, and others, may be important but their roles are complex and involve other attributes of the environments as well.

We can eliminate a certain aspect of the time-stability hypothesis. The time aspect of this hypothesis is inferred to involve long periods of geologic time. The data from the foraminiferal fossil record suggest, however, that as major physico-chemical changes took place on the sea floor in response to changes in the southern hemisphere, species originations increased (Douglas and Woodruff, 1981). There appears to be little time lag between these phenomena, suggesting that long geologic periods of time are not required, at least for foraminifera. Times required for speciation are on the order of thousands, not millions, of years. Indeed, Grassle and Sanders (1973) made just such a prediction with respect to the time-stability hypothesis.

However, in a general sense, the time-stability hypothesis must have some validity. Certainly, time is required for evolutionary change, and the environment must remain relatively stable (predictable) within that time period. The fossil record of foraminifera from the deep sea, thus, does not disprove the time-stability hypothesis because the oceanographic instabilities recorded there took place over too great a time period relative to the dynamics of speciation.

Although this hypothesis may predict the conditions necessary for the development of numerous species, it does not address the problem of what mechanisms might be important in the speciation process. The other three hypotheses concerning trophic resource regimes, spatial heterogeneity, and disturbance do address the problem of how species may accommodate themselves within a predictable environment. In the testing and development of these kinds of hypotheses, natural history observations are important (Bernstein et al., 1978), and morphology and life history bear strongly on such observations. We cannot disprove any of these hypotheses, but each predicts (Grassle and Sanders, 1973) certain morphologic and life-history characteristics that we can examine.

Disturbance may play a role in speciation processes in the deep sea, but the life-history and morphologic characteristics of longevity, small brood size, a predominance of older individuals in populations, the persistence of patch structure, and the survival of delicate structures such as tubes, indicate that this mechanism is not of primary importance in speciation. A continual disturbance model would predict just the opposite characteristics.

Spatial heterogeneity seems to be very important (Bernstein and Meador, 1979; Jumars, 1975, 1976). The hypothesis predicts that substrate utilization should be great and diverse. Numerous examples of spatial utilization of a variety of permanent and ephemeral structures (foraminiferal tests, manganese nodules, fecal pellets) are known in the deep sea (Jumars, 1976). Size is important in the use of space. Small or dwarf organisms may more effectively use unique substrates and microhabitats. The small limpets discussed below are characteristically associated with wood, turtle grass, eel grass, and a variety of other plant material of terrestrial or shallow marine origin. Squid beaks form local accumulations in some parts of the deep sea (Hessler, 1974), and there is a small family of minute gastropod limpets that feeds consistently on squid beaks and produces characteristic homing scars.[3] Yet another minute gastropod limpet seems to be consistently associated with old polychaete tubes. Dead (or alive) foraminiferal tests are commonly found with epibionts including, usually, other foraminifera. Because of the tendency to sort material taxonomically for distribution following a deep-sea sampling program, it is possible that many small species that occupy microhabitats on larger host species escape notice.

A variety of microhabitats occur in Antarctica as well that are used by small organisms. Lipps and Delaca (1980) record how algal canopies are subdivided by a whole microcommunity, with a corresponding increase in diversity. Another unique microhabitat in the shallow Antarctic is the thick mat of siliceous spicules of the sponge-dominated benthic communities. Where it occurs, the mat is usually more than a meter deep and provides a substrate for a diverse and rich epifauna and infauna (Dayton et al., 1970). Microhabitat utilization is common, however, in many different environments.

One aspect of both the Antarctic and deep-sea realms is resource limitation, either on a continuous (deep sea) or on a predictable seasonal (Antarctic) basis. Where energy is limiting, organisms should predictably shift to efficient methods to capture and use it. In the deep sea and seasonally in Antarctica, the quality and quantity of food, available chiefly as detritus, is low; and, because of low temperature, some building materials, notably calcium carbonate, are metabolically expensive to secrete and maintain. Under these fundamental energy restrictions, imposed by the combination of cold and short

[3]C. S. Hickman, unpublished.

supply, dwarfing occurs primarily in those organisms that secrete calcium carbonate, and it is most common infaunally. Gigantism, on the other hand, occurs primarily in organisms that live epifaunally or are planktonic, and it is achieved through one of a series of low-energy constructional schemes for becoming large in a nontraditional sense. In other words, many of the so-called giants are actually caloric dwarfs of deceptive appearance.

As with gigantism, the causes of true dwarfing are not clear for any environments in which they have been studied. Classical attempts to explain dwarfing in marine faunas have invoked unfavorable environmental conditions or resource limitation (Hallam, 1965; Tasch, 1953) as direct causes of physiological stunting.

More recently, however, dwarfing has been linked with progenesis as an explanation of diminutive size (Ghiselin, 1974; Gould, 1977; Snyder and Bretsky, 1971; Swedmark, 1968) in a variety of marine contexts. Progenesis can be expected as an adaptive explanation of small size where selection favors either rapid or early reproductive maturity (Ghiselin, 1974; Gould, 1977) or small body size *per se* (Gould, 1977; Thorson, 1965). However, truncated ontogeny is decidedly not what we find in Antarctica and the deep sea, nor are the normal morphological correlates of progenesis (retention of juvenile characteristics) associated with the calcareous dwarfs in these habitats.

If anything, the majority of the evidence suggests delayed maturation and slow growth as correlates of small size in dwarfed bivalves. Sanders and Allen (1977) and Allen (1979) found that the minute bivalve *Tindaria callistiformis* shows no gonadial development during the first 30 to 40 years of its life and suggested that it may not become sexually mature until 100+ years.

Dwarfing in the deep sea and Antarctica requires a great deal of additional study in order to document the ways in which small size is accomplished, particularly in view of the apparent differences between dwarfs in these environments and the kinds of dwarfing that has been studied better in other kinds of organisms and environmental settings.

Mammalogists have taken considerable interest in the evolution of dwarf lineages in Quaternary faunas, where insular populations of elephants, deer, hippopotami, and Australian marsupials provide elegant examples of within-lineage size reduction (Foster, 1964; Hooijer, 1967; Marshall and Corruccini, 1978; Sondaar, 1977; Thaler, 1973). In mammals, this size decrease has been regarded primarily as a "strategy" (Rosenzweig, 1968) or density-dependent response to extreme resource (food) limitation (Marshall and Corruccini, 1978).

For organisms that secrete calcium carbonate in low-temperature regimes, size limitation more likely may result from calcium metabolism problems than from food limitation alone. Sanders *et al.* (1965) have made the obvious suggestion that small body size is one way of adapting to low productivity in the deep sea. However, bivalves in the shallow Antarctic that have access to high levels of summer productivity seem to remain small within those low temperature regimes. Those organisms that are able to achieve giant size in regions of sustained low temperatures and at least seasonal periods of resource starvation are almost all organisms that require no calcium or that use it in small amounts.

A second line of evidence against food limitations as an explanation of dwarfing comes from small gastropods in eutrophic deep-sea settings.[4] Great quantities of terrestrial plant material accumulate in near-shore hadal trenches to provide a rich food

_____

[4] *Ibid.*

source for a variety of unusual deep-sea organisms (George and Higgins, 1979; Wolff, 1979). Particularly striking is the large number of minute (shells less than 5 mm in maximum length) "limpets" that have evolved in a number of families (some as yet undescribed) to exploit these resources. Although food is abundant, individuals do not attain large size. However, high population densities may be attained, a further suggestion that food is not limiting. In some of these minute deep-sea gastropods there is a well-developed periostracum or uncalcified organic matrix that shields the underlying calcium carbonate shell from contact with seawater. A detail of the periostracum of a deep-sea cocculinacean gastropod is illustrated in Fig. 14-13.

Our analysis of morphology suggests that morphologic similarity does not necessarily have a taxonomic basis, but instead results from adaptation to similar environmental regimes. As others have suggested (Douglas and Woodruff, 1981), we believe that many of the morphologic similarities, at least on a gross scale and probably even on a minor scale, result from convergence in adaptation. Evidence suggests that alternative ways of coping with resource limitation (food and, secondarily, skeletal materials) and microhabitat specialization contribute significantly to species diversification and maintenance in the deep sea and in the Antarctic. Further work must be done to test these hypotheses, and we suggest that detailed morphologic interpretation, analysis of convergence, and determination of microhabitat preference and ways and kinds of resource utilization would be fruitful lines of study.

Fig. 14-13.  Detail of the periostracum, an uncalcified organic construction, covering the shell of a deep-sea cocculinacean limpet. Bar is 100 $\mu$m.

## CONCLUSIONS

The deep-sea and shallow Antarctic marine environments share a number of physical and biotic characteristics. The similarities have suggested to different workers a series of conflicting hypotheses for the origin, age, and evolutionary processes, although some of the hypotheses infer a close relationship between the two environments. Reevaluation of data combined with recently acquired evidence from the fossil record, geology, and paleoceanography disprove some of the hypotheses. The evidence supports the hypothesis that the two faunas have been evolving primarily in place, with minor immigration, at least since the Mesozoic. Species living in each environment today are not especially ancient, certainly not relics, and many of their morphological similarities are the result of convergence rather than close taxonomic affinity.

Morphological analysis of some of the peculiarities of deep-sea and Antarctic organisms helps establish gigantism, dwarfing, and certain bizarre constructions as common responses to resource limitation (food and skeletal materials). Six constructional mechanisms for attaining and maintaining large size at low energetic cost include: elongation of appendages and sense organs, extreme flattening, mineralized lattice construction, stalk elongation, agglutinization, and big bag construction. The physical giants in these environments are actually caloric dwarfs. Associated with the morphological peculiarities are other adaptations for efficient utilization of resources including lowered metabolic rates, low reproductive output, brooding, and use of skeletal materials other than calcium carbonate. Specializations for efficient use of a diversity of unique microhabitats in these environments is an emerging pattern that requires further detailed study.

## ACKNOWLEDGMENTS

We thank Joel Hedgpeth for supplying us with his original and painstaking drawing of the giant pycnogonid, *Colossendeis colossea*. Ellen Bailey, University of California, Davis, executed Figs. 14-1, 14-2, and 14-9, and Mary Taylor, University of California, Berkeley, drew Figs. 14-4, 14-5, 14-7, 14-8, 14-10, 14-11, and 14-12. Our work on deep sea and Antarctic faunas has been supported by NSF grants DEB 80-20992, DPP 76-17231, DPP 77-21735, and GV 31162.

## REFERENCES

Allen, J. A., 1979, The adaptations and radiation of deep-sea bivalves: *Sarsia,* v. 64, p. 19–27.

Arnaud, P. M., 1977, Adaptations within the Antarctic marine benthic ecosystem, *in* Llano, G. A., ed., *Adaptations within Antarctic ecosystems:* Proc. 3rd SCAR Symposium on Antarctic Biology, Washington, D.C., 1974, Smithsonian Inst. and Gulf Publ. Co., Houston, p. 135–157.

Barham, E. G., 1971, Deep-sea fishes lethargy and vertical orientation, *in* Farquhar, G. B., ed., *Proceedings of an International Symposium on Biological Sound Scattering in the Ocean:* Maury Center for Ocean Science, Washington, D.C., 1971, p. 100–118.

——, 1979, Giant larvacean houses—observations from deep submersibles: *Science,* v. 205, p. 1129–1131.

Barnes, A. T., Quetin, L. B., Childress, J. L., and Pawson, D. L., 1976, Deep-sea macroplanktonic sea cucumbers—suspended sediment feeders captured from deep submergence vehicle: *Science,* v. 194, p. 1083–1085.

Benson, R. H., 1975, The origin of the psychrosphere as recorded in changes of deep-sea ostracode assemblages: *Lethaia*, v. 8, p. 69–83.

Bergman, C., 1847, Uber die Verhaltnisse der Warmeokonomie der Tiere zu ihrer Grosse: *Gottinger Studien*, pt. 1, p. 595–708.

Bernstein, B. B., Hessler, R. R., Smith, R., and Jumars, P. A., 1978, Spatial dispersion of benthic foraminifera in the abyssal central North Pacific: *Deep-Sea Res.*, v. 23, p. 401–416.

——, and Meador, J. P., 1979, Temporal persistence of biological patch structure in an abyssal benthic community: *Mar. Biol.*, v. 51, p. 179–183.

Bone, D. G., 1972, Aspects of the biology of the Antarctic amphipod *Bovallia gigantea* Pfeffer at Signy Island, South Orkney Islands: *British Antarctic Survey Bull.*, no. 27, p. 105–122.

Bonner, J. T., 1968, Size change in development and evolution, *in* Macurda, D. B., Jr., ed., *Paleobiological aspects of growth and development, a symposium:* Paleontological Society Mem. 2, p. 1–15.

Bouchet, P., and Waren, A., 1979, Planktotrophic larval development in deep-water gastropods: *Sarsia*, v. 64, p. 37–40.

Brady, H., and Martin, H., 1979, Ross Sea region in the Middle Miocene—a glimpse into the past: *Science*, v. 203, p. 437–438.

Bregazzi, P. K., 1972, Life cycles and seasonal movements of *Cheirimedon femoratus* (Pfeffer) and *Tryphosella kergueleni* (Miers) Crustacea:Amphipoda); *British Antarctic Surv. Bull.*, no. 30, p. 1–34.

Burton, M., 1929, Porifera II. Antarctic sponges: *Nat. Hist. Rep. Antarct. Terra Nova Exped., 1910*, v. 6, no. 1, p. 393–458.

Clarke, A., 1979, On living in cold water—*K*-strategies in Antarctic benthos: *Mar. Biol.*, v. 55, p. 111–119.

——, 1980, A reappraisal of the concept of metabolic cold adaptation in polar marine invertebrates: *Biol. J. Linn. Soc.*, v. 14, p. 77–92.

Clarke, A. H., 1962, Annotated list and bibliography of the abyssal marine molluscs of the world: *Bull. National Mus. Canada*, no. 181, 114 p.

Daniels, R. A., and Lipps, J. H., in press, Distribution and ecology of fishes of the Antarctic Peninsula: *J. Biogeography*.

Dayton, P. K., and Hessler, R. R., 1972, Role of biological disturbance in maintaining diversity in the deep sea: *Deep-Sea Res.*, v. 19, p. 199–208.

——, and Oliver, J. S., 1977, Antarctic soft-bottom benthos in oligotrophic and eutrophic environments: *Science*, v. 197, p. 55–58.

——, Robilliard, G. A., and Paine, R. T., 1970, Benthic faunal zonation as a result of anchor ice at McMurdo Sound, Antarctica, *in* Holdgate, M. W., ed., *Antarctic Ecology, Vol. 1:* London, Academic Press, p. 244–258.

DeBroyer, C., 1977, Analysis of the gigantism and dwarfness of Antarctic and sub-Antarctic gammaridean Amphipods, *in* Llano, G. A., ed., *Adaptations within Antarctic ecosystems:* Proc. 3rd SCAR Symposium on Antarctic Biology, Washington, D.C., 1974, Smithsonian Inst. and Gulf Pub. Co., Houston, p. 327–334.

DeLaca, T. E., Karl D. M., and Lipps, J. H., 1981, Direct use of dissolved organic carbon by agglutinated benthic foraminifera: *Nature*, v. 289, p. 287–289.

——, Lipps, J. H., and Hessler, R. R., 1980, The morphology and ecology of a new large agglutinated Antarctic foraminifera (Textulariina:Notodendroididai nov.): *Zool. J. Linn. Soc.*, v. 69, p. 205–224.

Dell, R. K., 1972, Antarctic benthos: *Adv. Mar. Biol.*, v. 10, p. 11–216.

——, and Fleming, C. A., 1975, Oligocene-Miocene bivalve mollusca and other macrofossils from Sites 270 and 272 (Ross Sea), DSDP, Leg 28: *Initial Reports of the Deep Sea Drilling Project,* U.S.G.P.O., Washington, D.C., v. 28, p. 693–700.

DeWitt, H. H., 1971, Coastal and deep-water benthic fishes of the Antarctic: *Amer. Geograph. Soc. Antarctic Map Folio Ser.,* no. 15, p. 1–10.

Douglas, R. G., and Woodruff, F., 1981. Deep-sea foraminifera, *in* Emiliani, C., ed., *The Sea, Vol. 7,* New York, Wiley-Interscience Pub., p. 1233–1328.

Everson, I., 1977, Antarctic marine secondary production and the phenomenon of cold adaptation: *Phil. Trans. Roy. Soc. Lond. (B),* v. 279, p. 55–66.

Foster, J. B., 1964, Evolution of mammals on islands: *Nature,* v. 202, p. 234–235.

George, R. Y., and Higgins, R. P., 1979, Eutrophic hadal benthic community in the Puerto Rico Trench: *AMBIO Special Rpt.,* no. 6, p. 51–58.

Ghiselin, M., 1974, *The Economy of Nature and the Evolution of Sex:* Berkeley, Univ. California Press, 346 p.

Gould, S. J., 1975, On the scaling of tooth size in mammals: *Amer. Zool.,* v. 15, p. 351–362.

——, 1977, *Ontogeny and phylogeny:* Cambridge, Mass., Belknap Press, 501 p.

Grassle, J. F., 1977, Slow recolonization of deep-sea sediment: *Nature,* v. 265, p. 618–619.

——, 1978, Diversity and population dynamics of benthic organisms: *Oceanus,* v. 21, no. 1, p. 42–49.

——, and Sanders, H. L., 1973, Life histories and the role of disturbance: *Deep-Sea Res.,* v. 20, p. 643–659.

Hallam, A. J., 1965, Environmental causes of stunting in living and fossil marine benthonic invertebrates: *Palaeontology,* v. 8, p. 132–155.

Hedgpeth, J. W., 1969, Introduction to Antarctic zoogeography: *Amer. Geogr. Soc. Antarctic Map Folio Ser.,* no. 11, p. 1–9.

——, 1971, Perspectives of benthic ecology in Antarctica, *in* Quam, L. E., ed., *Research in the Antarctic:* Washington, D.C., Amer. Assoc. Adv. Sci., p. 93–136.

Hayes, D. E., Frakes, L. A., Barrett, P. J., Burns, D. A., Chen, P.-H., Ford, A. B., Kaneps, A. G., Kemp, E. M., McCollum, D. W., Piper, D. J. W., Wall, R. E., and Webb, P. N., 1975, Leg 28: *Initial Reports of the Deep Sea Drilling Project,* U.S.G.P.O., Washington, D.C., v. 28, p. 1–1017.

Hertlein, L. G., 1969, Fossiliferous boulder of early Tertiary age from Ross Island, Antarctica: *Antarctic J. U.S.,* v. 4, p. 199–200.

Hessler, R. R., 1974, The structure of deep benthic communities from central oceanic waters, *in* Miller, C., ed., *The Biology of the Oceanic Pacific:* Oregon State Univ. Press, Corvalis, p. 79–93.

——, and Jumars, P. A., 1974, Abyssal community analysis from replicate box cores in the central North Pacific: *Deep-Sea Res.,* v. 21, p. 185–209.

——, and Sanders, H. L., 1967, Faunal diversity in the deep sea: *Deep-Sea Res.,* v. 14, p. 65–78.

——, and Thistle, D., 1975, On the place of origin of deep-sea isopods: *Mar. Biol.,* v. 32, p. 155–165.

Hickman, C. S., 1976, Bathyal gastropods of the Family Turridae in the Early Oligocene Keasey Formation in Oregon, with a review of some deep-water genera in the Paleogene of the eastern Pacific: *Bull. Amer. Paleontol.,* v. 70, p. 1–119.

——, 1980a, Paleogene marine gastropods of the Keasey Formation in Oregon: *Bull. Amer. Paleontol.,* v. 78, p. 1–112.

——, 1980b, Gastropod radulae and the assessment of form in evolutionary paleontology: *Paleobiology*, v. 6, p. 276–294.

Hooijer, D. A., 1967, Indo-Australian insular elephants: *Genetica*, v. 38, p. 143–162.

Jumars, P. A., 1975, Environmental grain and polychaete species' diversity in a bathyal benthic community: *Mar. Biol.*, v. 30, p. 253–266.

——, 1976, Deep-sea species diversity—does it have a characteristic scale? *J. Mar. Res.*, v. 34, p. 217–246.

Kauffman, E. G., 1976, Deep-sea Cretaceous macrofossils: Hole 317A, Manihiki Plateau: *Initial Reports of the Deep Sea Drilling Project*, U.S.G.P.O., Washington, D.C., v. 33, p. 503–535.

Kennett, J. P., 1977, Cenozoic evolution of Antarctic glaciation, the Circum-Antarctic Ocean, and their impact on global paleoceanography: *J. Geophys. Res.*, v. 82, p. 3843–3860.

——, 1978, The development of planktonic biogeography in the southern ocean during the Cenozoic: *Mar. Micropaleontol.*, v. 3, p. 301–345.

Knudsen, J., 1970, The systematics and biology of abyssal and hadal bivalvia: *Galathea Rpts.*, v. 11, p. 7–236.

——, 1979, Deep-sea bivalves, *in* van der Spoel, S., and others, eds., *Pathways in Malacology:* Bohn, Scheltema and Holkema, Utrecht, p. 195–224.

Koltun, V. M., 1970, Sponges of the Arctic and Antarctic—a faunistic review: *Symp. Zool. Soc. Lond.*, v. 25, p. 285–297.

Kussakin, O. G., 1973, Peculiarities of the geographical and vertical distribution of marine isopods and the problem of deep-sea fauna origin: *Mar. Biol.*, v. 23, p. 19–34.

Lemeche, H., Hansen, B., Madsen, F. J., Tendal, O. S., and Wolff, T., 1976, Hadal life as analyzed from photographs: *Vidensk. Meddr. Dansk. Naturh. Foren.*, v. 139, p. 263–336.

Lipps, J. H., and DeLaca, T. E., 1976, Shallow-water marine associations, Antarctic Peninsula: *Antarctic J. U.S.*, v. 11, p. 12–20.

——, and DeLaca, T. E., 1980, Shallow-water foraminiferal ecology, Pacific Ocean, *in* Field, M. E., *et al.*, eds., *Quaternary Depositional Environments of the Pacific Coast:* Pacific Coast Paleogeography Symposium 4, Pacific Section, Soc. Econ. Paleontol. Mineral., Los Angeles, p. 325–340.

——, Ronan, T. E., Jr., and DeLaca, T. E., 1979, Life below the Ross Ice Shelf: *Science,* v. 203, p. 447–449.

McDougall, K., 1980, Paleoecological evaluation of late Eocene biostratigraphic zonations of the Pacific coast of North America: *Soc. Econ. Paleontol. Mineral. Paleontol. Monogr.*, no. 2 (*J. Paleontol.*, v. 54, no. 4, suppl.), 75 p.

Madsen, F. J., 1961, On the zoogeography and origin of the abyssal fauna in view of the knowledge of the Porcellanasteridae: *Galathea Rpt.*, v. 4, p. 177–218.

Marshall, L. G., and Corruccini, R. S., 1978, Variability, evolutionary rates, and allometry in dwarfing lineages: *Paleobiology*, v. 4, p. 101–119.

Menzies, R. J., and George, R. Y., 1967, A re-evaluation of the concept of hadal or ultra-abyssal fauna: *Deep-Sea Res.*, v. 14, p. 703–723.

——, George, R. Y., and Rowe, G. T., 1973, *Abyssal Environment and Ecology of the World Oceans:* New York, Wiley, 488 p.

Nicol, D., 1966, Size of pelecypods in Recent marine faunas: *Nautilus*, v. 79, p. 109–113.

——, 1967, Some characteristics of cold-water marine pelecypods. *J. Paleontol.*, v. 41, no. 6, p. 1330–1340.

——, 1970, Antarctic pelecypod faunal peculiarities: *Science,* v. 168, p. 1248–1249.

Oliver, P. G., 1979, Adaptations of some deep-sea suspension-feeding bivalves (*Limopsis* and *Bathyarca: Sarsia,* v. 64, p. 33–36.

Pearse, J. S., 1965, Reproductive periodicities in several contrasting populations of *Odontaster validus* Koehler, a common Antarctic asteroid: *Antarctic Res. Ser.,* no. 5, p. 39–85.

Picken, G. B., 1979a, Non-pelagic reproduction of some Antarctic prosobranch gastropods from Signy Island, South Orkney Islands: *Malacologia,* v. 19, p. 109–128.

——, 1979b, Growth, production and biomass of the Antarctic gastropod *Laevilacunaria antarctica* Martens 1885: *J. Exp. Mar. Biol. and Ecol.,* v. 40, p. 71–79.

Rokop, F., 1974, Reproductive patterns in the deep-sea benthos: *Science,* v. 186, p. 743–745.

Rosenzweig, M. L., 1968, The strategy of body size in mammalian carnivores: *Amer. Midl. Nat.,* v. 80, p. 299–315.

Sanders, H. L., 1968, Marine benthic diversity—a comparative study: *Amer. Nat.,* v. 102, p. 243–282.

——, and Allen, J. A., 1973, Studies on deep-sea Protobranchia (Bivalvia); Prologue and the Pristiglomidae: *Bull. Mus. Comp. Zool., Harv.,* v. 145, p. 237–262.

——, Hessler, R. R., and Hampson, G. R., 1965, An introduction to the study of deep-sea benthic faunal assemblages along the Gay Head–Bermuda transect: *Deep-Sea Res.,* v. 12, p. 845–867.

Savin, S. M., 1977, The history of the earth's surface temperature during the past 100 million years: *Ann. Rev. Earth Planet. Sci.,* v. 5, p. 319–355.

Scheltema, R. S., 1972, Reproduction and dispersal of bottom dwelling deep-sea invertebrates—a speculative summary, *in* Braver, L. W., ed., *Barobiology and the Experimental Biology of the Deep Sea:* Univ. N. Carolina Press, p. 58–66.

Schlanager, S. O., and Jenkyns, H. C., 1976, Cretaceous oceanic anoxic events—causes and consequences: *Geol. en Mijnbouw,* v. 55, p. 179–184.

Seager, J. R., 1979, Reproductive biology of the Antarctic opisthobranch *Philene gibba* Strebel: *J. Exp. Mar. Biol. and Ecol.,* v. 41, p. 51–74.

Shackleton, N. J., and Kenneth, J. P., 1975, Paleotemperature history of the Cenozoic and the initiation of Antarctic glaciation—oxygen and carbon isotope analyses in DSDP Sites 277, 279, and 281: *Initial Reports of the Deep Sea Drilling Project,* U.S.G.P.O., Washington, D.C., v. 29, p. 743–755.

Smith, K. L., Jr., and Hessler, R. R., 1974, Respiration of benthopelagic fishes—*in situ* measurements at 1230 meters: *Science,* v. 184, p. 72–73.

——, and Teal, J. M., 1973, Deep-sea benthic community respiration—an *in situ* study at 1850 meters: *Science,* v. 179, p. 282–283.

Synder, J., and Bretsky, P. W., 1971, Life habits of diminutive bivalve mollusks in the Maquoketa Formation (Upper Ordovician): *Amer. Jour. Sci.,* v. 271, p. 227–251.

Sondaar, P. Y., 1977, Insularity and its effect on mammal evolution, *in* Hecht, M. K., Goody, P. C., and Hecht, B. M., eds., *Major Patterns of Vertebrate Evolution:* New York, Plenum, p. 671–707.

Soot-Ryen, T., 1951, Antarctic pelecypods: *Sci. Results of the Norwegian Antarctic Expedition 1927-1928,* v. 32, p. 1–46.

Swedmark, B., 1968, The biology of interstitial Mollusca: *Symp. Zool. Soc. London,* no. 22, p. 135–149.

Tasch, P., 1953, Causes and paleoecological significance of dwarfed fossil marine invertebrates: *J. Paleontol.*, v. 27, p. 356–444.

Tendal, O. S., 1972, A monograph of the Xenophyophoria (Rhizopoda, Protozoa): *Galathea Rpt.*, v. 12, p. 7–103.

——, and Hessler, R. R., 1977, An introduction to the biology and systematics of Komokiacea (Textulariina, Foraminiferida): *Galathea Rpt.*, v. 14, p. 165–194.

Thaler, L., 1973, Nanisme et gigantisme insularies: *La Recherche*, v. 4, p. 741–750.

Thiel, H., 1975, The size structure of the deep-sea benthos: *Int. Revue Ges. Hydrobiol. Hydrogr.*, v. 60, p. 575–602.

Thorson, G., 1936, The larval development, growth, and metabolism of the Arctic marine bottom invertebrates compared with those of other seas: *Med. om Gronland*, v. 100, no. 6, p. 1–155.

——, 1950, Reproductive and larval ecology of marine bottom invertebrates: *Biol. Rev.*, v. 25, p. 1–45.

——, 1965, A neotenous dwarf-form of *Capulus ungaricus* (L.) (Gastropoda, Prosobranchia) commensalistic on *Turritella communis* Risso: *Ophelia*, v. 2, p. 175–210.

Thurston, M. H., 1972, The Crustacea Amphipoda of Signy Island, South Orkney Islands: *British Antarctic Surv. Sci. Rpt.*, no. 71, 127 p.

Tipper, R., 1968, Ecological aspects of two wood-boring mollusks from the continental terrace off Oregon: Ph.D. thesis, Oregon State Univ. 137 p.

Turekian, K. K., Cochran, J. K., Kharkar, D. P., Cerrato, R. M., Vaisnys, J. R., Sanders, H. L., Grassle, J. F., and Allen, J. A., 1975, Slow growth rate of deep-sea clam determined by $^{228}$Ra chronology: *Proc. Nat. Acad. Sci. (USA)*, v. 72, p. 2829–2832.

Turner, R. D., 1973, Wood-boring bivalves, opportunistic species in the deep sea: *Science*, v. 180, p. 1377–1379.

Valentine, J. W., 1973, *Evolutionary Paleoecology of the Marine Biosphere:* Englewood Cliffs, N.J., Prentice-Hall, 511 p.

Vogel, S., 1974, Current-induced flow through the sponge, *Halichondria: Biol. Bull.*, v. 147, p. 443–456.

——, and Bretz, W. L., 1972, Interfacial organisms—passive ventilation in velocity gradients near surfaces: *Science*, v. 175, p. 210–211.

Webb, P. N., 1974, Micropaleontology, paleoecology and correlation of the Pecten Gravels, Wright Valley, Antarctica, and description of *Trochoelphidiella onyxi* n. gen., n. sp.: *J. Foraminiferal Res.*, v. 4, p. 184–199.

——, Ronan, T. E., Jr., Lipps, J. H., and DeLaca, T. E., 1979, Miocene glaciomarine sediments from beneath the southern Ross Ice Shelf, Antarctica: *Science*, v. 203, p. 435–437.

Welton, B. J., and Zinsmeister, W. J., 1980, Eocene Neoselachians from the La Meseta Formation, Seymour Island, Antarctic Peninsula: *Nat. Hist. Museum of Los Angeles, Contrib. in Sci.*, no. 329, p. 1–10.

Wolff, T., 1961, Animal life from a single abyssal trawling: *Galathea Rpt.*, v. 5, p. 125–162.

——, 1962, The systematics and biology of bathyal and abyssal Isopoda Asellota: *Galathea Rpt.*, v. 6, p. 1–320.

——, 1977, Diversity and faunal composition of the deep-sea benthos: *Nature*, v. 267, p. 780–785.

——, 1979, Macrofaunal utilization of plant remains in the deep sea: *Sarsia,* v. 64, p. 117–136.

Woodruff, F., Savin, S. M., and Douglas, R. G., 1981, Miocene stable isotope record–a detailed deep Pacific Ocean study and its paleoclimatic implications: *Science,* v. 212, p. 665–668.

Zinsmeister, W. J., 1979, Biogeographic significance of the late Mesozoic and early Tertiary molluscan faunas of Seymour Island (Antarctic Peninsula) to the final breakup of Gondwanaland, *in* Gray, J., and Boucot, A. J., eds., *Historical Biogeography, Plate Tectonics and the Changing Environment:* Oregon State Univ. Press, Corvalis, p. 349–355.

——, and Camacho, H. H., 1980, Late Eocene Struthiolariidae (Mollusca:Gastropoda) from Seymour Island, Antarctic Peninsula and their significance to the biogeography of early Tertiary shallow-water faunas of the Southern Hemisphere: *J. Paleontol.,* v. 54, p. 1–14.

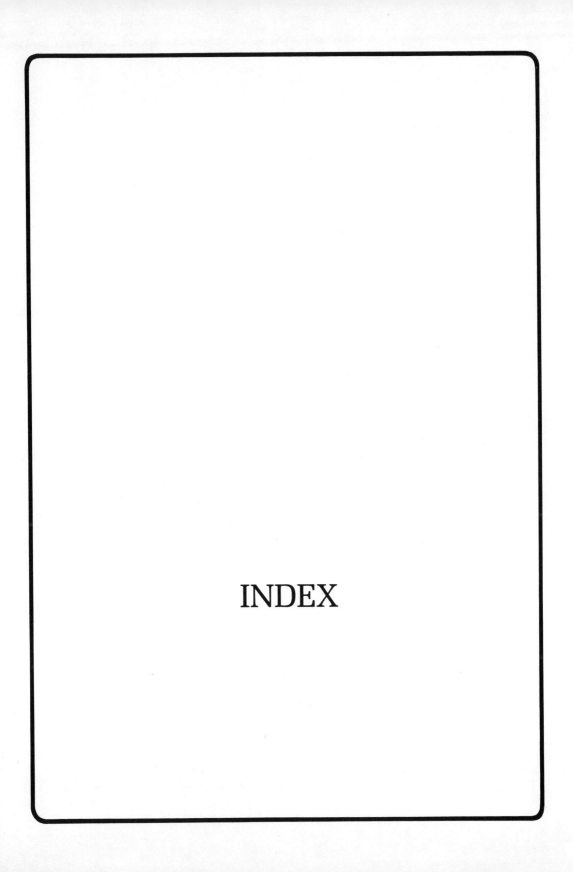

INDEX

# A

Abelson, P. H., 48
Abyssal habitats:
  dissolved organic material (DOM) in, 315–19
    concentration of DOM, 317
    metabolic requirements and energy flow, 316–17
    uptake of DOM by abyssal animals, 317–18
Abyssal plains, 5, 14–16
Adaptation (*See* Fishes, adaptations of, physiological and biochemical
Adelseck, C. G., 138
Agassiz, A., 270
Ages of faunas, 328, 345
Agglutinization, 339–41
Aging techniques, 230
Ahearn, G. A., 309
Albrecht, P., 94, 97–98
Alexander, M., 75
Ali, S. H., 205
Allen, J. A., 221, 230, 235, 343–44, 348
Alvarez, L. W., 43–44
*Alvin*, 180, 194
Ambler, J. W., 235
Amino acids:
  dissolved free (DFAA), 308
  transepidermal transport of, 302–19
    availability of substrate: levels of amino acid, 307
    *Dendraster excentricus,* 314
    general features of, in marine invertebrates, 305–6
    kinetics of, 308–15
    *Mytilus californianus,* 311–13
    nutritional requirements, 307
    techniques employed in the study of, 303–5
Amphipods, 225
  bacteria and, 195, 197
  respiration of, 297–98
  scavenging, 226, 234–35
Anaerobic bacteria, 185
Anaerobic muscle function, 274–75
Anaerobic respiration, 184–85
Andersen, M. E., 124
Anderson, J. W., 306
Anderson, N. H., 227
Anderson, T. F., 25
Angell, R. W., 22
*Anoplopoma fimbria,* 273
Antarctica:
  Eocene and oligocene glaciation, 30–31
  evolution of faunas in (*See* Evolution of faunas)

Antarctica (*cont.*)
  fossil record in, 330
Antarctic Convergence (Polar Front):
  carbon isotopes in, 37–38
Antarctic ice sheet:  Miocene changes in, 32
Apparent Michaelis constant ($K_m$), 261–64, 269
Aragonite, dissolution of, 165, 170–72
Armi, L., 233
Arnaud, P. M., 335–36, 344
Arrhenius, G. O. S., 124, 136
*Arthrobacter* sp., 206–7
Arthur, M. A., 29, 32–33, 38–40
Ascidians, 226, 231
Asteroids, Cretaceous/Tertiary changes and, 43
Atkinson, D. E., 312
Autoradiographic studies of amino acid transport, 303
Autotrophic prokaryotes, 186
Azam, F., 181

# B

Bacteria, 180–97, 231
  biochemical potential of, 181–83
  biomass of, 181–83
  commensal associations of, 187–88
  limits to distribution and activities of, 188–95
    nutrients, 192–95, 197
    pressure, 190–95
    temperature, 189, 193, 195
  low-nutrient, 195, 197
  manganese (*See* Manganese bacteria)
  metabolism, 184–87, 194
  nutrition, 183, 188–89, 192–95, 197
  respiration of, 298
  roles, postulated, 196–97
  size of, 181–82
  transport mechanisms of, 183, 193
Bacterioplankton, 192–93
*Baculella hirsuta,* 222
Bada, J. L., 308, 317
Bailey, S. A., 96
Bait, response to, 226, 234
Baldwin, R. J., 288, 294
Ballard, Robert D., 4, 186, 226, 240
Bamford, D. R., 309–10, 313
Banse, K., 270
Barham, E. G., 336, 342
Barnard, J. L., 229
Barnes, A. T., 296, 342
Barnes, S. S., 132, 140

Carbon isotopes (*cont.*)
negative δ$^{13}$C excursion at Cretaceous/
Tertiary boundary, 40–47
Pliocene/Pleistocene, 33–36, 47
positive δ$^{13}$C excursions in the Miocene
and Paleocene, 38–40
in planktonic foraminifera, 23–24
Carlucci, A. F., 85, 192–93, 298
Carnivores, 226–27 (*See also* Predators;
Scavengers)
Caswell, H., 245
Cavanaugh, C. M., 259
*Chaetozone* sp., 224
Chan, L. H., 59
Charnov, E. L., 242
Chemical reactions, deep-sea floor, 155–73
decomposition of organic matter, 160–65
diagenetic modeling, 158–60
Chemolithotrophy, 184–87, 197
manganese, 205
Chemosensory detection of carrion, 235–37
Chester, R., 138
Chia, F. S., 314
Childress, J. J., 262, 270, 272, 274–75, 294
Christensen, H. N., 309
Chukrov, F. V., 131
Chumley, J., 171
Chung, Y., 85
Cita, M. B., 32
Citrate synthase (CS), 271, 274
Clark, R. C., 94
Clarke, A., 335
Clarke, A. H., 334
Claypool, G., 161
Climap Project Members, 33
Cline, J. D., 80
Cobler, R., 136
Coccoliths, oxygen and carbon isotopes of,
25–27
Cochran, J. K., 63, 122, 124
Cody, M. L., 238
Cohen, Y., 75, 80, 84–85, 87, 90
Cole, S. A., 25
Colwell, R. R., 189
Comets, Cretaceous/Tertiary changes and,
44–45
Community structure, 219–49
patterns observed, 220–31
body size and areal density of fauna,
221–23
size-frequency distribution and
population dynamics, 229–31
species diversity, 227–29
taxonomic composition and trophic
structure, 224–27

Community structure (*cont.*)
processes implicated, 231–48
community level, 243–48
individual level, 231–39
population level, 239–43
Competitive exclusion, 244–45
Connell, J. H., 243, 245–46, 248
Conolly, J. R., 116
Continental margins, 16–17
Continental rise, 16
Continental shelves, 16–17
Continental slope, 16–17
Corals, 13–14
growth rates for, 69–70
Corliss, B. H., 31
Corliss, J. B., 68, 186, 315
Corner Rise, 13
Corpe, W. A., 187
Corruccini, R. S., 344, 348
*Coryphaenoides acrolepis,* 273
*Coryphaenoides armatus,* 273
Cosmogenic nuclides, 56–57
Costopulos, J. J., 306, 311
Craig, H., 22, 75, 83, 141
Cretaceous/Tertiary boundary, 40–47
asteroidal and cometary impacts, 43–45
carbon isotopes in, 40–43
temperature, 29, 43–47
Crisp, P. T., 95, 100–101
Cronan, D. S., 110, 112, 114, 126–27,
140, 143
Croppers, 241
Crowe, J. H., 304, 306, 311, 313
Crowely, P. H., 312
Crustaceans, 272
elongation of legs, 337
Crutzen, P. J., 74
Curran, P. F., 309
Curry, W. B., 22

# D

Daniels, R. A., 326, 328
Dastillung, M., 94, 97–98
David, D. J., 207
Dayton, P. K., 226–27, 241–46, 270,
326, 329, 339, 347
DeBroyer, C., 343
DeBurgh, M. E., 306
Decomposition of organic matter, 160–65
sedimentation rates and, 162–64

Farrington, J. W., 94, 308, 317
Fauchald, K., 226, 237
Faunas (*See* Community structure; *and specific types of fauna*
  ages of, 328, 345
  areal density of, 223
  body size of, 221–23
  origins of, 327–28, 345
  species diversity, 227–29, 243–45, 249
  taxonomic groups (*See* Taxonomic groups)
  trophic structure, 224–27
Fecundity, 230–31
Feeding guilds, 225–27, 241
Feeding strategies, 259, 273–75 (*See also* Deposit feeders; Foraging theory; Predators; Scavengers; Suspension feeders)
Felbeck, H., 259
Feller, R. J., 227, 234, 248
Fenchel, T. M., 183, 185, 193, 247, 316
Ferguson, J. C., 303, 314–15
Ferromanganese (or manganese) nodules, 106–45, 209
  bacteria and, 139–40, 187, 203, 207–9, 213
  buried, 110–14
  composition and abundance, relation between, 141–43
  concentration mechanisms, 135–41
  diagenesis of, 121–22
  elemental composition of, 126–30
  genesis of, 132–43
    activity of epibenthic microfauna on nodules, 140
    authigenic reactions in surficial sediments, 138–39
    bacterial activity in surface sediments and on nodules, 139–40
    concentration mechanisms, 135–41
    local sea-floor volcanism, 141
    precipitation from bottom waters, 136–37
    preferential incorporation of dissolved metals in specific iron/manganese oxyhydroxide minerals, 140–41
    sources of metals in, 132–35
    upward diffusion of metals dissolved in porewaters, 138
    vertical transport by settling particles, 136
  growth rates of, 123–26
    burial rates and, 111–12
  internal structure, 119–23
  laminae of, 123
  local distribution of, 107, 109
  mineralogy, 131–32
  organisms attached to, 118–19
  regional distribution of, 107
  shape of, 114–16

Ferromanganese (or manganese) nodules (*cont.*)
  size of, 116–17
  sources of metals in, 132–35
  surface texture, 117–18
  unresolved problems concerning, 143–45
Fersht, A., 261
Fewkes, R. H., 130, 132, 141
Feyerabend, P., 219
Fisher, F. M., 306
Fishes, adaptations of, physiological and biochemical, 257–76 (*See also* Community structure; Faunas)
  body composition, 269–75
  metabolism, 269–75
  pressure, 260–69, 276
  temperature, 260–69, 276
Fish trap respirometer (FTR), 285–86
Fleming, C. A., 330
Fleming, R. H., 75, 84
Fluorometric methods for studying amino acid transport, 303–4
Fogg, G. E., 85
Food intake, oxygen consumption and, 307
Food supply, 223 (*See also* Foraging theory; Trophic structure)
  adaptation to low, 259, 270, 273–74, 276
  resource limitation, 243–44
Food-type specialization, 244–45
Food web, 295–99
Foraging theory, optimal, 231–39, 248
Foraminifera:
  agglutinization in, 339–41
  fossil record, 330–31
  oxygen and carbon isotopes of, 27
  planktonic (*See* Planktonic foraminifera)
  size range of, 221–23
Ford, J., 206
Fornari, D. J., 126
Fossil record:
  in Antarctica, 330–31
  of the deep sea, 330–31
Foster, A. R., 121
Foster, J. B., 348
Frazer, J. L., 142
Free-vehicle grab respirometer (FVGR), 283–85
Free vehicle respirometer (FVR), 282–83
Friedman, D., 79
Friedrich, H., 225
Froelich, P. N., 134, 162, 185
Funnell, B. M., 20

# G

Gabe, D. R., 208
Gadgil, M. D., 242–43

Gage, J. D., 221, 231, 240
Gagner, C., 136
Gardner, W., 296
Gartner, S., 43
Gas chromatography for studying amino acid transport, 303
George, R., 346, 349
Ghiorse, W. C., 203, 205-6, 209
Ghiselin, M., 348
Gieskes, J. M., 80, 85
Gigantism and bizarre morphologies, 336-43, 348
    agglutinization, 339-41
    big bag construction, 342
    elongation of appendages and sense organs, 336-37
    extreme flattening, 337-38
    mineralized lattice construction, 338-39
    stalk elongation, 339
Glaciation of western Antarctica during the Eocene and Oligocene, 30-31
Glasby, G. P., 110, 113-14, 123
Goldan, P. D., 74
Goldberg, E. D., 63, 160
Goldhaber, M. B., 97, 163
Goodney, D. E., 25-26, 41-42
Goody, R. M., 74
Gordon, C. M., 240
Gordon, L. I., 75, 80, 84-85, 87, 90
Gould, S. J., 348
Grassle, F., 5
Grassle, J. F., 68, 228, 231, 238, 240-41, 244, 270, 296, 335-36, 343, 346-47
Greenslate, J. L., 109, 118, 136, 140, 143
Greenwalt, D., 240
Gregory, S., 160
Griggs, G. B., 240
Grizzle, P. L., 98
Growth rates of deep-sea organisms, 66-70
Guilds, 225-27, 241 (*See also* Deposit feeders; Predators; Scavengers; Suspension feeders)
Guinasso, N. L., Jr., 63, 160, 170
Gundlach, H., 141
Gupta, B. L., 317

# H

Haedrich, R. L., 223, 293, 297
Hahn, J., 74-75, 80, 85
Hajash, A., 133
Hajj, H., 205
Hallam, A. J., 348
Hamilton, R. J., 94

Hammen, C. S., 307, 312
Hamner, P., 237
Hamner, W. M., 237
Hansen, H. J., 22
Haq, B. U., 36, 38
Harada, K., 123
Harper, J. L., 243
Harris, T. G., 212
Hartman, B. A., 95-96, 102
Hastings, J. W., 187
Hawkins, A. D., 236
Hay, W. W., 48
Hayes, D. E., 330
Healy-Williams, N., 22
Heath, G. R., 109-11, 113, 124, 136, 141, 163
Hedges, J. I., 226
Hedgpeth, J. W., 326
Heezen, Bruce C., 5, 126, 233
Heirtzler, J. R., 5
Hemipelagic sediments, anoxic diagenesis of, 133-35
Henrichs, S. M., 308, 317
Herbivores, 226-27
Herd-forming urchins, 238-39
Herman, Y., 33
Herring, P. J., 192
Hertlein, L. G., 330
Hessler, R. R., 113, 143, 220-21, 226-27, 237, 241-46, 270, 286, 288, 293, 299, 326-29, 334-35, 341, 343, 347
Heye, D., 124, 143
Hickman, C. S., 330, 336
Higgins, R. P., 349
Hinga, K. R., 139, 285, 288
Hirsch, P., 205, 209
Hochachka, P. W., 261-62, 274-75
Hodson, R. E., 181, 184, 193
Holland, H. D., 133
Hollister, Charles D., 5, 233, 237
Holm-Hansen, O., 190
Honjo, S., 136, 138, 168, 259
Hooijer, D. A., 348
Hopkins, 33
Horn, D. R., 107, 110, 117, 126
Horn, H. S., 246
Hot water springs, 13
Hsü, K. J., 32, 43-46
Hughes, M. J., 138
Hungate, R. E., 185
Hurd, D. C., 172
Hurlbert, S. H., 228
Huston, M., 245
Hydrocarbons (*See* Petroleum contamination)
Hydrothermal systems, leaching of newly formed oceanic crust by, 132-34

Hydrothermal vents, 203, 259
Hypsometric curve, 5

# I

Immigration, potential for, 231
Iron, bacterial redox reactions and, 184–85
Isaacs, J. D., 226, 241, 299
Isopods, extreme flattening in, 337–38
Interparity, 242–43
Iwamoto, T., 292, 297

# J

James, D., 309
Jannasch, H. W., 56, 180, 185, 187, 193–95,
    281, 286, 288, 294, 298–99, 315
Jenkyns, H. C., 38–40, 332
Johannes, R. E., 308, 311
Johnson, P. W., 181, 192
Johnson, T. C., 173
Johnston, I. A., 266
Jones, D. S., 230
Jones, J. H., 80
Jorgensen, B. B., 183, 185, 193
Jørgensen, C. B., 233, 302, 307
Jørgensen, N. O. G., 306, 313
Jumars, P. A., 113, 143, 220–21, 226–27,
    229, 237, 241–46, 248, 329, 343, 347
Jung, W. K., 205
Junge, C., 74–75, 85

# K

Kahn, M. I., 22–23
Kaneko, T., 189
Kanwisher, J., 281
Kaplan, I. R., 80, 94–95, 97–98, 100–101,
    161, 163
Karas, M. C., 109, 145
Karl, D. M., 259
Kauffman, E. G., 330
Kauppi, P., 246
Keaney, J., 43
Keigwin, L. D., Jr., 36, 38
Keir, R. S., 168–70
Kemeny, J. G., 246
Kennedy, J. S., 236
Kennett, J. P., 25, 29–33, 36, 41, 116, 143,
    332
Killingley, J. S., 23
Kim, J., 190–91
Kitchell, J. A., 238

Klinkhammer, G. P., 134–35, 141
$K_m$, 261–64, 269
Knauer, G. A., 136
Knudsen, J., 335, 345
Koblentz-Mishke, O. J., 85
Kohlmeyer, J., 316
Koide, M., 63, 160
Kollatukudy, P. E., 94
Koltun, V. M., 339
Krishnaswami, S., 63, 122, 124
Krogh, August, 302
Kroopnick, P., 23–24, 33, 38–40
Krumbein, W. E., 207
Ku, T. L., 59, 123–25
Kuhn, T. S., 219
Kussakin, O. G., 326–27
Kutzova, R. S., 209
*Kuznetzovia,* 209

# L

Lactate, 275
Lactate dehydrogenase (LDH), 262–65,
    267–69, 271–72, 274–76
Lactic acid metabolism, 274
Laidler, K. J., 267
Lakatos, I., 219
Lalou, C., 124
LaPlace, P., 307
LaRock, P. A., 209
Late Cretaceous (*See* Cretaceous/Tertiary
    boundary)
Late Miocene, carbon isotopes in, 36–38
Lavoisier, A., 307
Lawson, D. S., 172
LDH (*See* Lactate dehydrogenase)
Lee, C., 308, 317
LeMasurier, W. E., 30
Lemche, H., 341
*Leptothrix (Sphaerotilus) discorphorus,* 205
Lerman, A., 164
Levinton, J. S., 237, 241
Li, Y. H., 160, 167
Life-history tactics, 240–41, 243, 248
Lindroth, K. J., 25
Lipps, J. H., 326, 328, 340, 347
Little, C., 317
Liu, Kon-Kee, 77, 80, 84–86
Liu, S., 75
Loblich, A. R., 45
Lonsdale, P., 226
Lopez, G. R., 237
Loutit, T. S., 36
Low, P. S., 266

Murray, J., 106, 115–16, 119, 141, 171
Murray, J. W., 203
Muscle enzymic activities, 262, 272–75
    (*See also* Lactate dehydrogenase)
*Mytilus californianus,* amino acid transport in,
    311–13
*Mytilus Seamount,* 13–14

## N

Nakhshina, Y. P., 203
Nannofossils, calcareous, oxygen and carbon
    isotopes of, 25–27
Nealson, K. H., 139, 187, 189, 206–7, 209
Nelson, D. E., 125
New England seamount chain, 13–14
*Nezumia sclerorhynchus,* 230
Nicol, D., 344
Nicotinamide adenine dinucleotide (NADH),
    264
Ninhydrin technique for studying amino acid
    transport, 303
Nishida, S., 123
Nissenbaum, A., 100
Nitrate reduction, bacterial, 184–85, 187
Nitrifying bacteria, 186
Nitrogen cycle, 75, 186
Nitrous oxide, 74–90
  apparent production of, 87–88
  in the atmosphere, 74–75
  method of measuring, 77–79
  mole fractions in the ocean air, 81–83, 89
  $\Delta N_2O/\Delta(NO_3^- + NO_2^-)$ relationship, 88–89
  ozone layer and, 74–75
  in subsurface water and basin water, 85–86
  supersaturation of, 75, 89–90
  in surface water, 83–85
Nozaki, Y., 61–62, 160
Nuclides (*See* Radionuclides)
*Nucula cancelleta,* 231
Nutrition (*See* Dissolved organic material,
    nutritional role of; Food supply; Trophic
    structure):
  bacterial, 183, 188–89, 192–95, 197
Nygaard, M. H., 270

## O

Oaks, J. A., 306
Ocean basin, 5
Ocean floor:
  depth of, 4–5
  origin of, 7–10
  shape of, 5–6

Ockelmann, K. W., 235
*Ocypode quadrata,* 235
Oil pollution (*See* Petroleum contamination)
O'Keefe, J. A., 31
Oligocene temperatures, 30–31
Oliver, P. G., 326, 344
Opdyke, N. D., 21, 33
Optimal foraging theory, 231–39, 248
Orians, G. H., 236
Origin of faunas, 327–28, 345
Osman, R. W., 245
Otto, J. C. G., 207–8
Oxidation, bacterial, 184–87
  manganese (*See* Manganese oxidation,
    bacterial)
Oxygen consumption, 275 (*See also*
    Respiration)
  depth of occurrence and, 272
  food intake and, 307
Oxygen isotopes, 20
  of benthic foraminifera, 27
  of calcareous nannofossils, 25–27
  Eocene/Oligocene changes in, 30–31
  Miocene changes in, 31–33
  of planktonic foraminifera, 21–23
  Pliocene/Pleistocene changes, 33
Ozone layer, nitrous oxide in the atmo-
    sphere and, 74–75

## P

Paine, R. T., 248
Paleoceanography, 19–48
  carbon isotopes in planktonic
    foraminifera, 23–24
  oxygen and carbon isotopes of calcareous
    nannofossils, 25–27
  oxygen isotopes of planktonic foraminifera,
    21–23
Paleocene:
  carbon isotopes in, 38–43
  temperature, 29–30
Paleotemperatures, 332
  history of, 27–33, 332–33
    Eocene/Oligocene, 30–31
    Late Cretaceous/Paleocene, 29, 43–47
    Miocene, 31–33
    Pliocene/Pleistocene, 33
  oxygen isotopes of planktonic
    foraminifera and estimates of, 21–22
Pamatmat, M. M., 288, 307
Parker, P. L., 23
Parsons, T. R., 192
Particle-mixing rates in deep-sea sediments,
    60–66

Particulate organic carbon (POC),
  bacteria and, 192–94, 197
Paul, A. Z., 113, 118, 143
Payne, R. R., 116
Pearcy, W. G., 235
Pearse, J. S., 314, 335
Pearse, V. B., 314
Pearson, N. E., 236
Pelagic animals, muscle enzymic activity of,
  273
Pelagic sediments, oxic diagenesis of, 135
Peng, T. H., 63, 83–84, 160
Péquignat, E., 313
Perch-Nielsen, K., 43
Perfil'ev, B. V., 209
Peterson, M. N. A., 168
Petroleum contamination, 94–102
*Phormosoma placenta,* 238–39
Photosynthesis, 157–58
Physiographic maps of ocean floor, 5–6
Picken, G. B., 335
Pierotti, D., 74, 79, 81, 83, 85, 89
Pillow lava, 11–12
Piper, D. Z., 126
Plankton:
  definition of, 281
  in food web, 295–99
  respiration of, 293–94, 297–98
  respirometry, 286–88
Planktonic foraminifera:
  carbon isotopes in, 23–24
  depth habitats of, 22, 25–27
  oxygen isotopes of, 21–23
Plate tectonics, 8, 10
Platt, J. R., 219
Pleistocene:
  carbon isotope history, 33–35, 47
  temperatures, 33
Pliocene:
  carbon isotopes in, 36
  temperatures, 33
Polloni, P., 221, 223, 296
*Polycarpa delta,* 231
Polychaetes, 224–29
Popper, K. R., 219
Population control, predation and, 241–45,
  248
Population density, 223
  of benthopelagic animals, 296–97
Population dynamics, size-frequency distribu-
  tion and, 229–31, 241
Population growth rates, 229–30
  predation rates and, 245
  *r-K* selection theory, 239–41, 243, 245
Populations, persistence of, 239–40

Porewaters, 155–56
  upward diffusion of metals dissolved in, 138
Pramer, D., 193
Predator-prey relationships, 245
Predators (predation), 229, 235, 259, 273,
    274 (*See also* Scavengers)
  population control and, 241–45, 248
Pressure:
  adaptation to, 258–69, 276
  bacterial growth and, 188–95
Price, B. A., 83, 87
Price, N. B., 126–30
Pritchard, P. H., 187
Psychrophilic bacteria, 190
Pteropod ooze, 170–71
Pteropods, sedimentation of, 170–72
Pütter, August, 302
Pycnogonids, gigantism of, 336
Pyke, G. H., 231–32, 238
Pyruvate, 262–64
Pyruvate kinase (PK), 271–72, 274–75


# R

Raab, W., 129–30, 138
Radioactive decay, 57–58
Radionuclides, naturally occurring, 56–71
  cosmogenic nuclides, 56–57
  growth rates of deep-sea organisms and,
    66–70
  particle-mixing rates in deep-sea sediments
    and, 60–66
  supply and removal of, 58–60
Rarefaction procedure, Hurlbert, 228
Rasmussen, K. J., 236
Rasmussen, R. A., 74, 79, 81, 83, 85, 89
Rattails, 259, 273
Rau, G. H., 226
Reck, R. A., 75
Redox reactions, bacterial, 184–85
  manganese (*See* Manganese bacteria)
Reed, W. E., 95
Rehoboth Seamount, 13
Reichelt, J. L., 184
Reid, A. M., 236
Reid, R. G., 236
Reidel, W. R., 20
Renard, A. F., 106, 115–16, 119, 141
Reproduction, 231, 240, 242
Resource limitation, 243–44
Resource partitioning, 243–45
Respiration:
  of amphipods, 297–98
  anaerobic, 184–85

Respiration (*cont.*)
    anaerobic enzymic activity an, 274–75
    of bacteria, 184–85, 298
    in benthic boundary layers (*See* Benthic
        boundary layer, respiration in)
    of benthopelagic animals, 297
    depth of occurrence and, 270–71
    of plankton, 293–94, 297–98
    of sediment community, 289–92, 296
Respirometry (respirometers):
    bell jar, 282–83
    of benthopelagic animals, 285–86
    fish trap (FTR), 285–86
    free vehicle (FVR), 282–83
    free-vehicle grab (FVGR), 283–85
    of plankton, 286–88
    of sediment community, 282–85
    slurp gun (SGR), 286–88, 294
    submersible-placed grab (DSGR), 283–84
Rex, M. A., 229, 245
Rice, A. L., 233
Richerson, P., 244
Riley, 89–90
Risk, M. J., 238
*r-K* selection theory, 239–41, 243, 245, 248
Roberts, F. S., 246
Rokop, F., 335
Rokop, F. J., 230–31
Roman, M. R., 259
Root, R. B., 225
Roper, C. F. E., 342
Rosenzweig, M. L., 348
Rosson, R. A., 134, 140, 206, 212
Rotty, R. M., 48
Rowe, G. T., 220, 223, 231, 293, 296–97
Ruby, E. G., 189, 192
Ryan, W. B. F., 32

# S

Sablefish, 273
Sachs, H. M., 20
Saito, T., 29, 40
Sampling devices, 220, 229
    pressure apparatus, 190–91
Sanders, H. L., 68, 221, 227, 230, 240–41,
        243–44, 270, 328, 335, 343–44, 346–48
Saunders, R. L., 293
Savin, S. M., 21–22, 25–27, 29–33, 38, 40, 43,
        332
Scavengers, 226, 229, 241, 248, 259
    foraging theory and, 234–37
    hovering height of, 236–37
Schaffer, H. A., 77, 87
Schaffer, W. M., 242–43

Scheltema, R. S., 335
Schink D. R., 63, 160, 170, 173
Schinske, R. A., 303
Schlanger, S. O., 38–40, 332
Schlichter, D., 306
Schoener, T. W., 244
Scholle, P. A., 38–39
Schultz, S. G., 309
Schutt, C., 207–8
Schwartzlose, R. A., 226, 299
Schwarz, J. R., 190
Schweisfurth, R., 205, 212
Schwertmann, V., 138
Sclater, F. R., 136
Scott, 58
Sea floor (*See* Ocean floor)
Sea-floor spreading, magnetic anomaly
        stripes and, 8
Seager, J. R., 335
Sea level, 17
Sea squirts, 226
*Sebastolobus alascanus,* 263–65
*Sebastolobus altivelis,* 263–65
Sedimentation rates, decomposition of organic
        matter and, 162–64
Sediment community:
    definition of, 280–81
    in food web, 295–99
    respiration of, 289–92, 296
    respirometry, 282–85
Sediments, 13–14
    bacteria in, 183, 193–94
        anaerobic respiration, 184–85
        manganese in, 207–9
    particle-mixing rates in, 60–66
    thicknesses of, 7
Seifert, W. K., 101
Seki, H., 231, 316
Self, R. F. L., 237, 241, 248
Sepers, A. B. J., 302
Sessions, M. H., 113, 143
Seyfried, W. E., Jr., 133
SG-1, 206, 210
Shackleton, N. J., 21–22, 26–27, 30–33, 45, 47
Shells, growth rates of, 66–70
Shick, J. M., 304, 318
Shimp, S. L., 193
Shokes, R., 99–100
Sholkovitz, E. R., 80, 85
Shorey, H. H., 236
Shulenberger, E., 229
Siebenaller, J., 317
Siebenaller, J. F., 264–65, 267
Sieburth, J. McNeill, 181–82, 187–89, 193,
        197, 316